Plate Heat Exchangers

International Series on Developments in Heat Transfer

Objectives

The Developments in Heat Transfer book Series publishes state-of-the-art books and provides valuable contributions to the literature in the field of heat transfer. The overall aim of the Series is to bring to the attention of the international community recent advances in heat transfer by authors in academic research and the engineering industry.

Research and development in heat transfer is of significant importance to many branches of technology, not least in energy technology. Developments include new, efficient heat exchangers, novel heat transfer equipment as well as the introduction of systems of heat exchangers in industrial processes. Application areas include heat recovery in the chemical and process industries, and buildings and dwelling houses where heat transfer plays a major role. Heat exchange combined with heat storage is also a methodology for improving the energy efficiency in industry, while cooling in gas turbine systems and combustion engines is another important area of heat transfer research.

To progress developments within the field both basic and applied research is needed. Advances in numerical solution methods of partial differential equations, high-speed, efficient and cheap computers, advanced experimental methods using LDV (laser-doppler-velocimetry), PIV (particle-image-velocimetry) and image processing of thermal pictures of liquid crystals, have all led to dramatic advances during recent years in the solution and investigation of complex problems within the field.

The aims of the Series are achieved by contributions to the volumes from invited authors only. This is backed by an internationally recognised Editorial Board for the Series who represent much of the active research worldwide. Volumes planned for the series include the following topics: Compact Heat Exchangers, Engineering Heat Transfer Phenomena, Fins and Fin Systems, Condensation, Materials Processing, Gas Turbine Cooling, Electronics Cooling, Combustion-Related Heat Transfer, Heat Transfer in Gas-Solid Flows, Thermal Radiation, the Boundary Element Method in Heat Transfer, Phase Change Problems, Heat Transfer in Micro-Devices, Plate-and-Frame Heat Exchangers, Turbulent Convective Heat Transfer in Ducts, Enhancement of Heat Transfer and other selected topics.

Plate Heat Exchangers

Design, Applications and Performance

L. Wang
Siemens Industrial Turbines, Sweden

B. Sundén
Lund Institute of Technology, Sweden

&

R.M. Manglik
University of Cincinnati, USA

WIT PRESS Southampton, Boston

L. Wang
Siemens Industrial Turbines, Sweden

B. Sundén
Lund Institute of Technology, Sweden

R.M. Manglik
University of Cincinnati, USA

Published by

WIT Press
Ashurst Lodge, Ashurst, Southampton, SO40 7AA, UK
Tel: 44 (0) 238 029 3223; Fax: 44 (0) 238 029 2853
E-Mail: witpress@witpress.com
http://www.witpress.com

For USA, Canada and Mexico

WIT Press
25 Bridge Street, Billerica, MA 01821, USA
Tel: 978 667 5841; Fax: 978 667 7582
E-Mail: infousa@witpress.com
http://www.witpress.com

British Library Cataloguing-in-Publication Data

A Catalogue record for this book is available
from the British Library

ISBN: 978-1-85312-737-3
ISSN: 1369-7331

Library of Congress Catalog Card Number: 2002111282

Printed in Great Britain by Athenaeum Press Ltd.

Contents

Preface

Heat exchangers are important, and used frequently in the processing, heat and power, air-conditioning and refrigeration, heat recovery, transportation and manufacturing industries. Such equipment is also important in electronics cooling and for environmental issues like thermal pollution, waste disposal and sustainable development. Various types of heat exchangers exist. In textbooks of heat transfer, commonly a brief chapter is provided for the introduction of heat exchangers and elementary theory of design, rating and sizing are presented. There also exist many books on heat exchangers either as textbooks or edited volumes. However, most such books treat a variety of heat exchanger types or specific problems and do not specialize in any particular heat exchanger type. Therefore, a lack of comprehensive and in-depth textbooks on specific heat exchangers exists.

The present book concerns plate heat exchangers (PHEs), which are one of the most common types in practice. The overall objectives are to present comprehensive descriptions of such heat exchangers and their advantages and limitations, to provide in-depth thermal and hydraulic design theory for PHEs, and to present state-of-the-art knowledge.

The book starts with a general introduction and historical background to PHEs, then discusses construction and operation (PHE types, plate pattern, etc.) and gives examples of PHEs in different application areas. Material issues (plates, gaskets, brazing materials) and manufacturing methods are also treated. The major part of the book concerns the basic design methods for both single-phase and two-phase flow cases, various flow arrangements, thermal-hydraulic performance in single-phase flow and for PHEs operating as condensers and evaporators. Fouling problems are also discussed and in a section on extended design and operation issues, modern Research and Development (R & D) tools like computational fluid dynamics (CFD) methods are discussed. Unique features for PHEs are discussed throughout.

Extensive R & D activities are carried out at companies and universities worldwide and originally this book was intended as an edited volume reflecting current research and state-of the-art. However, as time elapsed and the lack of a comprehensive textbook was identified, the objectives were changed.

We believe this book will be useful as both a textbook at various educational levels and as a reference source book for PHEs.

We are grateful to the companies providing us with a lot of information on their products and their R & D works. We also appreciate the cooperation and patience provided by the staff at WIT Press and for their encouragement and assistance in producing this book.

Lieke Wang, Raj M. Manglik and Bengt Sundén

CHAPTER 1

Basic features and development of plate heat exchangers

1.1 Introduction

The heat exchange process between two or more streams at different temperatures occurs in numerous industrial, commercial, and domestic applications and is usually effected in some type of a heat exchanger. A varied set of heat exchangers, spanning almost the entire spectrum of shape, size, transfer mode, and other features can be found in the process, power, petroleum, transportation, air conditioning, refrigeration, cryogenics, and biomedical industries, among others. They can be broadly classified as either direct contact or indirect contact (or transmural) type of heat exchangers. In the former, the fluids transferring heat are in direct contact and not separated by a wall. Owing to the absence of a wall, closer approach temperatures are attained, and the heat transfer is often also accompanied with mass transfer. The cooling tower is an example of a direct contact type heat exchanger. In transmural heat exchangers, a wall (tubular, plate, or some other non-circular geometry) separates the hot and cold fluids streams, and heat exchange between them takes place across this interface. Shell-and-tube and plate (or plate-and-frame) heat exchangers are examples of indirect contact type exchangers; a rotary regenerator is an example of the indirect contact exchanger where heat is exchanged in a transient manner. A more detailed classification of heat exchangers can be made on the basis of their construction features, modes of transfer, and heat duty specifications, and several different schemes have been outlined by Walker [1], Sukhatme and Devotta [2], Hewitt et al. [3], Shah and Sekulić [4], Kuppan [5], Kakaç and Liu [6], and others.

The competitive pressures of the global market and the growing urgency for energy conservation and reduction of environmental degradation have placed a greater emphasis on the use of high efficiency heat exchangers. Although the traditional shell-and-tube heat exchanger is still quite commonly used in numerous industrial applications, and its performance can be significantly improved by means of enhancement devices [7, 8], it tends to have large hydraulic diameters and surface area to volume ratios. The engineering cognizance of this problem has led to the development of a variety of different high performance heat exchangers, with varying degrees of compactness, that are employed in practice [3, 4, 6, 9, 10]. A compact heat exchanger has been defined as one with a heat transfer surface area density of above $700\,m^2/m^3$ on at least one of the fluid sides [11]. They not only provide a smaller size footprint, but their specific construction features also promote enhanced thermal-hydraulic performances and increased energy efficiencies, with significant materials, capital, and operating costs savings.

Although the plate heat exchanger (PHE) is categorized at the lower end of the compactness spectrum [11], it offers many distinct advantages and unique application features over some of the other highly compact exchangers. These include flexible thermal sizing (plates can simply be added or removed to meet varying batch-processing heat-load demands), easy cleaning for sustaining extreme hygienic conditions (necessary for food, pharmaceutical, and dairy product processing), close approach temperature pure counter-flow operation (needed for cryogenic applications and mitigation of thermal degradation of some process fluids), and enhanced heat transfer performance. As such, the design of PHEs tends to be highly specialized in nature, with a variety of different plates and their surface corrugations, flow arrangements, performance characteristics, and construction and application features, and these are given special attention in the extended discussions of this book.

1.2 Historical background

The earliest development and usage of PHEs were in response to increasingly stringent statutory requirements from foodstuffs, particularly dairy products, in the late nineteenth century. At the beginning of the 1880s, there was growing public awareness that diseases, such as tuberculosis, were spread by 'raw' or untreated milk. This initiated the early experiments with milk pasteurization, which involved heating the milk to a certain temperature that does not influence the taste, holding it at this temperature for a short time, and then immediately cooling it. This process requires the heat transfer equipment to be thermally very efficient and, more importantly, be easily cleaned (which had to be conducted daily). It was indeed difficult to meet these operational requirements in most of the early heat transfer equipments that were used for pasteurization of milk, which in time led to the development of PHEs.

What is perhaps the very first patent for a PHE, which pre-dates this activity by a few years, was granted to a German by the name of Albrecht Dracke in 1878 [12]. In this patent, the cooling of one liquid by another, with each flowing in a layer on opposite sides of a series of plates, was proposed. The subsequent years of the late 1800s and early 1900s saw considerable inventive and developmental activity that resulted in several other patents covering different forms of PHEs. Figure 1.1 gives a schematic illustration of one such patent for a PHE and its plate pack, which was issued to two Germans (Langen and Hundhausen, Patent No. 58504) on 5 April 1890 [13], and that offered some new 'improvements'.

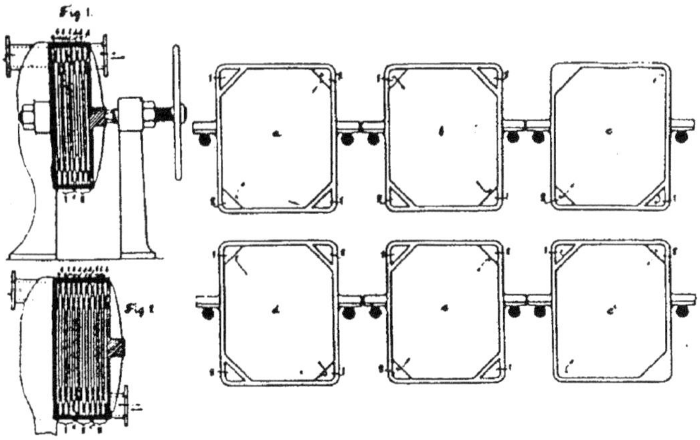

Figure 1.1: Patent No. 58504 for an early PHE issued on 5 April 1890 [13].

Figure 1.2: The PHE invented by Dr Richard Seligman (courtesy of APV).

However, PHEs were not commercially exploited until the 1920s, when Dr Richard Seligman, the founder of APV International in England, invented the first operational PHE (Fig. 1.2) in 1923. The device was called plate pasteurizer, which was destined to revolutionize the 'thermal curing' work in dairies to much the same extent as had the separator in its time. Almost a decade later, around 1930, Bergedorfer Eisenwerk of Alfa Laval in Sweden (AB Separator at that time) developed a similar commercial PHE, the company's first such product (Fig. 1.3) [14]. Although the basic concept and operation of a PHE has changed little since then, its overall design and construction has progressed significantly in order to accommodate larger throughput capacities, higher working temperatures, and larger working pressures, among other factors. These changes have helped expand the applications for PHEs from the original milk pasteurization to a very broad range of industrial heat exchanger needs.

Figure 1.3: The PHE developed by Bergedorf (courtesy of Alfa Laval).

This activity came to the fore in the 1970s and thereafter, when it gained added impetus from the energy crisis. Several new designs for plate surface embossing and their manufacturing techniques were developed, and a representation is illustrated in Fig. 1.4. One of the driving motivations was to develop new plate patterns for effective heating/cooling of process fluids, with cost-effective manufacturing processes and improved structural integrity. Also, the late 1970s saw a concerted effort to conserve energy by improving the thermal-hydraulic performance of heat exchangers. An outcome of this in PHE design was the use of chevron-type corrugated plates in a 'mixed' flow arrangement, where a combination of low, medium, or high chevron angle plates are used [15]; as shown later, this can promote increased thermal effectiveness with a reduced pressure drop penalty. A relatively recent and novel extension of this approach has been to segment the plate surface into four quadrants, with each quadrant having a different chevron angle corrugation [16]. The chief objective here is to match better the pressure drop and heat transfer rate constraints in a two-fluid stream application, reduce overall flow maldistribution, and optimize the heat transfer area.

Today, there are many well-known manufacturers around the world, commercially offering a wide variety of PHEs for a very broad spectrum of different applications. A representative listing of some manufacturers includes Alfa Laval Lund AB (http://www.alfalaval.com), APV (http://www.apv.com), SWEP

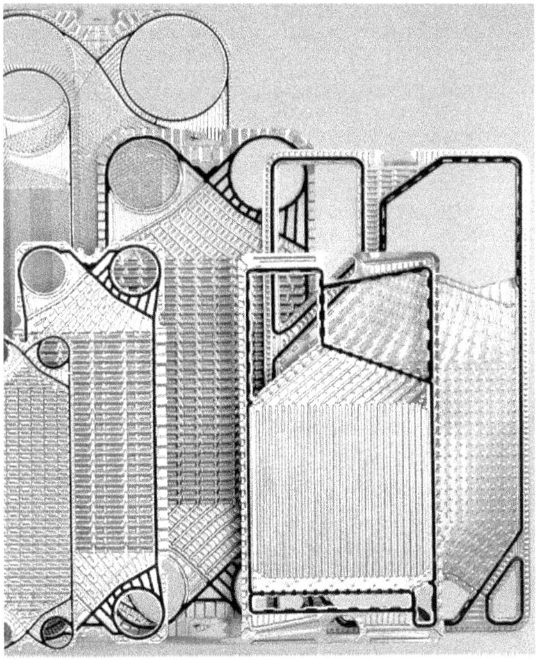

Figure 1.4: Typical PHE plates with different surface corrugation patterns and port geometries (courtesy of GEA Ahlborn Gmbh).

(http://www.swepphe.com), Tranter (http://www.tranterphe.com), GEA (http://www.gea-ag.com), Sondex A/S (http://www.sondex.com), Hisaka Works (http://www.hisaka.co.jp), among many others. This is by no means a comprehensive list or an endorsement of some specific companies. As the markets for PHE applications expand, it would likely attract numerous manufacturers to the field.

1.3 Basic principle

A cut-away, disassembled illustration, which highlights the primary structural components of a typical modern plate or plate-and-frame heat exchanger (PHE), is given in Fig. 1.5. The primary elements of a PHE include the heat transfer and flow separation plates, gaskets, the supporting end frames (a fixed and a movable cover plate or frame) that house inlet/outlet fluid port nozzles, horizontal plate carrying and guiding bars, the support column, and end-frame tightening bolts.

The main heat exchange core consists of thin, rectangular, pressed sheet-metal plates, which are sandwiched between full peripheral gaskets and assembled in a pack and bolted in a frame. The edge gaskets between each pair of embossed plates confine each fluid stream within the prescribed inter-plate flow channels and prevent their intermixing as well as any fluid leakage out of the plate pack to the surroundings. The plates hang from a top carrying bar and are located in place with

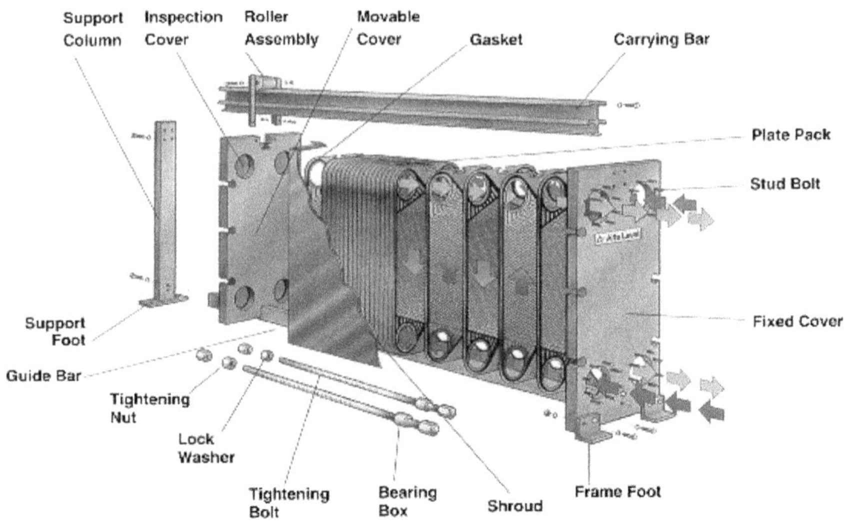

Figure 1.5: An exploded view of a PHE (courtesy of Alfa Laval).

proper alignment of the stack by the lower bar. In most units, the carrying bars are supported and held apart at the other end of the frame by a supporting column. The plate pack that is sandwiched between the fixed frame plate and the movable cover or pressure plate is clamped together by lateral bolts that are tightened to a predetermined overall length. It should be pointed out that the first pattern-embossed plate in the core pack has gaskets around all four holes to prevent the fluids from leaking out between the first plate and the frame as they enter or leave the PHE via the nozzle connections in the frame.

As shown in Fig. 1.5, the hot and cold fluid streams flow in alternate channels separated by the embossed or corrugated plate. Each pair of adjacent plates in the PHE forms this channel, and the fluid media, entering and leaving via inlet/outlet ports, are routed through alternate inter-plate channels by the gasketing at the periphery of each plate and their pack arrangement. Heat is thus transferred from the warm fluid via the dividing wall to the colder fluid in a pure counter-flow arrangement, which lends to the high effectiveness of the PHEs.

1.4 General characteristics

Because of their structural features, PHEs offer a number of advantages over the traditional shell-and-tube heat exchangers. These include the following:

1. The plate surface corrugations readily promote enhanced heat transfer by means of several mechanisms that include promoting swirl or vortex flows, disruption and reattachment of boundary layers, small hydraulic diameter flow passages, and increased effective heat transfer area. The heat transfer

Figure 1.6: Schematic representation of recirculating and swirling fluid motion caused by plate surface corrugations (courtesy of Alfa Laval). (a) Washboard plate (in-phase parallel corrugations) channels with lateral swirl in surface troughs; (b) Chevron or Herringbone plate (cross-corrugated) flow channels with longitudinal helical swirl.

coefficients so obtained are significantly higher than those in shell-and-tube heat exchangers for comparable fluid conditions. The dominant mechanism among these is the production of either lateral vortices in the corrugation troughs or helical swirl flows along the plate length, depending upon the plate corrugation arrangement (in-phase parallel waviness or cross-corrugations) as shown in Fig. 1.6, which result in well-mixed flows with high convective heat transfer coefficients.

2. PHEs usually have a much smaller thermal (and physical) size due to high heat transfer coefficients. For the same effective heat transfer area, the weight and volume of PHEs are approximately only 30% and 20%, respectively, of those of shell-and-tube heat exchangers. A typical pictorial depiction of this size advantage of a PHE over a shell-and-tube heat exchanger for the same heat duty is shown in Fig. 1.7, and this makes it particularly attractive for marine and offshore oil- and gas-rig applications.
3. Owing to their high heat transfer coefficients and true counter flow arrangement, PHEs are able to operate under very close approach temperature ($\sim$1°C) conditions. As a result, up to 90% heat recovery can be attained, which is a significantly higher thermal performance compared to the 50% recovery for shell-and-tube heat exchangers. PHEs are therefore highly suited for use in the heat recovery from rather low-grade heat sources.
4. In PHEs, each fluid medium is confined to channels between individually gasketed plates. The space between gaskets is vented to atmosphere, thereby eliminating the possibility of any cross-contamination of fluids.

Figure 1.7: An illustrative depiction of the typical size differences between a PHE and a shell-and-tube heat exchanger for a given heat-load application (courtesy of Alfa Laval).

5. PHEs are very suitable for high viscosity fluids, which tend to flow in the laminar regime in most heat exchangers because of their swirl-producing characteristics; helical vortices are often generated at rather low Reynolds numbers in cross-corrugated flow passages. This mitigates fouling as well in many cases.
6. The plate-pack in a gasketed PHE can be very easily disassembled and reassembled. Thus, their inspection, cleaning, and gasket replacement can be very conveniently carried out, and, in particular, very hygienic conditions can be maintained for dairy and food processing applications. Additionally, the easy assembly/disassembly provides a great flexibility for altering the thermal size of the PHE by simply adding or removing some plates to meet the changing heat load requirements in a process plant.
7. Owing to the thin channels created between the two adjacent plates, the volume of fluid contained in PHEs is small. It, therefore, reacts to the process condition changes in rather short time transients and is thereby easier to control.
8. It is possible to heat or cool two or more fluids within the same unit by simply installing intermediate divider sections between the heat transfer plates. This can reduce the structural complexity of the process plant.
9. Plates with different surface patterns can be combined in a single PHE. Different multi-pass arrangements can also be configured. This flexibility enables better optimization of operating conditions for PHEs.
10. Because PHEs generally have low hold-up volume and less weight, low costs are always achieved for handling, transportation, and foundations.
11. Flow-induced vibration, noise, and erosion-corrosion due to fluid impingement on heat transfer surface are eliminated in PHEs.
12. Only the plate edges are exposed to the atmosphere. Thus, the heat loss is negligible and no insulation is generally required.

These and some other advantageous features of a gasketed PHE, in comparison to the shell-and-tube heat exchanger, are summarized in Table 1.1. These have made

Table 1: Comparison between PHEs and shell-and-tube heat exchangers.

	Gasketed PHE	Shell-and-tube heat exchanger
Temperature cross	Impossible	Possible
Approach ΔT	~1°C	~5°C
Multiple duty	Possible	Impossible
Piping connections	From one direction	From several directions
Heat transfer ratio	~3–5	1
Operating weight ratio	1	~3–10
Hold-up volume	Low	High
Space ratio	1	~2–5
Welds	None	Welded
Sensitivity to vibrations	Not sensitive	Sensitive
Gaskets	On every plate	On each bonnet
Leakage detection	Easy to detect	Difficult to detect
Access for inspection	On each side of plate	Limited
Disassembly time	~15 min	~60–90 min
Repair	Easy to replace plates and gaskets	Requires tube plugging = decreased capacity
Thermal size modification	Easily achieved by adding or removing plates	Difficult
Fouling ratio	~0.1–0.25	1

PHEs very competitive, when compared to other types of compact and non-compact heat exchangers as well, for a variety of different applications.

The main disadvantage, however, is that PHEs are limited to applications with relatively low fluid pressure and temperature conditions. This is essentially due to the restrictions imposed by the gasket material, which cannot withstand the high pressure or temperature or the corrosiveness of the fluid, and thus suffers from leakage problems; the thin plate material also lends to some operating pressure restrictions. In current practice, the standard gasketed plate-and-frame heat exchangers have been generally considered for operating pressures of up to about 25 bar; somewhat higher pressures can be achieved using heavy-duty frames. Likewise, the upper limit of 160°C is found in most PHE applications, though gaskets made of some special materials have been claimed to operate up to temperatures of 400°C [17, 18]. To overcome this disadvantage imposed by operating pressure and temperature restrictions, several variant types of PHEs have been developed and they are described in the next chapter.

References

[1] Walker, G., *Industrial Heat Exchangers*, 2nd edition, Hemisphere, New York, NY, 1990.

[2] Sukhatme, S.P. and Devotta, S., Classification of Heat Transfer Equipment, in *Heat Transfer Equipment Design*, ed. R.K. Shah, E.C. Subbarao, and R.A. Mashelkar, Hemisphere, New York, NY, pp. 7–22, 1988.

[3] Hewitt, G.F., Shires, G.L., and Bott, T.R., *Process Heat Transfer*, Begell – CRC Press, Boca Raton, FL, 1994.

[4] Shah, R.K. and Sekulić, D.P., Heat Exchangers, in *Handbook of Heat Transfer*, 3rd edition, ed. W.M. Rohsenow, J.P. Hartnett, and Y.I. Cho, McGraw-Hill, New York, NY, chapter 17, 1998.

[5] Kuppan, K., *Heat Exchanger Design Handbook*, Marcel Dekker, New York, NY, 2000.

[6] Kakaç, S. and Liu, H., *Heat Exchangers: Selection, Rating and Thermal Design*, 2nd edition, CRC Press, Boca Raton, FL, 2002.

[7] Bergles, A.E., Techniques to Enhance Heat Transfer, in *Handbook of Heat Transfer*, 3rd edition, ed. W.M. Rohsenow, J.P. Hartnett, and Y.I. Cho, McGraw-Hill, New York, NY, chapter 11, 1998.

[8] Manglik, R.M., Heat Transfer Enhancement, in *Heat Transfer Handbook*, ed. A. Bejan and A.D. Kraus, Wiley, New York, NY, chapter 11, 2002.

[9] Kays, W.M. and London, A.L., *Compact Heat Exchangers*, 3rd edition, McGraw-Hill, New York, NY, 1984.

[10] Shah, R.K., Kraus, A.D., and Metzger, D., *Compact Heat Exchangers: A Festschrift for A. L. London*, Hemisphere, New York, NY, 1990.

[11] Shah, R.K., Classification of Heat Exchangers, in *Heat Exchangers – Thermo-Hydraulic Fundamentals and Design*, ed. S. Kakaç, A.E. Bergles, and F. Mayinger, Wiley, New York, 1981.

[12] Magnusson, B. and Samuelson, B., *The Story of the Alfa Laval Plate Heat Exchanger and the Spiral Heat Exchanger*, Norstedts Tryckeri, Stockholm, 1985.

[13] Clark, D.F., Plate Heat Exchangers, *The Chemical Engineer*, no. 285, pp. 275–285, May 1974.

[14] Magnusson, B., *The Origins and Evolution of the Alfa Laval Plate Heat Exchanger*, Norstedts Tryckeri, Stockholm, Sweden, 1985.

[15] Marriott, J., Performance of an Alfaflex Plate Heat Exchanger, *Chemical Engineering Progress*, vol. 73, no. 2, pp. 73–78, 1977.

[16] Moon, M.G., Invention Paves Way for Third Generation Heat Exchangers, *Australian Refrigeration, Air Conditioning and Heating*, vol. 43, no. 9, pp. 21–25, 1989.

[17] Wadekar, V.V., Improving Industrial Heat Transfer – Compact and Not-So-Compact Heat Exchangers, *Journal of Enhanced Heat Transfer*, vol. 5, pp. 53–69, 1998.

[18] Manglik, R.M. and Muley, A., *Heat Transfer and Pressure Drop Characteristics of Plate-and-Frame Heat Exchangers: A Literature Review*, Report No. TFL-Int-1, University of Cincinnati, Cincinnati, OH, 1993.

CHAPTER 2

Construction and operation

2.1 Gasketed heat exchanger

As briefly described in the previous chapter, a PHE primarily consists of thin, rectangular, pressed sheet metal plates that are sandwiched between full peripheral gaskets and clamped together in a frame as described in Fig. 1.5. The frame has a fixed end-cover plate, fitted with the connecting ports, which is bolted together with a movable cover plate to hold the embossed plates in between; the top and bottom carrying bars allow proper alignment of the plate stacks. The front and side views of a typical fully assembled PHE are depicted in Fig. 2.1, with the inlet and outlet port connections for both fluid streams shown on the fixed end-cover plate side. This configuration is generally referred to as a U-arrangement; an alternative type is the Z-arrangement, where one set of ports is on the movable cover plates. The typical flow arrangements in the two schemes are shown in Fig. 2.2.

The ports on the end frames can have different types of connections, and the three most common nozzle options are illustrated in Fig. 2.3. Inlet and outlet fluid stream portholes are also punched into the corners of each corrugated plate in the stack, and in the assembled frame ports are aligned to form flow distribution headers for the two fluid streams. The peripheral gaskets provide the flow channel space between the adjacent plates, as well as appropriate flow paths for the hot and cold fluids in alternate channels.

2.1.1 Corrugated plate patterns

The thermal-hydraulic performance of PHEs is strongly influenced by the plate surface corrugation patterns in the plate-pack they are fitted with. A plate is normally produced by stamping or embossing a corrugated (or wavy) surface pattern on metallic sheets. Also, on one side of each corrugated plate, special gasket grooves are stamped or embossed along the edges of the plate and around the ports so as to 'seal' the inter-plate channels and provide alternating port fluid inlet/outlet access. Plates are embossed or stamped in many different patterns [1], as well as

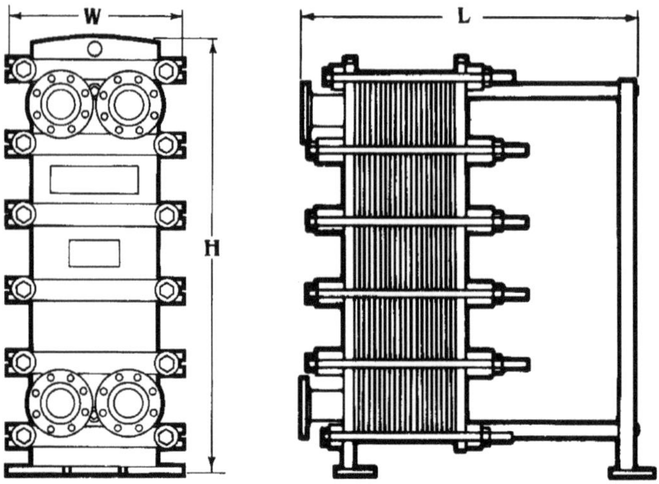

Figure 2.1: Front and side views of an assembled U-arrangement type PHE (courtesy of Tranter).

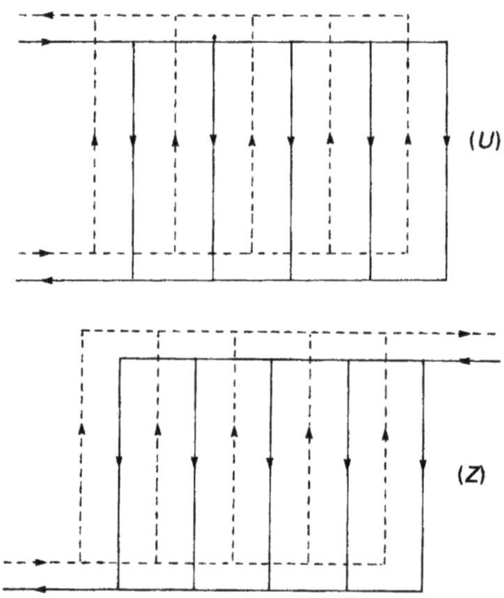

Figure 2.2: Typical flow arrangements of hot and cold fluid streams in PHEs with U- and Z-type port connections.

port geometries, and representations of their variety are illustrated in Fig. 1.4. The various surface geometries have generally resulted from a parallel, competitive commercial development of PHEs, and their respective thermal-hydraulic performance advantage is somewhat unclear [1, 2]. When assembled in a stack, the corrugations on the adjoining plates abut to form narrow interrupted flow passages, and

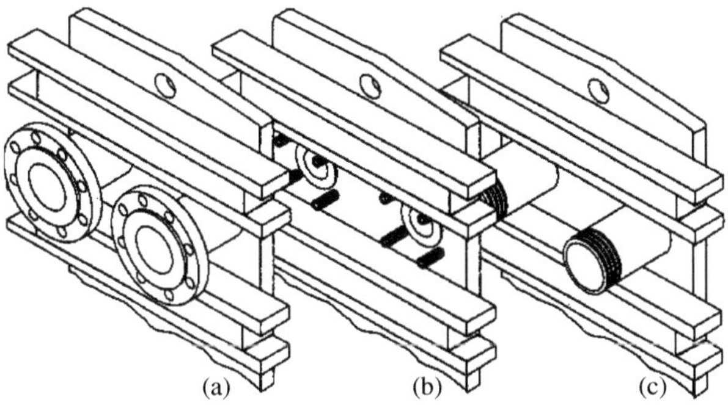

Figure 2.3: Commonly used port nozzle connections (courtesy of Tranter): (a) flanged (standard), (b) studded, and (c) threaded pipe.

Figure 2.4: Vintage *paraflow*® with gunmetal plate (courtesy of APV).

the inter-plate, inter-corrugation flow paths so formed are largely responsible for promoting enhanced convective heat transfer coefficients and decreased fouling characteristics. The corrugations also increase the effective surface area for heat transfer as well as plate rigidity, and the multiple metal-to-metal contact points between adjacent plates lend greater mechanical support to the stack.

The modern plate geometries have evolved since 1940s. The early heat transfer plates were very different, and Figs 2.4 and 2.5 show two plate patterns that were prevalent during the 1930s [3]. A decade later in the 1940s, there were several more manufacturers in the market and many different plate sizes and forms, each having its own characteristics. Figure 2.6 illustrates four plates produced by different manufacturers in that period.

As pointed out by Shah and Focke [4], more than 60 different plate-surface corrugation patterns have been developed worldwide during the past century. A representation of the different categories among these that are employed in present-day applications is shown in Fig. 2.7. Of these, the chevron plates are perhaps the more prevalent [2, 5] and typical features of their surface pattern and inter-plate cross-corrugated flow channels are depicted in Fig. 2.8. These flow passages enhance

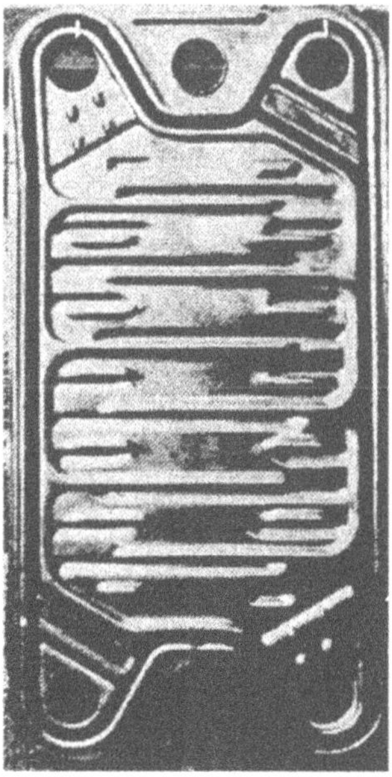

Figure 2.5: A vintage P3® stainless steel plate (courtesy of Alfa Laval).

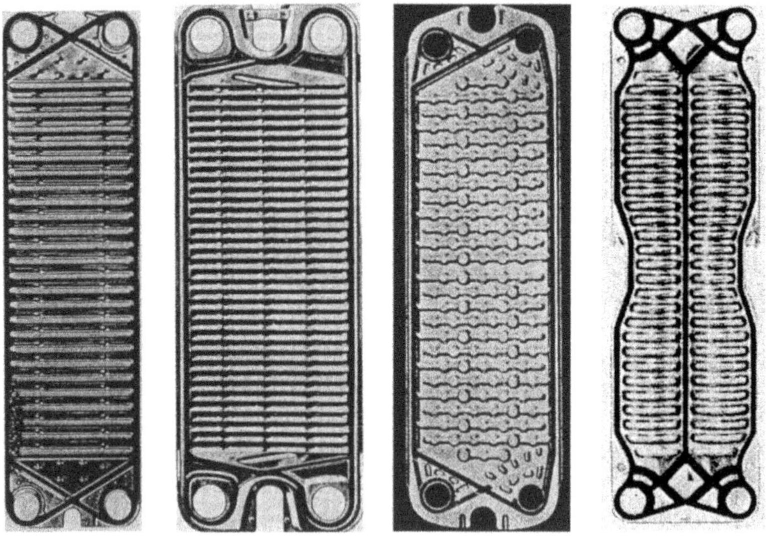

(a) (b) (c) (d)

Figure 2.6: Plates with different surface corrugation patterns produced during the 1940s by different manufacturers: (a) APV HM, (b) Alfa Laval P12, (c) Schimdt Sigma 20, and (d) Ahlborn.

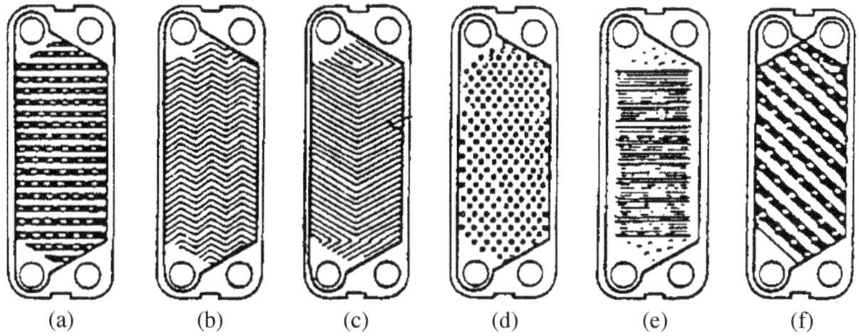

Figure 2.7: Typical categories of different plate-surface corrugation patterns [4]: (a) washboard, (b) herringbone or zig-zag, (c) chevron, (d) protrusions and depressions, (e) washboard with secondary corrugations, and (f) oblique washboard.

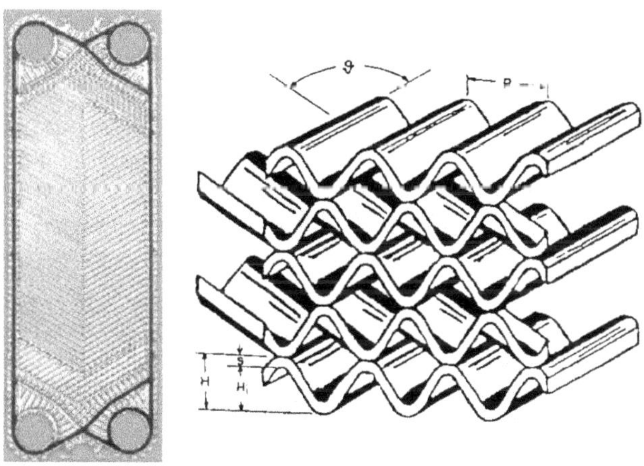

Figure 2.8: A PHE plate with chevron-type corrugation pattern (courtesy of GEA Ahlborn) and the associated inter-plate flow channels [6].

the heat transfer coefficient by increasing the effective surface area, disrupting boundary layers, and promoting swirl flows [2, 7]. The last also tends to decrease the fouling resistance by increasing the level of fluid mixing and agitation, though this generally leads to a higher pressure drop penalty.

2.1.2 Geometrical characterization of chevron-type plates

The last decade has seen a considerable expansion of the application of chevron-type plates [1, 2, 5] and they have generally been perhaps the most successful design offered by the majority of manufacturers [8–12]. A description of their important geometrical dimensions is given in Fig. 2.9. As seen in this figure, the plate surface is divided into a top and bottom flow distribution region and a core heat transfer

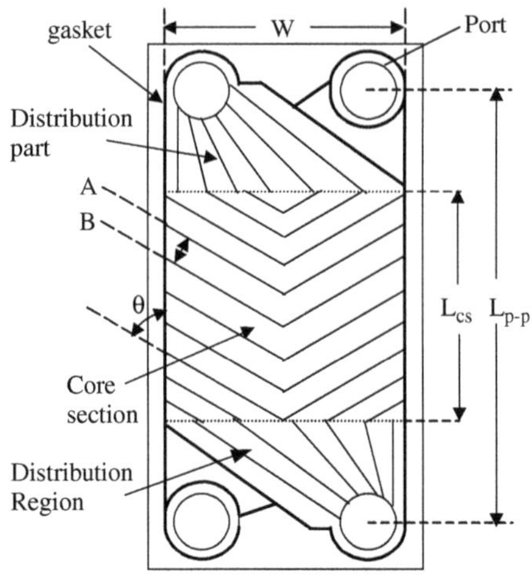

Figure 2.9: Geometrical attributes of a chevron plate-corrugation pattern.

section (L_{cs}). The two distribution regions are designed to evenly distribute the fluid flowing in from the inlet port to the core region and then collect and channel it to the outlet port. They are usually designed to minimize flow maldistribution over the core heat transfer section, so that the thermal and hydraulic performance does not degrade. However, both the core and flow distribution regions are active heat transfer surfaces.

To describe the various geometry and surface corrugation features of a chevron plate (Figs 2.8 and 2.9), several different nomenclatures have been used in the literature, including manufacturers' catalogues [1, 4, 5, 8–12]. A common set of definitions is adopted in this book to circumvent the associated confusion, and these are as follows: corrugation wavelength P, inter-plate channel height H_i, amplitude of corrugation waviness ($0.5H_i$), included chevron angle θ, plate thickness S, inside width between gasket edges W, length ($L = L_{p-p}$), and effective projected area ($W \times L$). For the fluid flow and convective heat transfer calculations in the inter-plate flow channels, an equivalent diameter D_e or a hydraulic diameter D_h is used. The equivalent diameter D_e is defined as

$$D_e = \frac{4WH_i}{2(W + H_i)} \approx 2H_i \tag{2.1}$$

and the expression for hydraulic diameter D_h is

$$D_h = \frac{4 \times \text{minimum free flow area}}{\text{Wetted perimeter}} = \frac{2H_i}{\phi} \tag{2.2}$$

where ϕ is the effective heat transfer surface area enlargement factor and is defined as

$$\phi = \frac{\text{Developed area}}{\text{Projected area}} \tag{2.3}$$

It may be noted that $D_\mathrm{h} = (D_\mathrm{e}/\phi)$ and that although both are used in the literature as the length scale for characterizing the Reynolds number, Nusselt number, and friction factor, it is preferred to use equivalent diameter D_e. This has the advantage of scaling the convection performance to that of plain parallel plates and isolating the effects of the area enlargement factor ϕ on the heat transfer enhancement [2, 4, 7].

2.2 Evolution of plate heat exchangers

Since its inception in the 1920s for commercial usage, primarily in the dairy industry, the traditional plate-and-frame heat exchanger has evolved over the last several decades and variant models of PHEs have been developed. Although some of these modifications have been driven by new strategies for making more compact equipment, others have focused on overcoming some of the disadvantages mentioned in the previous chapter and expanding the applications spectrum of PHEs. This evolution has typically manifested either as an altered structure or construction of the PHE or as variations in the plate surfaces corrugation patterns. In the former category during the last several decades, brazed, semi-welded, fully welded, wide-gap, double-wall PHEs, among many others, where the gasketing has been eliminated, have been developed.

2.2.1 Brazed plate heat exchanger

The brazed plate heat exchanger (BPHE), as schematically shown in Fig. 2.10, is essentially made up of a pack of thin corrugated stainless steel plates that are brazed together using copper as a brazing material to form a self-contained unit. Brazing eliminates the need of either frames or gaskets and results in a very compact

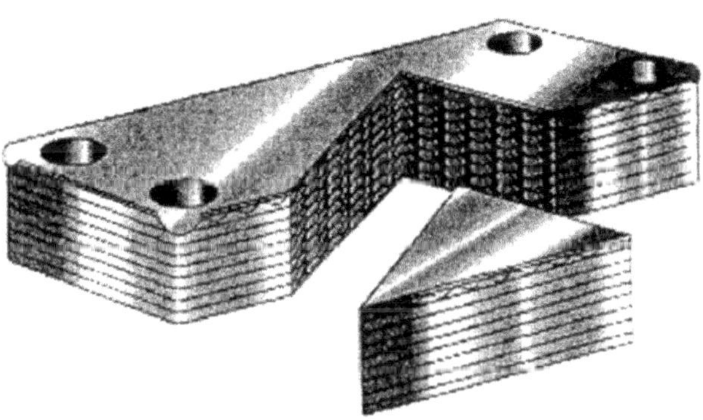

Figure 2.10: The brazed PHE (courtesy of Alfa Laval).

exchanger. Also, instead of copper, where its use presents a compatibility problem with a process stream (e.g. ammonia), nickel or some other brazing material is used.

Because the plates are brazed to each other and there are no frames or gaskets, BHEs can handle higher pressure and temperatures than plate-and-frame heat exchangers, e.g. situations with pressures up to 30 bar and temperatures up to 400°C. They are also characterized by very low weight due to the absence of frames. However, the exchanger length is usually less than 1 m because of the brazing furnace size limitations, where their capacity is restricted to a single BPHE unit. Typical applications include heating and cooling (sensible or with phase change) in the process industry, evaporation and condensation in refrigeration systems, and other HVAC installations.

2.2.2 Semi-welded plate heat exchanger

By welding heat exchanger plates in pairs, to make what are commercially called twin plates, a semi-welded PHE is configured by assembling them in a plate-and-frame pack with gaskets only in the plate channels that handle the alternate fluid stream (Fig. 2.11). This design is especially useful for handling relatively corrosive media, which flows in the welded twin-plate channels. The only gaskets in contact with this medium are two circular porthole gaskets between the welded plate pairs that are typically available in highly resistant elastomer and non-elastomer materials. The channels containing the non-corrosive, non-aggressive, secondary heating or cooling fluid medium flows are sealed using traditional elastomer gaskets.

The semi-welded PHEs can withstand pressures up to 30 bar on the welded twin-plate fluid side, though it should be pointed out that frames are still needed to hold the plate pack. The relatively higher pressure operation extends its applications to include evaporation and condensation in refrigeration and air-conditioning systems, among others.

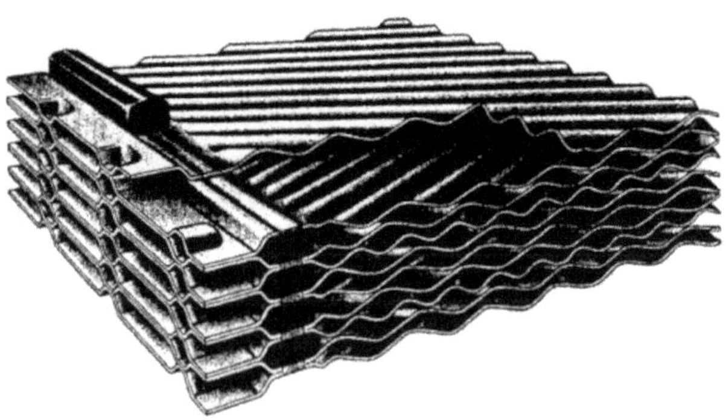

Figure 2.11: The semi-welded PHE (courtesy of Alfa Laval).

2.2.3 Fully welded plate heat exchanger

The fully welded PHE depicted in Fig. 2.12 is a gasket-free version, where a completely welded plate pack is bolted between the two end plates in a conventional frame. By joining the plates at their edges and eliminating the gaskets, the structural integrity of the plate pack is significantly enhanced, and so are the operating temperature and pressure limits of the gasketed PHEs. The laser welds are applied in two spatial dimensions along the edges in the plane of the plates. This allows the plate pack to expand and contract along its length as temperature and pressure changes take place, thereby making the pack more fatigue resistant. Consequently, they are particularly attractive for applications where the heat transfer or thermal processing undergoes rapid changes in temperature and/or pressure. However, fully welded PHEs, unlike gasketed and semi-welded models, lose the flexibility of either expanding or decreasing their surface area by adding or removing plates for meeting varying heat load requirements. Also, they cannot be cleaned readily by mechanical methods, and only chemical cleaning methods can be employed.

The fully welded PHEs are intended for thermal processes with severe duty requirements that often involve handling of highly aggressive or corrosive fluids. They can withstand temperatures up to 350°C and pressures up to 40 bar. Typical applications include exchangers for desuperheating in heat recovery systems, refrigeration interchangers, and heaters of organic chemicals such as solvents, vegetable oil, steam, and batch reactors, among others.

Figure 2.12: Cutaway section of a fully welded PHE (courtesy of Alfa Laval).

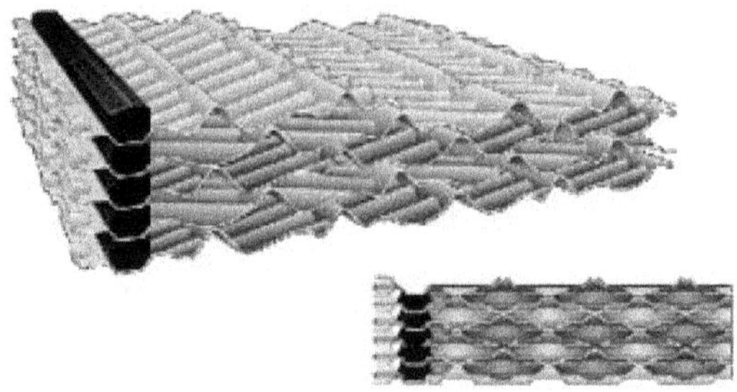

Figure 2.13: Sectional view of inter-plate passages in a wide-gap PHE (courtesy of Alfa Laval).

2.2.4 Wide-gap plate heat exchanger

PHEs with wide-gap plate packs (Fig. 2.13) provide larger free-flow area channels for handling fluids containing fibres or coarse particles and high-viscous fluids, which normally clog or cannot be satisfactorily treated in other types of PHEs and still retain some of enhanced thermal-hydraulic performance characteristics. The plate-surface corrugations and gaskets are designed such that in the inter-plate channels the flow cross-section has a maximum gap of up to 16 mm. The plate corrugations still provide an effective area enlargement and promote swirl flows to effect high heat transfer coefficients; the wider flow gap, however, tends to reduce the pressure drop penalty.

Typical applications include heating of raw, limed, and mixed juice in sugar mills, cooling and bleaching of plant filtrate in pulp and paper mills, and sanitization of fibrous food product slurries.

2.2.5 Double-wall plate heat exchangers

The double-wall PHE is designed for use with either a reacting media or when product contamination between the two fluid streams must be avoided 'fail safe'. Double plates, sealed by conventional gaskets, replace the single plate that normally separates the two fluid media. In the event of the media reacting with and corroding the surface of the double-wall plates, the leakage is directed in the passages between the double plates (Fig. 2.14). This essentially minimizes the possibility of inter-fluid contamination, and the leakage also becomes easily visible on the outside of the heat exchanger. The more common applications of this type of PHE include, among others, heating and cooling of drinking water, pharmaceutical media, lubricating oil, and transformer oil.

2.2.6 Diabon graphite plate heat exchanger

The Diabon graphite PHE employs graphite plates, shown in Fig. 2.15, which were developed for thermal processing of media that is normally too corrosive for plates

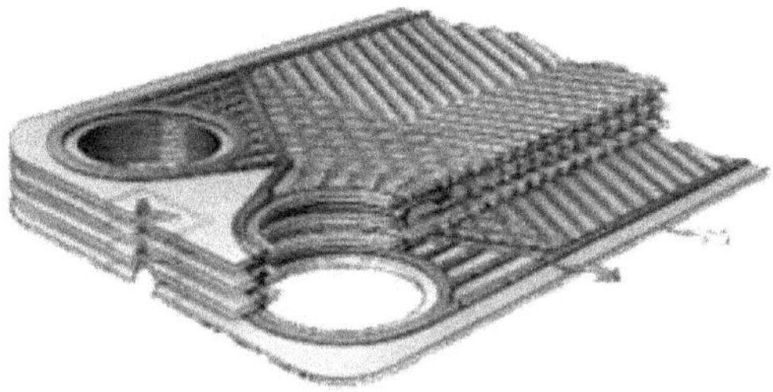

Figure 2.14: Cross-section of a double-wall plate pack (courtesy of Alfa Laval).

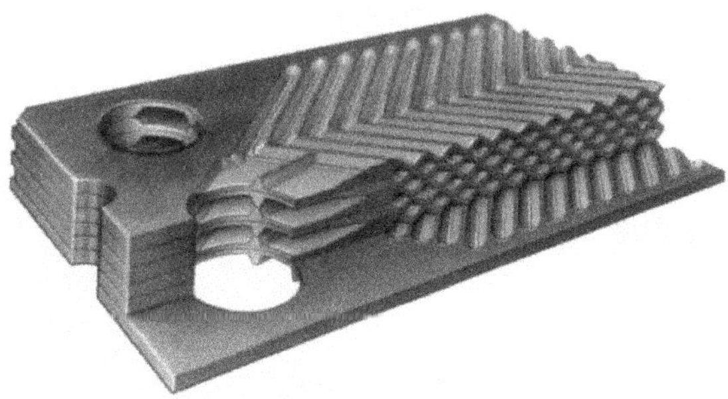

Figure 2.15: Cross-section of the Diabon F 100 graphite PHE (courtesy of Alfa Laval).

made of exotic metals and alloys. The Diabon F 100, or NS1 graphite, is a composite material made up of graphite and fluoroplastics. The material is compressed into the shape of surface corrugated plates, and the formed plates are fitted with thin, flat, corrosion-resistant gaskets. Besides being corrosion resistant and capable of withstanding high temperatures, graphite plates offer good heat transfer characteristics in combination with low thermal expansion and high-pressure operation.

Some of the prevalent applications of PHEs with graphite plates include heating of pickling baths, surface treatment of metals, hydrochloric acid production, and flue-gas waste-heat recovery.

2.2.7 Minex plate heat exchanger

The design of the Minex PHE, shown in Fig. 2.16, is a 'miniaturization' of the conventional plate-and-frame heat exchanger. The end frames and carrying or guide bars are eliminated, and tightening bolts are located with the outline dimensions of the heat transfer plates instead. This configuration makes it possible to combine a compact design with the PHE flexibility options of manual cleaning and ease of surface are expansion or reduction to meet changing application requirements. However, because only tightening bolts are used to keep the gasketed plate pack together, the small and compact Minex PHE is generally used in heat transfer duties where the required capacity is also very small. This type of PHE is intended for

Figure 2.16: The Minex PHE (courtesy of SWEP).

general-purpose, fixed load applications (e.g. sanitary water heating), and constitutes an attractive alternative to larger exchangers.

Several other types and modifications of PHEs are available commercially, and the above selections represent the more prominent sampling. While these new developments have overcome many of the deficiencies of the traditional plate-and-frame heat exchangers, the newer models that discard gaskets lose the advantages of easy cleaning and flexibility of adjusting or altering heat transfer area. Nevertheless, the family of PHEs available has enlarged considerably with these new developments, which provides more competitive functional and cost attributes, and expands their domain in modern industrial applications.

2.3 Operation and selection

The basic operation of a PHE, or majority of its variants, is no different than any other heat exchanger, including the shell-and-tube exchanger. Both the steady state and transient responses are dependent upon the size, hold-up volume, and operating conditions (flow rates and approach temperature difference of the two fluid streams). In principle, a single-pass PHE is a pure counter-flow exchanger, which along with its relative compactness and enhanced convective characteristics allows for a very small approach temperature operation. This is particularly useful in processing fluid media that are temperature sensitive or prone to thermal degradation.

Though they provide a relatively more compact and enhanced performance alternative to shell-and-tube and other traditional heat exchangers, the selection, application, and operation of gasketed PHEs are constrained by the operating temperature

(~160–250°C) and pressure (~25–30 bar) limits [13–15]. Also, gasket retention in its grooves and plate deformation due to high pressures of the fluid streams have to be considered in their selection. There are several other factors that impose operation constraints and should be factored in selecting PHEs, and some of these are enumerated as follows:

1 Complex inter-plate flow channels usually impart high shear rates and shear-sensitive media may thus be prone to degradation.
2 Possibility of flow maldistribution in handling highly viscous fluids or very low flow rates.
3 Plate manufacturing places an upper limit on their size (based on size of presses for stamping them), which in turn restricts applications requiring very high flow rates (typically in the process industry) as the pressure drop becomes excessive.
4 High-pressure drop also makes them unsuitable for air cooling, gas-to-gas heat exchange, and low operating pressure condensation applications.

Of course, regardless of these factors, and as pointed out in Chapter 1, PHEs offer a viable alternative to most types of exchangers. Their compact construction, ease of cleaning (particularly where very high hygienic conditions are to be imposed),

Table 2.1: Typical operating range of gasketed and brazed PHEs.

	Gasketed PHEs	Brazed PHEs
Maximum operating pressure	25 bar (30 bar with special construction)	30 bar
Maximum operating temperature	160°C (200°C with special gaskets)	225°C (minimum ~195°C)
Maximum flow rate	3600 m^3/h	140 m^3/h
Heat transfer coefficient	Up to 7500 $W/m^2 \cdot K$	Up to 7500 $W/m^2 \cdot K$
Heat transfer area	0.1–2200 m^2	0.02–60 m^2
Maximum connection size	450 mm	100 mm
Approach temperature difference	As low as 1°C	
Heat recovery	As high as 93%	
NTU	0.3–6.0	
Pressure drop	Up to 100 kPa per m channel length	
Number of plates	Up to 700	
Port size	Up to 435 mm	
Plate thickness	0.4–1.2 mm	
Plate size	0.3–3.5 m length	
Plate spacing	1.5–5.4 m	
Corrugation depth	1.5–5.4 mm	

flexibility of altering the thermal size, pure counter-current flow operation, and enhanced thermal-hydraulic performance more than compensate for any limitations in most cases.

Because there are many different types of PHEs and numerous manufacturers, it is difficult to precisely summarize the operating limits. The purpose here is to give a general overview; a summary of typical operating ranges for plate-and-frame and brazed heat exchangers is given in Table 2.1. This is based on the product lists from the following manufacturers: Alfa Laval Lund AB [8], SWEP [9], Tranter [10], and APV [11]. This, however, should not be construed as an endorsement of these particular manufacturers, and equipment that has comparable performance, reliability, and economic competitiveness from many other vendors is acknowledged.

Furthermore, based on the information given by one manufacturer (Alfa Laval), a general selection guide for types of PHEs and their specific compatibility with

Table 2.2: A selection guide of PHEs for different operating and service conditions.

	Gasketed	Brazed	Semi-welded	Fully welded	Wide gap	Double wall	Diabon graphite
Operating conditions							
Pressure [bar]	2.5	3.0	2.5	4.0	0.9	2.5	0.6
Temperature [°C]	30–200	−195–225	−30–200	−200–350	30–200	30–200	0–140
Type of service applicability							
Liquid–liquid	E	E	E	E	E	E	E
Gas–liquid	G–P[a]	E	E–P	E–P	G–P	G–P	E–P
Gas–gas	M–P	G–P	G–P	G–P	M–P	M–P	E–M
Condensation	G–P	E	E–M	E–M	E–M	E–M	E–M
Evaporation	G–P	E	E–M	E–M	E–M	E–M	E–M
Nature or type of fluid media							
Corrosive/aggressive	M–P	M–P	E	E	M	M	E
Viscous	E	M–P	E	G–M	E	E	M
Heat sensitive	E	E	E	E	E	E	E
Hostile reaction	M–P	M	G	G	M–P	E	G–M
Fibrous	P	P	P	P	E	P	P
Slurries	M–P	P	M	M	G	M	M
Fouling	G–M	M	M	M	G	M	M
Maintenance flexibility							
Mechanical cleaning	B	N	O	N	B	B	B
Repair	B	N	O	N	B	B	B
Modifications	B	N	O	N	B	B	B

Notes: E – excellent, G – good, M – moderate, P – poor, B – both sides of plate, O – one side of plate, N – no side of plate.

[a]Depending upon the operating pressure and gas/vapour density.

different operating conditions is presented in Table 2.2. Again, this is only a representative listing and not an endorsement.

References

[1] Manglik, R.M. and Muley, A., *Heat Transfer and Pressure Drop Characteristics of Plate-and-Frame Heat Exchangers: A Literature Review*, Report No. TFL-Int-1, University of Cincinnati, Cincinnati, OH, September 1993.
[2] Manglik, R.M., Plate Heat Exchangers for Process Industry Applications: Enhanced Thermal-Hydraulic Characteristics of Chevron Plates, in *Process, Enhanced and Multiphase Heat Transfer*, ed. R.M. Manglik and A.D. Kraus, Begell House, New York, NY, pp. 267–276, 1996.
[3] Clark, D.F., Plate Heat Exchanger Design and Development, *The Chemical Engineer*, no. 285, pp. 275–279, May 1974.
[4] Shah, R.K. and Focke, W.W., Plate Heat Exchangers and Their Design Theory, in *Heat Transfer Equipment Design*, ed. R.K. Shah, E.C. Subbarao, and R.M. Mashelkar, Hemisphere, Washington, DC, pp. 913–932, 1983.
[5] Martin, H.A., Theoretical Approach to Predict the Performance of Chevron-type Plate Heat Exchangers, *Chemical Engineering and Processing*, vol. 35, pp. 301–310, 1996.
[6] Ciofalo, M., Stasiek, J., and Collins, M.W., Investigation of Flow and Heat Transfer in Corrugated Passages – II. Numerical Simulations, *International Journal of Heat and Mass Transfer*, vol. 39, no. 1, pp. 165–192, 1996.
[7] Muley, A., Manglik, R.M., and Metwally, H.M., Enhanced Heat Transfer Characteristics of Viscous Liquid Low Reynolds Number Flows in a Chevron Plate Heat Exchanger, *Journal of Heat Transfer*, vol. 121, no. 4, pp. 1011–1017, 1999.
[8] Alfa Laval, Product Catalogue, Sweden (http://www.alfalaval.com).
[9] SWEP International, Product Catalogue, Sweden (http://www.swep-phe.com).
[10] Tranter Inc., Product Catalogue, USA (http://www.tranterphe.com).
[11] APV, Product Catalogue, UK (http://www.apv.com).
[12] GEA Ahlborn, Product Catalogue, Epsom, Surrey, UK.
[13] Kakaç, S. and Liu, H., *Heat Exchangers: Selection, Rating, and Thermal Design*, 2nd edition, CRC Press, Boca Raton, FL, 2002.
[14] Cooper, A. and Usher, J.D., Plate Heat Exchangers, in *Heat Exchanger Design Handbook*, ed. E.U. Schlünder, Hemisphere, Washington, DC, chapter 3, 1988.
[15] Gupta, J.P., *Fundamentals of Heat Exchanger and Pressure Vessel Technology*, Hemisphere, Washington, DC, 1986.

CHAPTER 3

Industrial applications

As previously indicated, starting from its dairy industry origins for milk pasteurization, the modern day applications of PHEs have expanded considerably to encompass a variety of different industries and processes. Much of this growth can be attributed to energy conservation imperatives in the latter part of the twentieth century. This led to the induction of heat transfer enhancement techniques and development of more compact exchangers [1–3]. PHEs, with their relative compactness and enhanced thermal-hydraulic performance, provide an additional attractive feature of flexibility in altering their thermal size (ease of addition and removal of plates). Their wide-ranging applications today cover refrigeration and air conditioning, energy (power) generation, marine power systems, pulp and paper production, food processing, and refinery and petrochemical systems, among others. In this chapter, a representative selection beginning with the original application for milk pasteurization is described in the ensuing sections.

3.1 Food processing

A general categorization of the food processing industry includes dairy products (e.g. milk, yoghurt, cream, and ice cream), brews and distilled products (e.g. wort, beer, wine, and alcohol), beverages (e.g. juice, carbonated drinks, tea, and coffee), and processed fruits and vegetables (e.g. purees, pastes, sauces, and jams). Here, pasteurization is perhaps one of the most important processes involved. Named after Louis Pasteur of France who discovered that spoilage organisms could be inactivated in wine by applying heat at temperature below its boiling point, it has been effectively applied to milk preservation, and it remains a cornerstone operation in the processing and mass marketing of milk products.

The principle of pasteurization essentially involves heating of the fluid media to a specific temperature for a specified period of time without allowing recontamination during the heat treatment process. The 'control' of this process is critical to both the effectiveness of pasteurization (maintaining the public health aspect) and the quality of the end product (taste, aroma, texture, etc.). The extent of pathogenic micro-organism inactivation depends on a combination of temperature

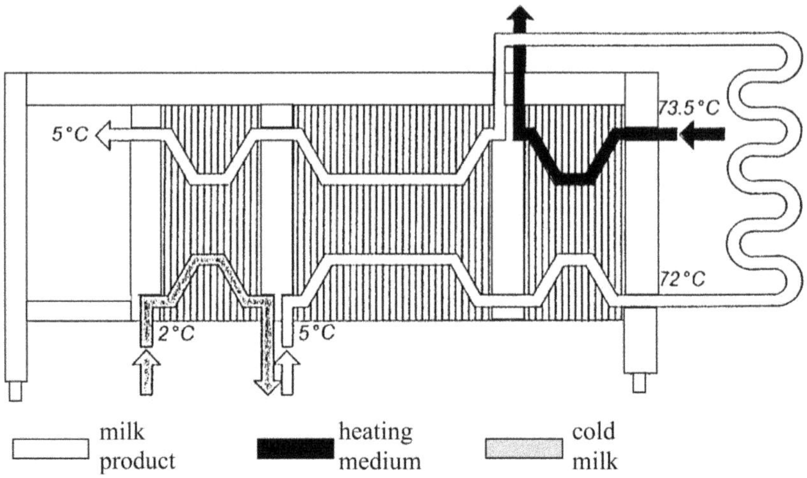

Figure 3.1: Process flow arrangement in a typical multi-pass PHE for milk pasteurization.

and holding time, and they must be highly regulated. For example, milk must be kept at 72°C for not less than 16 s. The continuous pasteurization method is perhaps the most common, where PHEs are widely used. A typical flow sheet is shown in Fig. 3.1, in which a large three-section PHE with dividing frames is used. Cold raw milk at about 5°C is drawn from a constant level tank (not shown in the figure) into the regenerative part of the PHE (middle section). Here it is warmed to $\sim 57 - 68$°C by the heat exchange with hot pasteurized milk flowing in the neighbouring channels. The warm raw milk is then continuously heated to at least 72°C in the right section of the exchanger by either vacuum steam or hot water. The milk, at pasteurization temperature and under pressure, then flows through the holding tube with a residence time of at least 16 s. Thereafter, the properly pasteurized milk flows through the regenerator part (middle section) where it gives up heat to the incoming raw milk and in turn is cooled to ~32–9°C. It finally passes through the cooling part of the exchanger (left section) where it is cooled to 5°C or below by means of chilled water, and is then ready for packaging.

There are several reasons why PHEs have been widely used historically as well as in present times for milk pasteurization. Perhaps the two most prominent factors are: (1) PHEs can be easily opened and thoroughly cleaned, and with plates made of high-grade stainless steel, this ensures maintenance of very stringent hygienic requirements and (2) the high heat transfer coefficients promoted by their corrugated inter-plate channels permit very close approach temperature difference (as low as 1°C) operation. Also, because the three parts can be assembled in one PHE, the overall exchanger unit is very compact thereby providing a significant reduction in installation space requirements as well as a capital cost advantage over other types of heat exchangers.

Pasteurization is also required in a variety of foods and beverage processing, including fruit juice, tomato paste, cream, whey, ice cream, beer, wine, etc., and PHEs are increasingly being used in most systems. In addition, today PHEs find numerous other food processing applications. Typical examples are sensible heating and cooling in pre-treatment and fermentation, drying, blending, freezing, sterilization, as well as phase change (boiling/evaporation) in cooking, forming pastes, crystallization, and polymerization.

3.2 Air-conditioning and refrigeration systems

In the past several decades, the high demands from residential comfort, commercial climate control, and food, biological, and biodegradable material preservation, etc., have greatly increased the general usage of heating, ventilation, air-conditioning, and refrigeration (HVAC&R) systems. Today air conditioning is found in most hotels, office buildings, industrial buildings, hospitals, and sports centres. In fact, their usage has been extended to temperature control in many industrial and manufacturing processes.

As a typical example, Fig. 3.2 gives a general schematic description for a central comfort cooling or building air-conditioning system. The main chiller segment is essentially a vapour-compression refrigeration cycle unit: compressed refrigerant vapour flows through the condenser (heat sink), the exiting liquid refrigerant then expands through the throttle valve, low-pressure liquid refrigerant then flows through the evaporator (heat source), and the outlet vapour is finally fed to the compressor again to complete the cycle. The secondary heat exchange units, both on the condenser and evaporator sides, usually operate with cold water or water–glycol solutions as working fluids.

A critical element in the effective performance of large chiller units is the removal of the significantly large condenser heat transfer rate. One of the following three basic condenser cooling alternatives are utilized to achieve this: an air-cooled system, water-cooled, closed-circuit cooling tower system, and a water-cooled, open

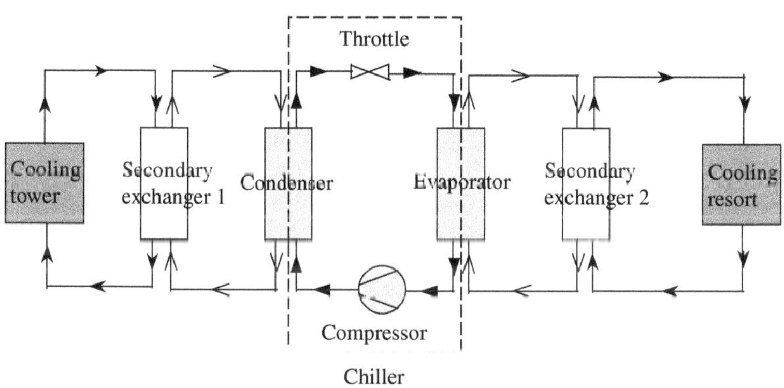

Figure 3.2: Flow schematic of a typical building air-conditioning system.

cooling tower system with 'open' cooling sources, e.g. sea or river water. Air coolers and closed-circuit cooling towers are large and expensive. Open cooling towers cost less but require a continuous water supply and are prone to fouling and shut downs when clean water is not available. This then often necessitates the use of closed-circuit coolers or air coolers. However, incorporating an indirect system with a secondary heat exchanger circuit as a 'divider' or 'filter' between the two systems provides a viable and attractive alternative. This eliminates many of the problems in direct cooling systems. PHEs are often the best choice for the secondary exchangers, and the numerous advantages that secondary cooling systems provide include:

1 low system cost, with a relatively short pay back period on the capital;
2 material savings in the condenser as less expensive materials can be used;
3 less maintenance;
4 with an intermediate heat exchanger, chillers as well as cooling towers can be run at optimal temperatures;
5 an intermediate heat exchanger minimizes the use of water treatment;
6 isolates the main vapour-compression chiller unit from raw water source contamination;
7 intermediate heat recovery reduces the environmental thermal pollution.

This secondary heat exchange scheme can also be incorporated on the evaporator side, as shown in Fig. 3.2. Again, the use of a PHE to separate the two cold-side circuits provides for a high thermal efficiency scheme with rather close approach temperature difference (as low as $\sim$0.5°C) operation. In addition, the secondary circuits protect from high pressures, minimize use of expensive and thermally inefficient energy storage systems, minimize risk of leakage damage, and lend to better control and better coolant flow plumbing in the building.

Of late, PHEs (particularly brazed or welded types) are increasingly being employed as a condenser or evaporator directly in chiller units, as they have been found to provide a viable high-performance alternative to the traditional tube-fin type heat exchangers [4–7].

3.3 Service heating and cogeneration

Heating, in most cases, is needed to provide a comfortable indoor environment in residential, commercial, office, and public facility buildings. However, heating needs are now increasingly required for a variety of other services that include, among others, tap water, swimming pools, and greenhouses. All heating applications are generally divided into two categories, service heating and district heating, based on several major differences. Service heating systems normally have the heat source inside and it is supplied only to a single building. The heat source is normally not only a boiler, but it can also come from heat pumps and solar panels. The normal operating temperature is below 100°C and the pressure is 6 bar or below for these systems. In contrast, district heating systems distribute hot water or steam to multiple buildings (and, in some cases, to an entire township). A variety of heat sources can be employed, including geothermal, cogeneration plants, and waste heat

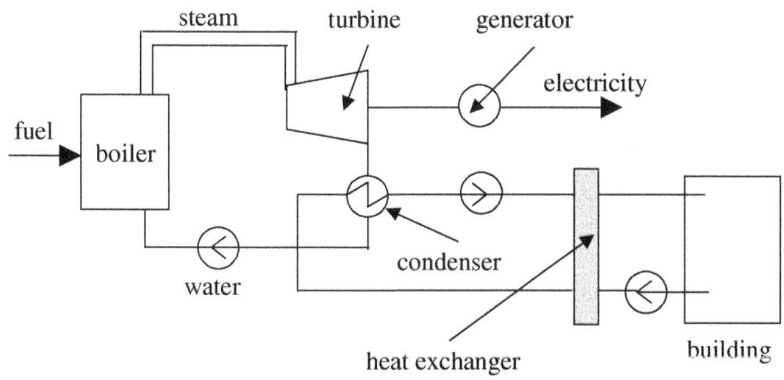

Figure 3.3: Process schematic of a typical combined heat and power or cogeneration plant.

recovery from industrial systems. Their normal operating temperature and pressure are, respectively, 100–150°C and 16 bar. With the variety of applications and system configurations, heat exchangers play an important role and PHEs offer attractive alternatives to traditional exchangers in terms of performance, reliability, and safety.

Consider the combined heat (for district heating) and power plant or a cogeneration system as an example, for which Fig. 3.3 gives a schematic description of its fundamental thermodynamic. This system essentially produces two end products: heat and electricity, and, as a result, it utilizes over 80% of the energy content of the fuel. In such a plant, the district heating water supply normally varies between 75°C and 135°C (depending upon the season of the year and climate conditions) and the returning water temperature normally varies between 40°C and 70°C. In some alternative configurations, low-pressure bleed steam from the turbine is also sometimes used for heating needs instead of the condenser cooling water exhaust. This steam may be used directly or routed through a secondary heat exchange system to produce hot water. In both types of systems, PHEs can serve both as the condenser and as the secondary heat exchanger. Their close approach temperature difference operation makes the system more energy efficient, and this economic incentive is further supplemented by the much smaller space needed for PHEs as compared to shell-and-tube heat exchangers [4–7].

3.4 Offshore gas and oil applications

Irrespective of the location of offshore oil/gas production platform and the type of well fluid to be processed, heating or cooling requirements are an essential element of the facility. This need for heat transfer equipment is dependent upon many factors such as the presence of gas compression or crude oil treatment on the platform. A critical set of requirements in their selection is that they should not only withstand corrosive media environments, but also be compact and lightweight. PHEs not only meet these operational constraints, but also provide the added flexibility of altering their thermal size in order to meet varying heat load requirements.

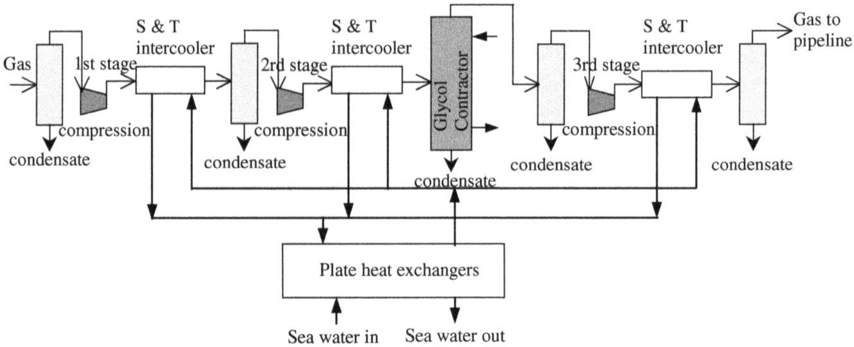

Figure 3.4: Closed-loop cooling scheme in a typical offshore gas compression plant.

A typical application for PHEs in offshore production platforms is in gas compression, which is necessary because of volumetric flow rate demands before it can be pumped through pipelines. The compression usually consists of several stages and a large amount of heat is generated during this process. In order to dissipate the generated heat, the use of a closed-loop cooling system that employs PHEs with titanium plates and seawater as the cooling medium has become increasingly popular in offshore platform installations. Figure 3.4 gives a process line diagram for a three-stage compression train in which the gas intercoolers are cooled by tempered water in shell-and-tube heat exchangers; the cooling water in turn is cooled by seawater in PHEs. This cooling system thus addresses the combined mechanical and metallurgical problem for heat exchangers in such an application by confining the high-pressure gas to the shell-and-tube exchanger and the corrosive seawater to the PHEs. Furthermore, in such a system, the loss of temperature differential because of a closed-circuit operation is effectively minimized by the high efficiency and small approach temperature difference operation of the PHE.

In offshore applications, PHEs are also used in the crude oil dehydration or desalting process, where the heat exchanger duties include crude emulsion-treated crude waste-heat recovery or 'interchange', crude emulsion heating, final cooling of the treated crude oil, and feed water-produced water heat recovery or interchange. Additionally, PHEs are also suitable for a wide range of small auxiliary applications on offshore platforms, which span various duties such as lube oil cooling, hydraulic oil cooling, freshwater cooling with seawater in conjunction with drilling, air compression, diesel engines, pumps, and HVAC systems, among others [4–7].

3.5 Marine applications

Heat exchangers play an important role in the overall safety, performance, and economy of any ship, whether it is a luxury yacht or a ferry (large or small), or a naval ship, or a tugboat, or the largest super tanker. In such sea-borne transportation systems, there are a large number of cooling and heating needs that require heat

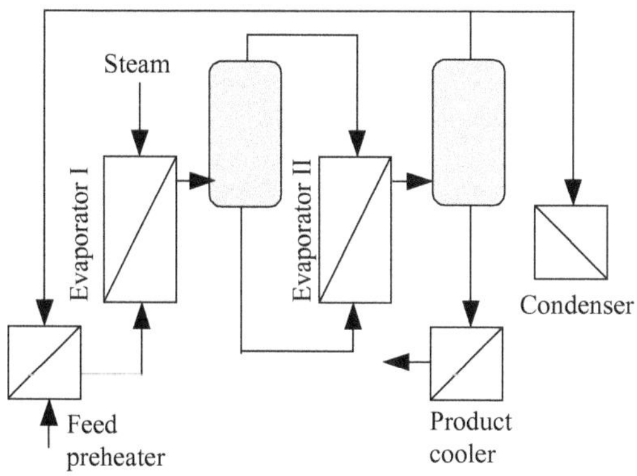

Figure 3.5: Schematic of an evaporation unit in a chemical processing plant.

exchangers. For example, 4-stroke engines are widely used in a variety of ships, ranging from small vessels to large ones with electric or diesel hydraulic systems, where engine cooling and gearbox lubrication oil cooling are routinely required. Some other examples include pre-heaters for heavy fuel oil (which tends to be dirtier and even contains some water condensate), seawater coolers for the engine cooling system, and evaporators and condensers in the central freshwater system, among many others.

Because of their relative compactness, significantly small space requirements, and high thermal-hydraulic efficiency, PHEs are found to be very attractive alternatives for conventional shell-and-tube heat exchangers in maritime applications. Furthermore, in order to reduce the risk from corrosive seawater usage for cooling needs, PHEs with their small approach temperature difference operation provide viable secondary closed-loop set ups; this, in principle, is similar to those employed in offshore oil and gas platform applications described in the previous section.

3.6 Chemical processing

Chemical processing generically encompasses many different types of industrial applications in petrochemicals, pharmaceuticals, oil or gas production and delivery, and manufacture of organic and inorganic chemicals, to name a few. In all of these applications, heat transfer plays a centrally important role and has a significant impact on the process efficiency, product delivery and reliability, and plant economics. Consequently, the selection of suitably efficient heat exchangers is very critical.

To illustrate the role of heat exchangers, and PHEs in particular, four typical thermal processes are described in this section, namely, evaporation, absorption and stripping, distillation, and reactor temperature control. These processes are representative of the chemical industry applications in general, and clearly highlight the advantages of selecting PHEs.

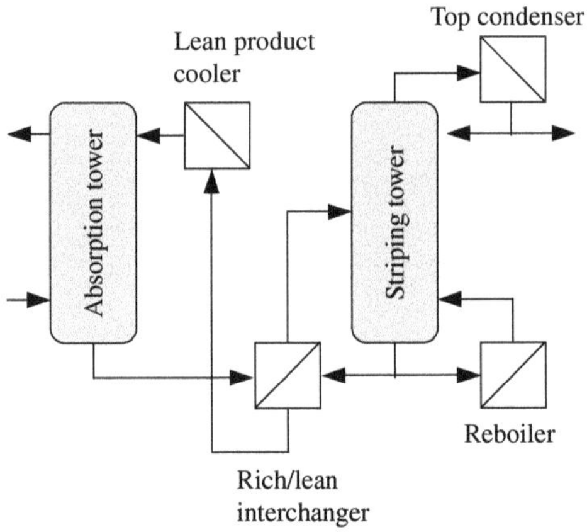

Figure 3.6: An absorption and stripping unit.

A typical flow sheet for an *evaporation unit* is depicted in Fig. 3.5. Although most evaporation systems have traditionally been based on shell-and-tube heat exchangers, PHEs have begun to be increasingly adopted as a better alternative in the last few decades. There are several reasons for selecting PHE evaporators. It is much less bulky than traditional tubular evaporators, which with a height of 10–12 m can be replaced by a plate evaporator of less than one-quarter the size. This results in substantially lower costs for foundations and piping. The smaller size also reduces hold-up volume and residence time, and the evaporation system can, therefore, be started and shut down quickly as well as attain a stable state in a relatively short time period. These factors help improve product quality and are especially important for products that are sensitive to heat. Their space-saving size and their flexibility of altering the thermal size – simply by adding or removing plates – also makes them quite effective as boosters, which work in parallel with existing evaporation systems to increase production. A further benefit of the plate evaporator's design is its relatively higher heat transfer efficiency. While tubular evaporators normally operate with a 5–10°C temperature difference between the two fluid media, plate evaporators can be effective with just 3–5°C temperature difference. This alloys the use of successive evaporators in series to achieve higher concentrations of the end product and at the same time use less steam and conserve energy. In addition to the savings in operating costs, plate evaporators also tend to have a lower capital cost than the tubular designs for the same duty – especially if exotic materials are used to handle corrosive fluids.

Plate evaporators today can be found in sugar factories, distilleries, paper mills, alkali and alumina plants, pharmaceutical production, and the manufacture of many inorganic salts and organic chemicals such as ethylene glycol, detergents, personal hygiene products, and glycerol.

Absorption and stripping, for which a typical process flow diagram is given in Fig. 3.6, are principally a way of recovering volatile organic compounds. While the economic incentive for this is to minimize the consumption of solvent, it is also increasingly a legal requirement to meet the newer more stringent environmental impact regulations. The efficiency of heat transfer is critical to the efficiency of the operation of the entire absorption and stripping system. The process requires lean absorption medium to be cooled before being fed to the top of the absorption tower and the rich absorption medium pre-heated before passing into the stripper. The system efficiency improvement essentially requires maximization of the heat recovery and minimization of the external heating and cooling utilities. Close approach temperature difference operation facilitates this and PHEs clearly have an advantage over traditional shell-and-tube heat exchangers. Moreover, with counter-current flows in PHEs temperature cross can be obviated, which is generally a problem in conventional shell-and-tube units. Aggressive media can also be handled conveniently but with some loss of flexibility in semi-welded and fully welded PHEs made of appropriate materials.

Figure 3.7 describes a typical *distillation* unit. This system consists of a condenser and a reboiler placed, respectively, at the top and bottom of the distillation column. For the condenser duty, PHEs are very attractive because they are small and light, and are thus easy to be mounted on top of a distillation column, saving space and reducing installation costs. Top mounting also entails reflux by gravity, where no pumps are required and the pressure drop is low because of the short flow length so that it is possible for vacuum condensation. Similar to the condenser, PHEs are suitable as reboilers mounted at the bottom as well, taking up minimum space and requiring minimum piping. The use of PHEs as condenser and reboiler further allows easier control of the distillation process, with more flexible and easier maintenance.

In modern chemical plants, the *reactor temperature control* process (shown in Fig. 3.8) can pose some serious problems, such as slow reaction times, poor product quality, waste of raw material, extensive energy consumption, slow change-over between different batch processes, etc. The net effect then is higher production

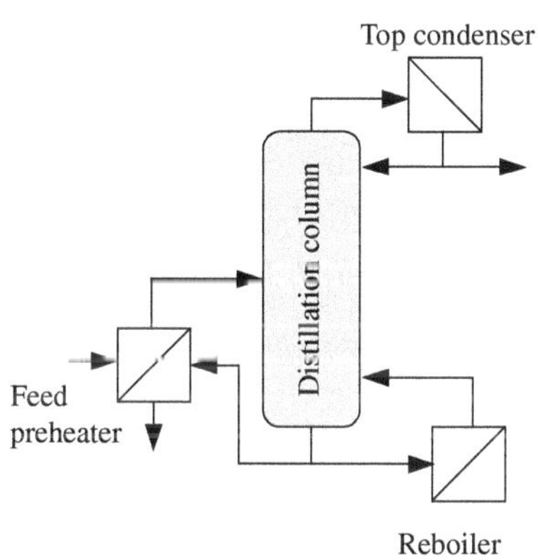

Figure 3.7: Process diagram of a distillation system.

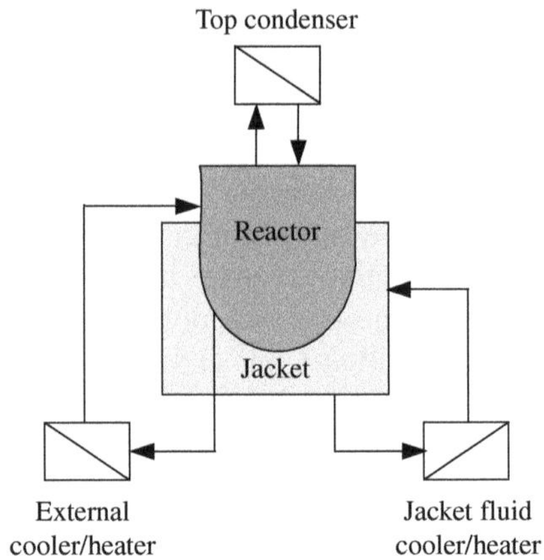

Figure 3.8: A reaction temperature control system.

costs and reduced market competitiveness of the output product. Efficient heat transfer can, however, reliably mitigate these problems. Jacketed heating and cooling depicted in Fig. 3.8 is the most common way to control reactor temperatures in both batch and continuous reactions. This is not very efficient because the heat transfer area is relatively small. A better way to implement jacket heating and cooling is to give each reactor vessel its own circuit for the heating and cooling fluids, and its own heat exchangers within that circuit. The relatively compact and space-saving PHEs are ideal for this situation. Furthermore, with the frequent heating and cooling temperature cycling in batch systems, conventional heat exchangers tend to fail and especially when large temperature swings are involved. PHEs are again a good alternative in such cases as they can consistently withstand thermal expansion and contraction with a predicted lifetime

3.7 Pulp and paper industry applications

As pulp and paper mills adopt closed water cycle systems, it has become more and more important for plant operators to focus on energy and water management. Inducting PHEs in such applications can provide viable solutions as well as meet new environmental requirements. Other factors that must be considered when selecting heat exchangers for a pulp and paper mill are the presence of fibres in one or both streams, high thermal efficiency or close approach temperature difference operation, space limitations, and ease of disassembly for regular scheduled maintenance. Standard PHEs readily meet all of these constraints, except for handling fibrous media. For those applications with fibres in one stream, the single-side wide-gap PHE (see Fig. 2.4) can be used; a double-sided wide-gap plate pack shown in Fig. 3.9 is used in cases where both fluid streams have fibrous content.

Typical applications involving PHEs in pulp and paper mills include cooling of bleaching plate filtrate by heating of process water (waste-heat recovery), heating

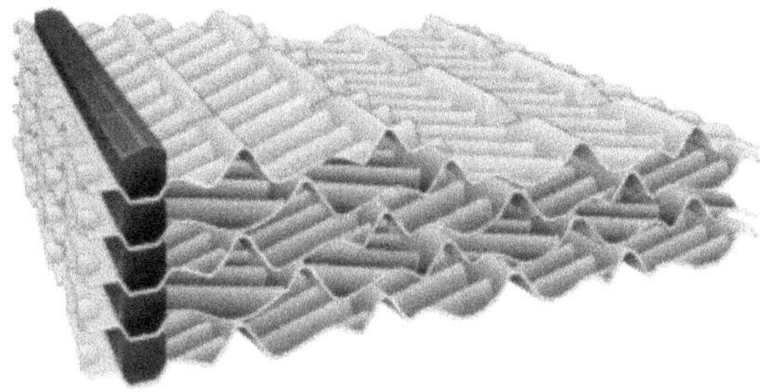

Figure 3.9: Double-sided wide-gap PHE plate pack (courtesy of Alfa Laval).

of intermediate black liquor by cooling of evaporation condensate, heating of wire pit white water by condensation of steam, and heating of fouled condensate before stripping by cooling of clean condensate after stripping, among others.

3.8 Solar energy applications

Solar energy is a clean and renewable energy source, which since its vogue in the 1970s post-mid-east oil embargo crisis has received considerable attention in recent times for its attractive viability in sustainable development [8–10]. One of the early solar heating configurations was the single-circuit system, where an internal heat exchanger was placed in the storage tank. The main disadvantage of this system is insufficient heat energy extraction that basically limits its usage for generating hot tap water only. An improved system, however, has been developed by including an external heat exchanger as schematically shown in Fig. 3.10. During sunlight hours, the photovoltaic (PV) module produces power, which runs a small circulating pump. Antifreeze is pumped through and heated in the solar collectors and is then fed to a reservoir in the heat exchanger module. Water coils in the reservoir absorb the heat from the solar fluid thereby heating the domestic water supply, which flows through these heat exchanger coils by natural thermo-siphon action: as the water is heated, it rises and returns into the top of the tank, drawing cold water from the bottom of the tank into the heat exchanger.

A comparison of the new system (Fig. 3.10) with the old single-circuit system reveals several advantages. First, the system performance is considerably enhanced. External heat exchangers can be configured so that the potable water circulates by natural convection or thermo-siphon action that allows excellent temperature stratification to be achieved in the storage tank. With the hot water remaining in the top of the tank, its usable supply then becomes readily and more rapidly available. Second, the thermodynamic efficiency of the system is improved with the external heat exchanger configuration. The rate of heat transfer is directly proportional to the temperature difference between the potable water being heated and the antifreeze from the solar collectors. With the external heat exchanger configuration, the storage tank coil is always surrounded by cold water, which results in higher temperature differences and improved thermal efficiency. Third, the overall cost of the new

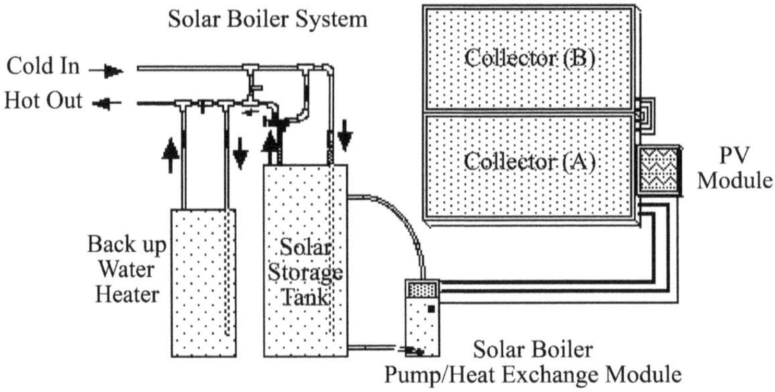

Figure 3.10: A solar energy system (courtesy of Solar-Works Inc.).

system is reduced owing to the long lifetime of the external heat exchanger in comparison with that of the solar tank; the external heat exchanger is salvageable in the event the solar tank develops a leak and needs replacement.

Obviously for such a system, external heat exchangers must be able to operate under very close approach temperature difference conditions because of the introduction of the extra secondary heat-exchange circuit. PHEs, with their additional advantages of relative compactness, easy maintenance, and low cost, are thus a natural and attractive choice for such usage.

3.9 Closing remarks

In closing, it must be pointed out that the applications involving PHEs are too numerous for all to be listed in this book. The examples briefly outlined in this chapter are quintessentially representative and serve as an introduction to readers. Many other industries, such as pharmaceuticals, electronics and electrical equipment, steel and metal industry, motor vehicle industry also use PHEs. However, in some of these cases special and customized exchangers may be needed.

References

[1] Bergles, A.E., Techniques to Enhance Heat Transfer, in *Handbook of Heat Transfer*, ed. W.M. Rohsenow, J.P. Hartnett, and Y.I. Cho, 3rd edition, McGraw-Hill, New York, chapter 11, 1998.

[2] Manglik, R.M., Heat Transfer Enhancement, in *Heat Transfer Handbook*, ed. A. Bejan and A.D. Kraus, Wiley, New York, chapter 14, 2003.

[3] Manglik, R.M. and Muley, A., *Heat Transfer and Pressure Drop Characteristics of Plate-and-Frame Heat Exchangers: A Literature Review*, Report No. TFL-Int-1, University of Cincinnati, Cincinnati, OH, September 1993.

[4] Alfa Laval Lund AB, Product Catalogue, Lund, Sweden.

[5] SWEP International, Product Catalogue, Landskrona, Sweden.
[6] Tranter, Product Catalogue, USA.
[7] APV, Prodcut Catalogue, UK.
[8] Saha, P.C., Sustainable Energy Development: A Challenge for Asia and the Pacific Region in the 21st Century, *Energy Policy*, vol. 31, no. 11, pp. 1051–1059, 2003.
[9] Darmstadter, J., Landsberg, H.H., Morton, H.C., and Coda, M.J., *Energy, Today and Tomorrow: Living with Uncertainty*, Prentice-Hall, Englewood Cliffs, NJ, 1983.
[10] Long, R.E., *Energy and Conservation*, H.W. Wilson, New York, NY, 1989.

CHAPTER 4

Materials and manufacturing

As is the case with most different types of heat exchangers, the material selection for PHEs and their manufacture are quite distinct. These are further dependent upon specific application needs such as heat duty, type of fluids handled, and mode of heat transfer (single-phase flows, boiling, and condensation). The three primary elements of a PHE that essentially govern its manufacture are the corrugated plates, gaskets, and end (pressure and frame) plates or the movable and fixed end-cover plates, of which the production of plates with different surface patterns is perhaps most critical and goes to the core of a PHEs thermal-hydraulic performance.

4.1 Plate material

The corrugated plates used in PHEs can be manufactured in essentially any metal or alloy that can be cold formed or welded. Stainless steel, titanium, nickel, aluminium, incoloy, hastelloy, tantalum, and monel are some examples. The cold stamping process used to press-form the surface patterns obviates the use of brittle materials such as high chromium steel and zirconium alloys, among others. The selection of plate material is primarily determined by fluid compatibility and heat duty; other factors such as thermal design optimization and manufacturers' preferences may also have a role. While special applications may require special plate material [1], they can be generally divided into the following four groups [1–4]:

1 Stainless steels, which include alloys such as 304, 316, 316 Ti, 254SMO, 904L, 317, 317LN, 6-XN, etc.
2 Nickel alloys, which include C-276, C-22, C-2000, G-30, D-205, 59, 31, 28, 3033, 825, 686, 625, 400, Nickel 200/201, etc.
3 Titanium and titanium alloys, which include the grades ASTM Gr1, Gr11, etc.
4 Other metal/metal alloys and non-metals, such as graphite, copper and copper alloys, tantalum, aluminium, etc.

Because the plates are rather thin, they are usually made of higher alloy grades so as to reduce corrosion and erosion. For example, while tubes made of 316 stainless steel may be adequate for handling acidic media in shell-and-tube exchangers, alloys containing 25 Ni, 20 Cr, 4 Mo, 2 Cu, or perhaps incoloy 825 would be needed for the plate material [1, 5]. Different manufacturers use different plate materials based on their specific design, application, and cost preferences. Brief descriptions of these features for some commonly used plate materials are given below:

AISI 304 This is perhaps the most inexpensive grade of austenitic stainless steels, and it offers good general corrosion resistance to a range of organic and non-organic fluid media, but exhibits poor resistance to sulfuric and hydrochloric acids. It does, however, develop crevice corrosion when chlorides are present in water.

AISI 316 A higher grade of stainless steel, which contains 2.5% molybdenum that makes it relatively more resistant to chloride-induced crevice corrosion. It can also be used with low concentrations (10–15%) sulfuric acid, though this depends upon the operating temperature.

Alloy 254SMO This stainless steel has a higher molybdenum content (6%) as compared to that in AISI 316, and thus has better resistance to chloride-induced pitting/crevice corrosion. It is commonly used in weak acid applications with higher chloride content, but should not be used for handling sea or brackish water (typically found in marine applications).

Alloy C-276 It is an expensive nickel alloy that is virtually resistant to chloride ions at low pH levels. It has good corrosion-resistant reliability with a variety of acids and has typically been used to handle sulfuric, hydrochloric, and wet phosphoric acids for a wide range of concentrations and temperature.

Titanium This is an excellent material for applications with chloride solutions and sea or brackish water. It can be typically used for sea water flows of up to 100–120°C. Though it is seldom the most economical choice, only the purest grade is used for most PHEs.

Palladium-stabilized titanium With the addition of very expensive palladium (0.15%), this titanium alloy has considerably higher corrosion resistance and is an attractive choice for many different applications that require this feature. Perhaps the most common use for palladium-stabilized titanium is for chloride solutions and sea water above 100–120°C, where commercially pure titanium tends to develop crevice corrosion.

A very general, and somewhat simplistic, selection guide based on fluid-plate material compatibility experience [3, 5, 6], is listed in Table 4.1. Also, for the thermal-hydraulic design of a PHE, the thermal resistance of the plate wall is an important consideration, and nominal thermal conductivity values for the more common plate materials are given in Table 4.2.

Table 4.1: Plate material selection guide for typical fluid media [3, 5].

Material	Fluid
Stainless steel	Water, cooling tower water, dilute chloride solutions (<200 ppm), copper sulfate solutions, food products, pharmaceutical media, brews, etc.
Nickel	Caustic (50–70%) solutions
Incoloy	Hydrogen gas/water vapour with mercury carryovers, and acids (≤70°C)
Hastelloy	Sulfuric and nitric acids
Titanium	Sea or brackish water, dilute acids (≤70°C), chloride solutions (>200 ppm), and chlorinated brines
Titanium–palladium alloy	Dilute nitric and sulfuric acids (10% concentration and ≤70°C)

Table 4.2: Thermal conductivity of some commonly used plate materials [7, 8].

Material	Thermal conductivity k [W/m · K] (at 300 K)
Stainless steel (304)	14.9
Stainless steel (316)	13.4
Alloy 254 SMO	13.0
Alloy C-276	10.6
Nickel	90.0
Nickel alloy	11.7
Hastelloy	10.6
Titanium	21.9
Graphite	5.7

4.2 Gasket material

The performance (safety and lifetime leakage-proof reliability) of plate-and-frame heat exchangers is highly dependent on the gaskets used and their material reliability [9]. Gaskets are typically one-piece moulded elastomers, and their specific material characteristics are selected on the basis of fluid compatibility, operating temperature, and pressure conditions. They essentially function as 'O' rings that are seated in the peripheral grooves on the corrugated plate surface, as shown in Fig. 4.1, and prevent intra- and extra-plate fluid leakage. The grooves are often tapered to ensure that a larger portion of the gasket surface area is in contact with the plate as

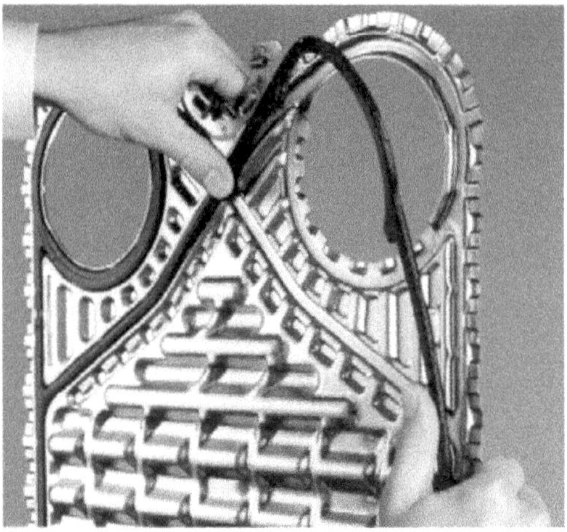

Figure 4.1: A typical gasket setting on a plate surface (courtesy of GEA Process Systems Ltd.).

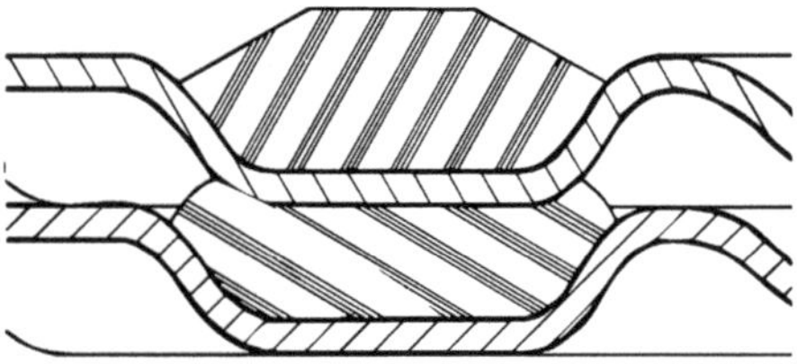

Figure 4.2: Tapered gasket grooves for larger plate-to-gasket surface area contact (courtesy of Tranter Inc.).

Figure 4.3: A section of a typical clip-on gasket and its assembly in plate surface groove (courtesy of Alfa Laval).

it gets compressed between plates (Fig. 4.2). Traditionally, gaskets have been glued onto the plate surface in their press-formed grooves. With newer advancements in elastomer materials and manufacturing capabilities, numerous PHE manufacturers have begun to supply glueless or adhesive-free and clip-on gaskets. A typical section of clip-on type is shown in Fig. 4.3.

Besides the operating conditions (type of liquid, concentration of additives, temperature, pressure, etc.), the compression characteristics to resist the stresses and forces that the flow of media generates are also critical. As such, gaskets are

Table 4.3: Listing of typical gasket materials, their limiting operating temperature, and applications [10].

Material	Maximum operating temperature [°C]	Typical applications
Rubber	70	Oxygenated solvents, acids, alcohols
Neoprene	70	Alcohols, alkalies, acids, hydrocarbon solvents
Styrene butadiene (SBR)	85	Aqueous solutions, alkalies, acids, oxygenated solvents
Nitrile	135	Dairy products, beverages, pharmaceutical and biochemical media, gasoline, oils, alkalies, organic solvents
Fluoro-elastomer	150	Oils
Butyl	155	Alkalies, acids, oils, aldehydes, ketones, phenols, ester
Ethylene propylene rubber (EPDM)	155	Alcohols, sodium hypochlorite, wide range of chemical media
Fluorinated rubber	180	Oils, gasoline, aqueous solutions, organic solvents
Silicone	180	Corrosive liquids

generally made from a variety of elastic and formable materials, such as rubber and its different polymerized forms. Each has limiting temperature threshold, and a nominal application and selection guide for some common materials is given in Table 4.3. Of the many rubber polymer types, one commonly used material is ethylene-propylene (EPDM), which is a hydrocarbon elastomer and is not oil-resistant. Another common polymer type has substituted hydrocarbons in which a hydrogen atom is replaced by a cyano group ($-C{\equiv}N$) to produce nitrile (NBR), or fluorine atoms giving fluoro-rubber (FKM, for which one trademark material is Viton from Dupont). In addition to polymers, rubber materials are composed of a number of other raw materials, such as vulcanization chemicals, fillers, and anti-degradation additives that facilitate handling and manufacturing. Thus, the selection and manufacture of gaskets is quite complex, which requires specialized expertise. In the current industry usage, the most common gasket materials are NBR, EPDM, and FPM, and brief descriptions of their properties are given below.

1 **NBR** is a general purpose rubber that is compatible with aqueous and fatty media environments, and it exhibits good chemical resistance to mineral oils,

vegetable oils, and fats. The operating temperature threshold of the sulfur-cured variety of this rubber is limited to approximately 110°C, while peroxide-cured NBR can be used up to 140°C. Standard grades of NBR (NBRS, NBRP, and NBRHTF) are formulated in accordance with the Food and Drug Administration (FDA) requirements, and are thus suitable for food and pharmaceuticals processing as well. A special hydrogenated nitrile rubber (HNBR) can be used up to approximately 160°C, which is mainly used in crude oil handling.

2 **EPDM** is another general purpose rubber, which finds a wide range of aqueous and chemical applications that include processing of many organic solvents, certain concentrations of acids, and strong alkalis. EPDM exhibits excellent resistance to steam and hot water and can be used up to 160°C. However, it is not suitable for most oils and fatty media applications. A special grade of EPDM (EPDMF) is also formulated in accordance with FDA requirements for usage under sanitary conditions.

3 **FKM** is a family of elastomers that offers good performance in handling acid, steam, and aqueous solutions. Although one grade (FKMT) can be used up to 180°C in certain applications, many types of FKM are unsuitable for use above 100°C. The standard grade (FKMG) material is resistant to water/steam up to a maximum temperature of 130°C. Another special grade (FKMS) with higher fluorine content is used for concentrated high temperature sulfuric acid. A combination of titanium plates and FKM gaskets can, however, cause corrosion, especially at elevated temperatures.

Generally, the cost of gasket material is directly proportional to its temperature, pressure, and corrosion resistance threshold; the higher the resistance of a gasket material, the higher its price. Gaskets typically last for at least one year if they operate at their maximum wet temperature mentioned above. Their lifetime is, of course, prolonged if the operating temperature of the application is lower than the stated maximum for the material. Normally the lifetime is increased by a factor of 2 for each 10°C temperature decrease below the material's upper operating limit. Some manufacturers have produced peroxide-cured gaskets instead of the standard sulfur-cured varieties, which tend to have improved compression resistance and high temperature capability.

4.3 Manufacturing

The primary elements in manufacturing plate-and-frame heat exchangers are the production of plates and end frames. Much of the plate production is focused on press-tool development and precision plate presses, and typical examples are shown in Fig. 4.4. The fabrication of frames generally requires that they have rugged durability and strength, with lower weight; heavier and bulkier frames, apart from their weight and cost, pose considerable handling and installation problems. Most suppliers adhere to ASME, AD-Merkblätter, BS 5500, Swedish Pressure Vessel Code, and other such international codes and standards.

(a)

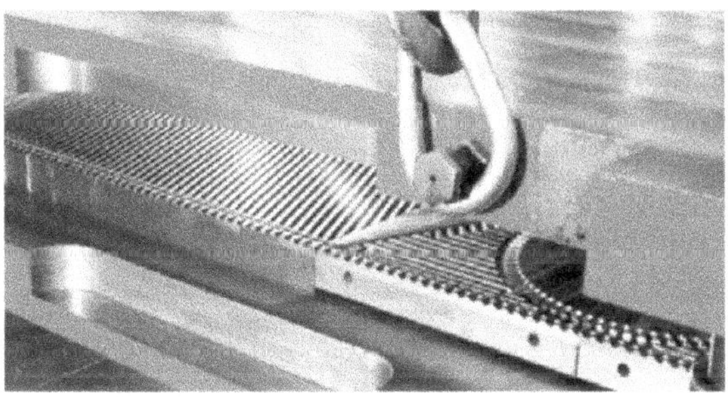

(b)

Figure 4.4: Press forming of corrugated plates: (a) a typical press tool or die with plate-surface corrugation patterns and (b) a computerized press (courtesy of GEA Process Systems Ltd.).

Brief overviews of the material flow, manufacture, and testing process involved in the production of plate-and-frame heat exchangers, brazed heat exchangers, semi-welded PHEs, and fully welded PHEs is given below.

4.3.1 Plate-and-frame heat exchangers

The raw material for the heat transfer plates – stainless steel, titanium, or any other plate material – is inspected before it enters the pressing shop. The coil is mounted on a reel and passes through a continuous cutting and pressing operation in the press line; the pressing essentially involves cold forming on a pressing-tool die with the desired surface corrugations in an NC press of the type illustrated in Fig. 4.4. During this process a code number is also stamped on the plate edge for its identification and specific material certification. Nozzle assemblies are carefully welded to the cast and machined end frames. Likewise, the tightening bolts, carrying and support (or guide) bars, support column, and other end-frame components are machined out of steel stock.

The inspected and certified gaskets are either glued or clipped on to the plates in the peripheral grooves before being placed in a plate pack. The plate-and-frame unit is then assembled in a frame to the customer's requirements (the number of plates, their size, surface geometry, and gaskets being dictated by the designed heat duty specifications). Next the completed unit is hydro-tested to a specified code and any other special requirements before final inspection and packing.

4.3.2 Brazed heat exchangers

After inspection of the raw materials (stainless steel and copper) for the channel plates, coils for each are set up in automated, computer-controlled press lines. The stainless steel and the copper feeds are then cut and pressed together simultaneously. The channel plates are stacked into plate packs containing the specified number of plates and then assembled between cover plates; the latter have the required nozzle connections located on them.

The assembled units are next stacked on specially designed brazing pallets. Individual BHEs are placed very carefully on the pallets, with each unit separated by graphite and specially designed webbing. The webbing material is typically one that reduces the impact on brazing and the consequent stresses due to the differences in thermal expansion of the plate and channel materials. The pallet stacks of BHEs are placed in a furnace where the units are vacuum brazed for about 8 hr. During this process, the brazing material collects at the contact points between the plates due to capillary forces, and thereby facilitates good brazed joints.

After the brazing process, each heat exchanger unit is marked and labelled with an engraved serial number (essential for identification, and material and operating design certification). The units are then tested for pressurized leakage and structural integrity, before being packed for delivery.

4.3.3 Semi-welded plate heat exchangers

The plates are manufactured in the same process as that for plate-and-frame heat exchangers briefly outlined above. Subsequently, twin-plates packs are produced by welding two plates in pairs. This is typically done by means of a laser welding method; it may be noted that most of the aforementioned materials, such as stainless steel 316, titanium, palladium-stabilized titanium, are suitable for laser welding.

Once again, appropriate gaskets are inspected before being glued or clipped on to the surface grooves on each twin-plate assembly. The semi-welded PHEs are then assembled in a frame by compressing a multiple twin-plate stack as per the specified operating heat transfer requirements in a manner similar to that for fully gasketed PHEs. The completed unit is then inspected and hydro-tested to manufacturing and application code requirements before final packing.

4.3.4 Fully welded plate heat exchangers

Unlike the semi-welded PHEs, the plate pack in this case is formed by welding plates together one by one in alternative grooves using the laser-welding process. The plate pack is then assembled between end-frame plates (pressure and frame plates), and compressed by lateral tightening bolts. Nozzle connections are located in the frame covers with linings welded to the plate pack. The completed unit is then hydro-tested to code requirements before final inspection and packing. Again, it should be pointed out that virtually all plate materials that are used in semi-welded PHEs are also suitable for fully welded PHEs.

References

[1] Gupta, J.P., *Fundamentals of Heat Exchanger and Pressure Vessel Technology*, Hemisphere, Washington, DC, 1986.

[2] Novak, L., Private communication, 2002.

[3] Raju, K.S.N. and Jagdish, C.B., Plate Heat Exchangers and Their Performance, in *Low Reynolds Number Flow Heat Exchangers*, ed., S. Kakaç, R.K. Shah, and A.E. Bergles, Hemisphere, Washington, DC, 1983, pp. 899–912.

[4] Saunders, E.A.D., *Heat Exchangers – Selection, Design and Construction*, Wiley, New York, 1988.

[5] Cowen, C.T., Choosing Materials of Construction for Plate Heat Exchangers – I, II, in *Process Heat Exchange*, ed. V. Cavaseno, McGraw-Hill, New York, 1980.

[6] Kakaç, S. and Liu, H., *Heat Exchangers – Selection, Rating, and Thermal Design*, 2nd edn, CRC Press, Boca Raton, FL, 2002.

[7] Touloukian, Y.S. and Ho, C.Y., eds., *Thermophysical Properties of Matter*, Plenum, New York, NY, 1972.

[8] Çengel, Y.A., *Heat Transfer*, 2nd edn, McGraw-Hill, New York, 2003.

[9] Kulesus, G., Select the Right Gasket for Plate Heat Exchangers, *Chemical Engineering*, vol. 99, no. 9, pp. 167–170, 1992.

[10] Manglik, R.M. and Muley, A., *Heat Transfer and Pressure Drop Characteristics of Plate-and-Frame Heat Exchangers: A Literature Review*, Report No. TFL-Int-1, University of Cincinnati, Cincinnati, OH, 1993.

CHAPTER 5

Basic design methods

5.1 Introduction

The thermal-hydraulic design of PHEs essentially follows the same general methodology employed for designing any other type of exchanger, and the overall algorithm is summarized in Fig. 5.1 [1]. As shown, the major design considerations include the following five blocks:

- Process/design or problem specifications
- Thermal and hydraulic design
- Mechanical/structural design, and operation and maintenance constraints
- Manufacturing considerations and cost
- Trade-off factors and system-based optimization.

The specification of process or design problem provides all the necessary information to design and optimize the exchanger for a particular application. This includes type of exchanger construction and materials used, types of fluids and their flow arrangement, heat load, and pressure drop constraints, among others. Heat exchanger thermal-hydraulic design involves the quantitative evaluation of heat transfer and pressure drop or an exchanger sizing and/or rating exercise. The mechanical design includes essential aspects of the mechanical or structural integrity of the exchanger under both steady-state and transient operating conditions, as well as compliance with the relevant local, national, and international codes and standards (e.g. TEMA, ASME Section VIII, API, and other pressure vessel and fatigue codes). Manufacturing evaluations and the requisite cost estimates have to be made then so that appropriate trade-offs can be considered in order to perform a system-based optimization. In this rather intricate heat exchanger design problem, only a part of the total design process consists of quantitative analytic evaluation. The large number of qualitative assessments and trade-offs that are involved, the design of a heat exchanger can be regarded as more of an *art* than a definitive science [1–4].

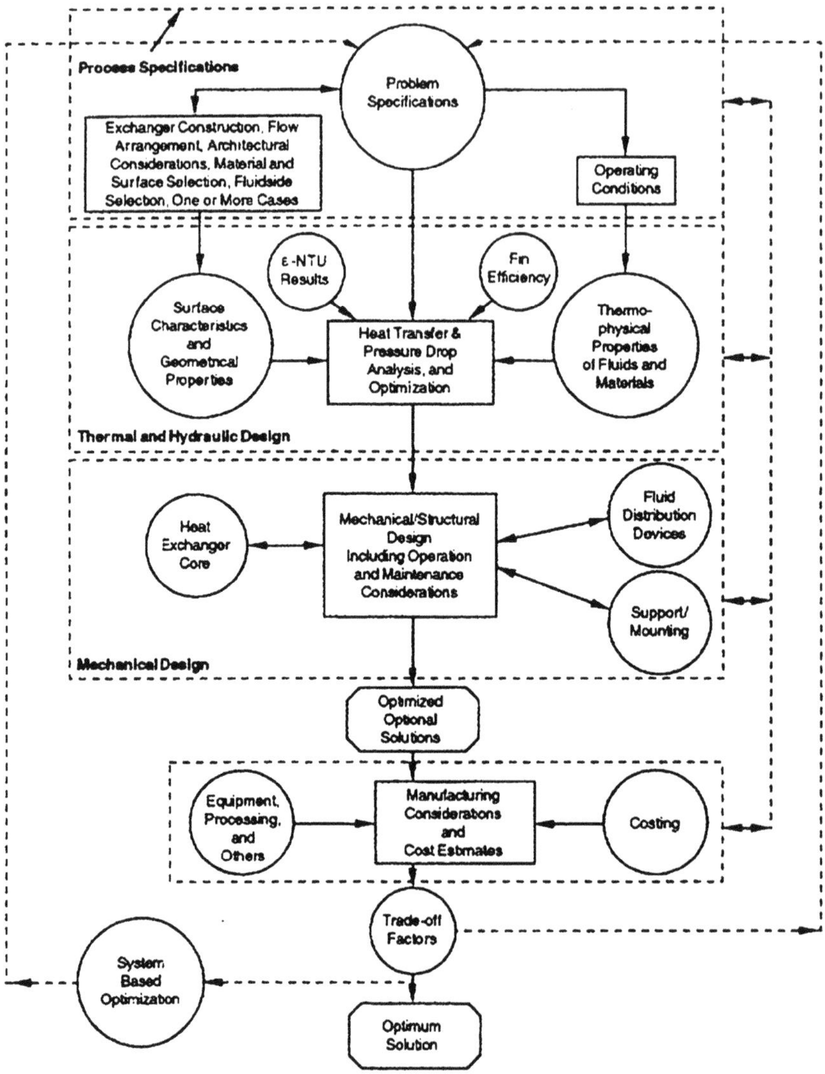

Figure 5.1: Heat exchanger overall design methodology (from Shah [1]).

In this chapter, the general thermal and hydraulic design methodology (step 2 in the list above and in Fig. 5.1) is presented; some constructional and manufacturing issues related to PHEs have been covered in Chapters 2 and 4, respectively. The two primary or fundamental problem specifications in the thermal and hydraulic design of heat exchangers are those of rating and sizing [4]. The rating problem is concerned with the determination of the heat load, fluids outlet temperatures, and pressure drop of each side in an existing PHE, where the following quantities are specified: type of construction and size with detailed dimensions, flow arrangement, thermal and hydraulic characteristics, and operating conditions of the fluid streams (flow rates, inlet temperatures, fouling factors, etc.). The sizing problem, on the other hand, requires the determination of construction type, flow arrangement, needed surface area (or size of exchanger) for a given set of fluid streams and their operating conditions (inlet/outlet temperatures and the flow rates), the specified heat load, and pressure drop constraints [4].

5.2 Basic energy balance and design equations

In all PHEs, the following three different primary flow arrangements for the hot and cold fluid streams are generally encountered: (1) parallel-flow arrangement with two fluid streams flowing in the same direction (Fig. 5.2); (2) counter-flow arrangement with two fluid streams flowing in opposite directions (Fig. 5.3); and (3) multi-pass arrangement where the path of at least one fluid stream is reversed through the flow length two or more times (Fig. 5.4). In this section, only parallel- and counter-flow arrangements are considered; an extended discussion of multi-pass arrangement and the related design issues are given separately in Chapter 6.

The fluid temperatures in PHEs generally vary along their flow path, even in the case of constant thermal resistance because of flow distribution and temperature

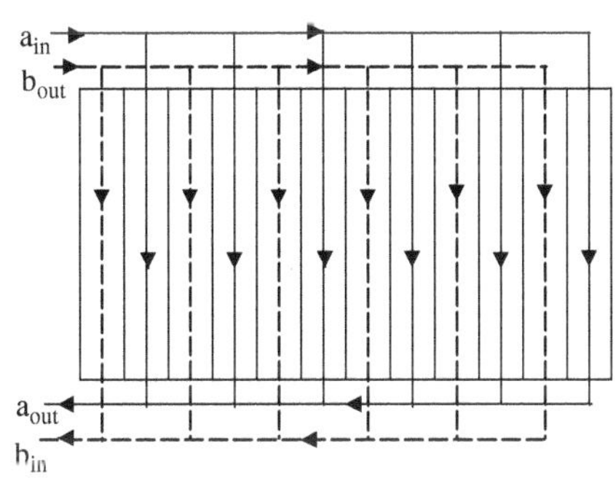

Figure 5.2: Parallel-flow arrangement in a two-fluid PHE.

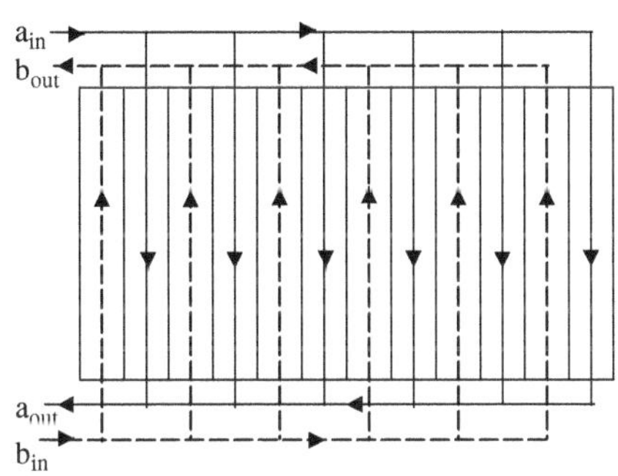

Figure 5.3: Counter-flow arrangement in a two-fluid PHE.

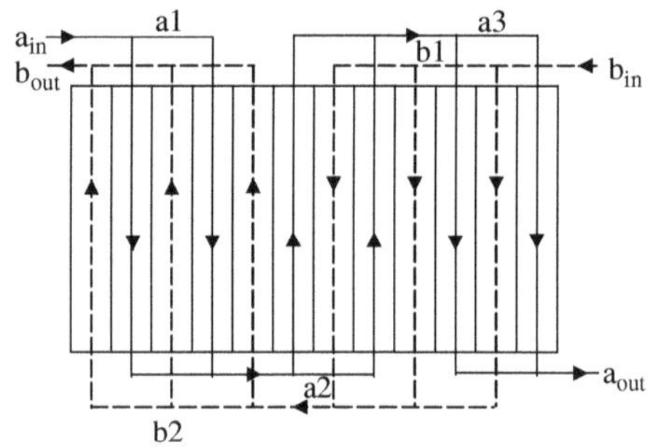

Figure 5.4: A typical multi-pass arrangement in a two-fluid PHE.

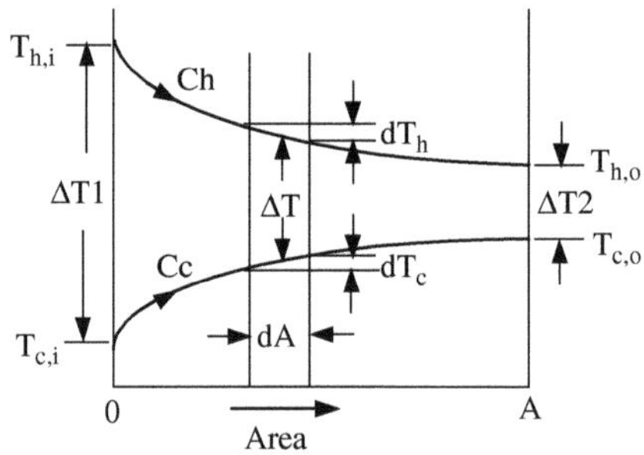

Figure 5.5: Temperature distribution in parallel-flow arrangement.

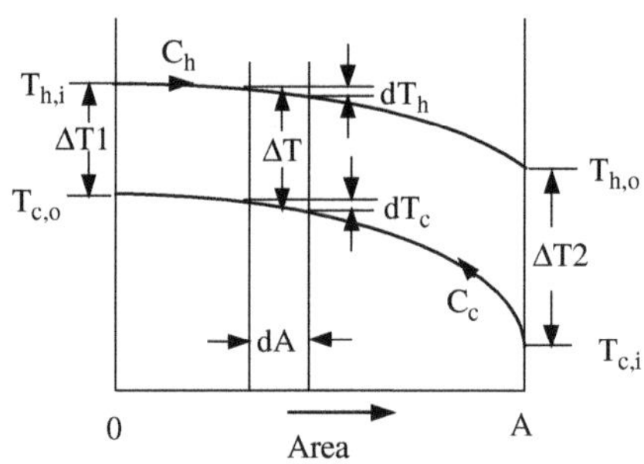

Figure 5.6: Temperature distribution in counter-flow arrangement.

gradient variations across the plates. The temperature variations in the most fundamental parallel- and counter-flow arrangements for single-phase flows of two fluid streams are schematically illustrated in Figs 5.5 and 5.6, respectively. Here the heat transfer surface area A is represented along the x-axis and the fluid-stream temperature along the y-axis. The essential aspect of the parallel-flow arrangement shown in Fig. 5.5 is that the final temperature of the cold fluid stream is always less than the outlet hot fluid stream temperature; in the limiting case of an infinitely large A, the two would be equal. In a counter-flow arrangement, on the other hand, because

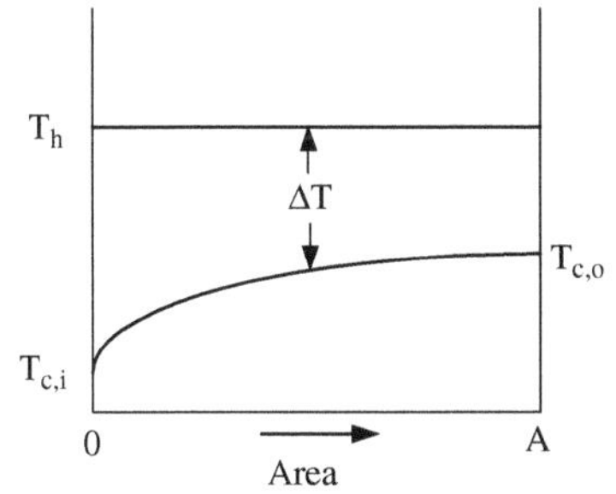

Figure 5.7: Temperature distribution in a condenser.

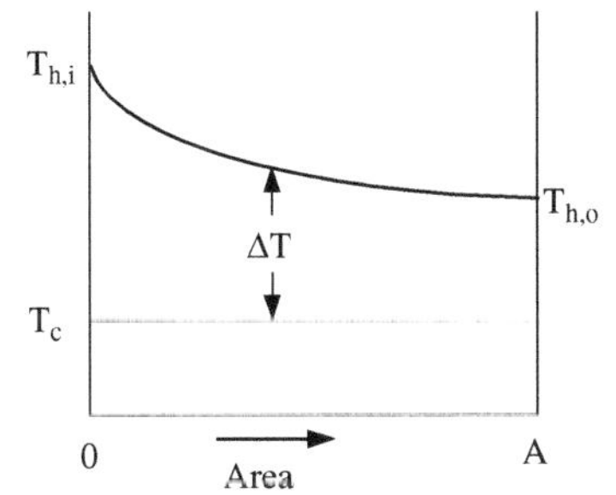

Figure 5.8: Temperature distribution in an evaporator.

of a favourable temperature gradient, the final cold fluid stream temperature may exceed the outlet temperature of the hot fluid stream. This thermodynamic advantage of a counter-flow arrangement lends to a smaller surface area requirement for a given heat load when compared with a parallel-flow arrangement.

When one of the two fluid streams undergoes phase change, the flow arrangement has no effect; the temperature distributions for a condenser and an evaporator, respectively, are illustrated in Figs 5.7 and 5.8. In the two cases, constant temperature vapour condensation or liquid evaporation along with single-phase liquid heating or cooling is depicted. It should be noted, however, that these conditions are based on the assumption that the phase change process occurs at constant pressure (zero pressure drop) and with constant overall heat transfer coefficient. If these assumptions cannot be approximated for a given design, then both the LMTD and ε–NTU methods are inapplicable and a stepwise design method for variable overall heat transfer coefficient must be used.

The application of the first law of thermodynamics over the control volume of any two-fluid heat exchanger provides the following overall energy balance:

$$Q = C_{\rm h}(T_{\rm h,i} - T_{\rm h,o}) \tag{5.1}$$

$$Q = C_{\rm c}(T_{\rm c,o} - T_{\rm c,i}) \tag{5.2}$$

where C is the fluid heat capacity flow rate or the product of its mass flow rate M and specific heat c_p. Given that the temperature difference between the hot and cold fluid streams varies with position in the heat exchanger (Figs 5.5 and 5.6), it can be more conveniently represented by a mean temperature difference ΔT_m such that the total heat load Q is given by

$$Q = UA\Delta T_m \tag{5.3}$$

where U is the overall heat transfer coefficient. Thus, Q and ΔT_m or required heat transfer surface A can be calculated from eqns (5.1) to (5.3) for the rating or sizing problem, respectively, once the overall heat transfer coefficient U is estimated for the two-fluid operating conditions.

The overall heat transfer coefficient in a PHE is a function of the convective heat transfer coefficients or the consequent resistances in the two fluid streams, their fouling resistances, and the thermal resistance due to conduction through plate thickness, as schematically illustrated in Fig. 5.9. Also, because the surface area remains unchanged, unlike the case in a shell-and-tube or tube-fin heat exchanger, the overall thermal resistance can be simply expressed as

$$\frac{1}{U} = \frac{1}{h_h} + \frac{1}{h_c} + \frac{\delta_p}{k_p} + R_{f,h} + R_{f,c} \tag{5.4}$$

In eqn (5.4), h_h and h_c are the heat transfer coefficients of the hot and cold fluid streams, respectively, δ_p is the plate thickness, k_p is the thermal conductivity of the plate material, and $R_{f,h}$ and $R_{f,c}$ are the fouling resistances on the plate surfaces on the hot and cold fluid stream sides, respectively. Fouling, which may manifest itself in a variety of forms (thin oxidation surface layer or a thick crust deposit, or corrosion, etc., all of which tend to be time dependent), is an important aspect in heat exchanger design, particularly for process industry applications, and is

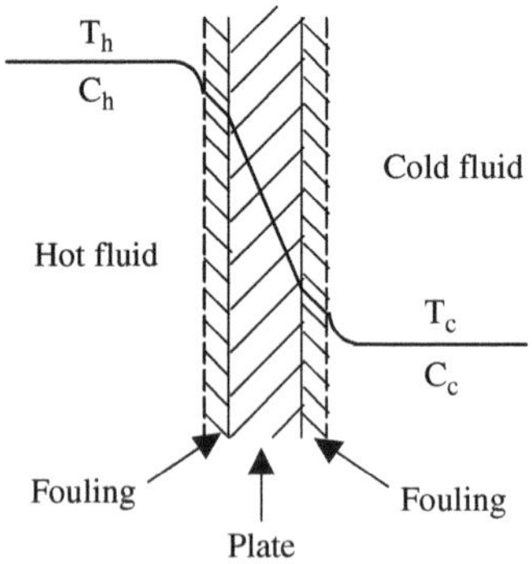

Figure 5.9: Elements of the overall thermal resistance of a single-pass two-fluid plate heat exchanger.

discussed separately in Chapter 9. The thermal resistance of the plate depends upon its material of construction and thickness, and the heat transfer coefficients of the hot and cold fluid streams depend upon the channel characteristics (plate pattern), flow conditions, fluid properties, etc.; these issues are explored at length in Chapters 7 and 8.

Once the overall heat transfer coefficient has been found, the only unresolved variables are the mean temperature difference ΔT_m, which is a function of the hot and cold fluid stream temperatures and their flow arrangement. Several different ways or methods to address this calculation, and the associated rating or sizing problem, have been proposed in the literature, and a few are described in the ensuing section.

5.3 Thermal design methods

To perform a thermal or heat transfer analysis of any heat exchanger, some of the prominent techniques include the LMTD, ε–NTU, and P–NTU methods, among others. Although all are subtle variations of essentially the same methodology, the last two are presented in dimensionless form and extended details of these and some other methods have been summarized by Shah and Mueller [5]. Both the LMTD and ε–NTU methods are applied widely in industrial practice, whereas the P–NTU method is often used for calculating the correction factor F for LMTD. These design methods are inherently based on the following simplifications and idealizations:

1 The heat exchanger operates under steady-state conditions.
2 All physical properties are constant in the PHE. Thus, the influence of temperature-dependent fluid property variations is ignored, and the overall heat transfer coefficient and number of heat transfer units are thereby considered constant along the flow length.
3 There is uniform temperature and fluid flow distribution in each pass of the exchanger, and that each fluid stream is perfectly mixed at any cross-section inside each flow channel.
4 Heat conduction is primarily perpendicular to the fluid flow in the inter-plate channels of the PHE, i.e. the longitudinal heat conduction in both plate and fluid is ignored.
5 Heat loss to the surrounding, external to the exchanger, is neglected; a well-insulated PHE as well as the dead air spaces between the cover and end plates ensures this condition.
6 Either there is no phase change (condensation or boiling) in any fluid stream, or if a fluid undergoes phase change, then it occurs under constant saturation temperature (constant pressure) and constant overall heat transfer coefficient conditions; the effective heat capacity rate for this then yields $C_{max} \rightarrow \infty$.

5.3.1 Logarithmic mean temperature difference method

The heat transfer rate across a differential area element dA along the exchanger length (see Figs 5.5 and 5.6) can be expressed as

$$dQ = U \cdot dA \cdot \Delta T = U \cdot dA \cdot (T_h - T_c) \tag{5.5}$$

This differential equation can then be integrated over the entire heat transfer surface area A along the length of the PHE. Here it may be noted that the overall heat transfer coefficient U is considered to be constant. Also, an energy balance over a differential area dA yields

$$dQ = -C_h dT_h = \pm C_c dT_c \tag{5.6}$$

where C_h and C_c are the respective hot and cold fluid stream heat capacity rates. Thus, for a counter-flow arrangement, we obtain

$$d(T_h - T_c) = dT_h - dT_c = dQ\left(\frac{1}{C_c} - \frac{1}{C_h}\right) \tag{5.7}$$

Substituting eqn (5.5) in eqn (5.7) results in

$$\frac{d(T_h - T_c)}{T_h - T_c} = U\left(\frac{1}{C_c} - \frac{1}{C_h}\right) dA \tag{5.8}$$

which can be integrated over the entire length of the PHE to yield

$$\ln \frac{T_{h,o} - T_{c,i}}{T_{h,i} - T_{c,o}} = \boldsymbol{UA}\left(\frac{1}{C_c} - \frac{1}{C_h}\right) \tag{5.9}$$

or

$$T_{h,o} - T_{c,i} = (T_{h,i} - T_{c,o}) \exp\left(\boldsymbol{UA}\left(\frac{1}{C_c} - \frac{1}{C_h}\right)\right) \tag{5.10}$$

Likewise for the parallel-flow arrangement, the following expression is obtained:

$$T_{h,o} - T_{c,o} = (T_{h,i} - T_{c,i}) \exp\left(\boldsymbol{UA}\left(\frac{1}{C_c} + \frac{1}{C_h}\right)\right) \tag{5.11}$$

Noting that $Q = C_h(T_{h,i} - T_{h,o}) = C_c(T_{c,o} - T_{c,i})$, the total heat load Q of the exchanger can be calculated as

$$Q = \boldsymbol{UA}\frac{(T_{h,i} - T_{c,o}) - (T_{h,o} - T_{c,i})}{\ln (T_{h,i} - T_{c,o}/T_{h,o} - T_{c,i})} = \boldsymbol{UA}\frac{\Delta T_1 - \Delta T_2}{\ln (\Delta T_1/\Delta T_2)} \tag{5.12}$$

where ΔT_1 is the temperature difference between the hot and cold fluid streams at one end, and ΔT_2 the temperature difference at the other end as shown in Figs (5.5) and (5.6).

Comparing eqn (5.12) with eqn (5.3) clearly suggests that the appropriate mean temperature difference between the hot and cold fluid streams over the entire length of the heat exchanger can be expressed as

$$\Delta T_{\mathrm{m}} = \frac{\Delta T_1 - \Delta T_2}{\ln(\Delta T_1/\Delta T_2)} \tag{5.13}$$

This expression is generally referred to as the logarithmic mean temperature difference, or LMTD or ΔT_{LMTD} or ΔT_{lm}. This is also applicable when the temperature of one of the fluid streams is constant or when one fluid undergoes constant pressure/temperature phase change as depicted in Figs 5.7 and 5.8. Furthermore, when $C_{\mathrm{h}} = C_{\mathrm{c}}$, the temperature difference is constant in the counter-flow arrangement, and $\Delta T_{\mathrm{LMTD}} = \Delta T_1 = \Delta T_2$, and is the same as the arithmetic mean temperature difference. Thus, if the temperature difference ΔT_1 is not more than 50% greater than ΔT_2, then the arithmetic mean temperature difference will be within 1% of ΔT_{LMTD} and may be used to simplify calculations. It should be noted that ΔT_{LMTD} represents the maximum temperature potential for heat transfer that can only be obtained in a counter-flow exchanger. Hence the surface area required for a prescribed heat load is smaller for a counter-flow arrangement than that for parallel-flow, cross-flow, or multi-pass arrangements.

It may be noted that the use of ΔT_{LMTD} lends only to an approximate analysis, as in practice, the overall heat transfer coefficient U is generally neither uniform nor constant. It varies considerably across the heat transfer surface, both longitudinally and laterally, and a numerical step-by-step integration of eqn (5.5) may be necessary, which is addressed separately in Section 5.5. In most standard or routine design work, however, a constant value is usually taken as representative of an average or mean overall heat transfer coefficient.

For a multi-pass flow arrangement in a PHE, the mathematical derivation of an expression for the mean temperature difference becomes more complex. The usual procedure is to modify the counter-flow log-mean temperature difference ΔT_{LMTD} by a correction factor, as in the following equation:

$$Q = (UA)F\,\Delta T_{\mathrm{LMTD,C\text{-}F}} \tag{5.14}$$

Here the correction factor F is the ratio of the true mean temperature difference to the logarithmic mean temperature difference for a counter-flow arrangement with the same approach temperature. It represents the degree by which the true mean temperature difference ΔT_{m} deviates from the counter-flow ΔT_{LMTD}, and should not be mistaken for the effectiveness of a PHE. For a given flow arrangement, the value of F depends upon the dimensionless temperature effectiveness P and the ratio of heat capacity rates of the two fluid streams R, and can be functionally represented as

$$F = \phi(P, R, \text{flowarrangement}) \tag{5.15}$$

where P and R are defined as

$$P = [(T_{c,o} - T_{c,i})/(T_{h,i} - T_{c,i})] = (\Delta T_c/\Delta T_{max}) \tag{5.16}$$

$$R = [(T_{h,i} - T_{h,o})/(T_{c,o} - T_{c,i})] = (C_c/C_h) \tag{5.17}$$

The correction factor F is less than 1 for all flow arrangements, except counter-flow and parallel-flow arrangements where it is equal to 1; in the latter case, $\Delta T_{LMTD,P-F}$ for a parallel-flow arrangement has to be evaluated. For PHEs, because of two end plates that do not have a role in the heat transfer process, the number of inter-plate channels can also have a significant influence on the *true* counter- or parallel-flow nature, and an extended discussion of this issue is given in Chapter 6. Furthermore, there are no explicit equations to predict correction factors for various flow arrangements, and the conventional way is to construct tables or graphs for possible arrangements through either analytical or numerical methods. For multi-pass PHEs, this has been worked out by Marriott [6], Bassiouny and Martin [7], Kandlikar and Shah [8–9], and Yang and Wang [10], among others, and detailed results are presented in Chapter 6.

5.3.2 The ε–NTU method

Though the LMTD (or ΔT_{LMTD}) method is widely employed for heat exchanger design, it is best suited for cases where all the terminal temperatures are known; the absence of such information requires a rather cumbersome iteration through logarithmic terms. This is particularly so in heat exchanger rating exercises and off-design operation performance estimation. The ε–NTU method circumvents this entirely and gives a very elegant method using dimensionless parameters that can be used with equal ease in both new design and performance rating problems of heat exchangers, including PHEs.

The heat exchanger effectiveness ε is defined as the ratio of the actual rate of heat transfer in a given heat exchanger to the maximum possible rate of heat exchange, which is only limited by the second law of thermodynamics, or

$$\varepsilon = (Q/Q_{max}) \leq 1 \tag{5.18}$$

where the expression for maximum possible heat load is

$$Q_{max} = C_{min}(T_{h,i} - T_{c,i}) \tag{5.19}$$

and where C_{min} is the smaller of the C_h and C_c. Thermodynamically, the maximum possible heat load represents Q that would be obtained in a counter-flow heat exchanger with infinite surface area, and the effectiveness ε can thus be calculated as follows:

$$\varepsilon = \frac{C_h(T_{h,i} - T_{h,o})}{C_{min}(T_{h,i} - T_{c,i})} = \frac{C_c(T_{c,o} - T_{c,i})}{C_{min}(T_{h,i} - T_{c,i})} \tag{5.20}$$

With this the rate of heat transfer in the heat exchanger can also be expressed as

$$Q = \varepsilon C_{\mathrm{min}}(T_{\mathrm{h,i}} - T_{\mathrm{c,i}}) \tag{5.21}$$

which provides the fundamentally basic relationship between the effectiveness, the smaller heat capacity flow rate, and the difference between the inlet temperatures (or the approach temperature difference) in the ε–NTU method. It replaces eqns (5.12) and/or (5.14) in the LMTD method and does not involve the outlet temperatures (unknowns in performance rating problems).

The thermal design problem in the ε–NTU method can functionally be stated as

$$\varepsilon = \phi(\mathrm{NTU}, R, \text{flow arrangement}) \tag{5.22}$$

Here NTU or the number of heat transfer units is defined as

$$\mathrm{NTU} = (\boldsymbol{UA}/C_{\mathrm{min}}) \tag{5.23}$$

The NTU is a measure of the heat transfer surface area requirements for a given heat duty or the size of the exchanger. The larger the value of NTU, the closer the heat exchanger approaches its thermodynamic limit. Furthermore, R is the ratio of the minimum and maximum heat capacity rates of the two fluid streams given by

$$R = (C_{\mathrm{min}}/C_{\mathrm{max}}) \tag{5.24}$$

In order to establish the exact relationship of eqn (5.23), consider, as an illustrative example, a PHE with *parallel-flow* arrangement. Equation (5.11) can be rewritten as

$$T_{\mathrm{h,o}} - T_{\mathrm{c,i}} = (T_{\mathrm{h,i}} - T_{\mathrm{c,o}})\exp\left[\frac{\boldsymbol{UA}}{C_{\mathrm{min}}}\left(\frac{C_{\mathrm{min}}}{C_{\mathrm{c}}} + \frac{C_{\mathrm{min}}}{C_{\mathrm{h}}}\right)\right] \tag{5.25}$$

Thus, by combining eqns (5.1), (5.2), and (5.20), and eliminating $T_{\mathrm{h,o}}$ and $T_{\mathrm{c,o}}$, the following expression can be obtained for the parallel-flow arrangement:

$$\varepsilon = \frac{1 - \exp[-\{1 + (C_{\mathrm{h}}/C_{\mathrm{c}})\}(UA/C_{\mathrm{h}})]}{(C_{\mathrm{min}}/C_{\mathrm{h}}) + (C_{\mathrm{min}}/C_{\mathrm{c}})} \tag{5.26}$$

In this equation, either of the two conditions $C_{\mathrm{h}} > C_{\mathrm{c}}$ or $C_{\mathrm{h}} < C_{\mathrm{c}}$ can be applied and the resulting expression for effectiveness ε can be written in the form

$$\varepsilon = \frac{1 - \exp[-(1 + C_{\mathrm{min}}/C_{\mathrm{max}})\boldsymbol{UA}/C_{\mathrm{min}}]}{1 + (1 + C_{\mathrm{min}}/C_{\mathrm{max}})} = \frac{1 - \exp[-(1 + R)\mathrm{NTU}]}{1 + R} \tag{5.27}$$

The foregoing derivation illustrates how the effectiveness for a given flow arrangement can be expressed in terms of two dimensionless parameters R and NTU. A similar mathematical development for the counter-flow arrangement yields the following effectiveness:

$$\varepsilon = \frac{1 - \exp[-(1 - R)\mathrm{NTU}]}{1 - R \cdot \exp[-(1 - R)\mathrm{NTU}]} \tag{5.28}$$

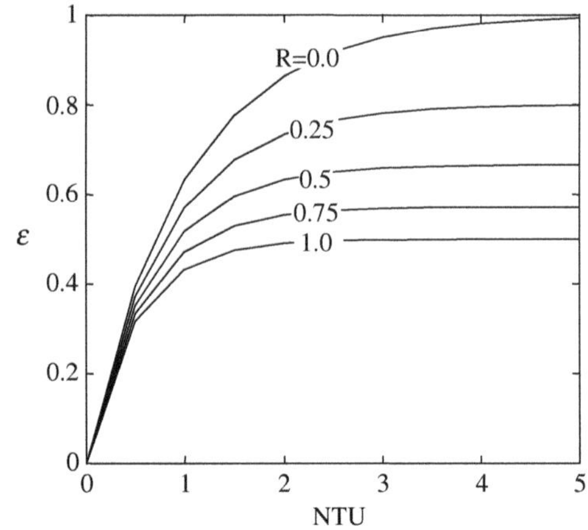

Figure 5.10: Variation in ε with NTU and R in a parallel-flow arrangement.

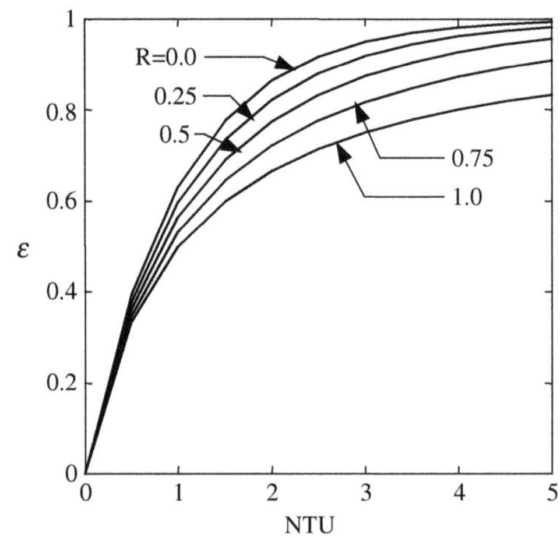

Figure 5.11: Variation in ε with NTU and R in a counter-flow arrangement.

The variations of ε with R and NTU for parallel- and counter-flow arrangements are graphed in Figs 5.10 and 5.11, respectively.

It may be noted here that with C_h or $C_c \to \infty$ in an evaporator or condenser, respectively, $R = 0$, and the effectiveness becomes independent of the flow direction and can be calculated from

$$\varepsilon = 1 - \exp(-\text{NTU}) \tag{5.29}$$

Similar expressions can be derived for multi-pass arrangements, though they are quite complex and the results are often presented in convenient graphs. This is reviewed in some detail in Chapter 6.

5.3.3 The *P*–NTU method

This method is in principle a variation of the ε–NTU method. While the latter distinguishes which flow stream has the smaller heat capacity rate, the former considers a temperature effectiveness *P* as a function of NTU and *R* that are explicitly defined for either the fluid 1 or fluid 2 side [11]. This nomenclature is adopted regardless of which side is the hot or cold stream, and the functional relationship along with the respective parameter definitions described below are based on the fluid 1 side that should be clearly identified in the sketch for each flow arrangement by the practitioner.

$$P_1 = \phi(\text{NTU}_1, R_1, \text{flow arrangement}) \tag{5.30}$$

$$Q = P_1 C_1 |T_{2,i} - T_{1,i}| \tag{5.31}$$

$$P_1 = [(T_{1,o} - T_{1,i})/(T_{2,i} - T_{1,i})] \tag{5.32}$$

$$R_1 = C_1/C_2 = [(T_{2,i} - T_{2,o})/(T_{1,o} - T_{1,i})] \tag{5.33}$$

$$\text{NTU}_1 = (UA/C_1) = (|T_{1,o} - T_{1,i}|/\Delta T_m) \tag{5.34}$$

Furthermore, the inter-relationship between these parameters, respectively, defined for the fluid 1 and fluid 2 sides can be shown as the following:

$$Q = P_1 C_1 |T_{2,i} - T_{1,i}| = P_2 C_2 |T_{1,i} - T_{2,i}| \tag{5.35}$$

$$P_1 = P_2 R_2 \tag{5.36}$$

$$P_2 = P_1 R_1 \tag{5.37}$$

$$\text{NTU}_1 = \text{NTU}_2 R_2 \tag{5.38}$$

$$\text{NTU}_2 = \text{NTU}_1 R_1 \tag{5.39}$$

$$R_1 = 1/R_2 \tag{5.40}$$

In practice, although this method is not often used for designing PHEs, it is usually employed in calculating the correcting factors required in the LMTD method. The correcting factor *F* is often functionally presented in terms of *P* and *R*, and some examples for various flow arrangements for multi-pass PHEs are discussed in Chapter 6.

5.3.4 Sizing and rating procedure

As stated earlier, the thermal analysis and design of heat exchangers primarily entail the sizing and rating problems. To solve these for PHEs, both the LMTD and the ε–NTU methods can be employed, although the latter tends to be more convenient.

In the sizing problem for a two fluid stream PHE, mass flow rates and inlet temperatures are usually given, along with outlet temperature of one fluid stream.

Its thermal analysis algorithm by the LMTD method consists of the following steps:

1 Calculate heat load Q and the unknown outlet temperature from eqns (5.1) and (5.2).
2 Calculate the temperature effectiveness P and the ratio of heat capacity rates R from eqns (5.16) and (5.17), respectively.
3 Specify flow arrangement, including number of passes and number of paths per pass, and then determine correction factor F from available $F - P$ curve (or formula) for the specified flow arrangement.
4 Calculate the hot and cold fluid side heat transfer coefficients based on the respective fluid flow conditions (see Chapters 7 and 8), estimate the fouling factors (if needed; Chapter 9), and then calculate the overall heat transfer coefficient U through eqn (5.4).
5 Determine the logarithmic mean temperature difference ΔT_{m} from eqn (5.13).
6 Finally, calculate the required heat transfer area A from eqn (5.14).

If the ε–NTU method is employed, the procedure has the following steps:

1 Calculate heat load Q and unknown T_{o} from eqns (5.1) and (5.2).
2 Calculate the heat exchanger effectiveness ε from eqn (5.20) and the heat capacity rates ratio R from eqn (5.24).
3 Specify the flow arrangement, including number of passes and number of paths per pass, and determine the number of heat transfer units NTU, the specific ε–NTU equation or a set of performance curves for that arrangement.
4 Calculate h_{h} and h_{c} based on the respective fluid flow conditions (Chapters 7 and 8), estimate the fouling factors (Chapter 9), and then calculate the overall heat transfer coefficient U through eqn (5.4).
5 Finally, calculate the required heat transfer area A from eqn (5.23).

In general, both these methods are effective for sizing PHEs and can be employed with ease.

In the rating problem for a two-fluid stream PHE, the mass flow rates and inlet temperatures of both fluids are known along with the flow arrangement (number of passes and number of paths per pass) and heat transfer area A (as this exercise seeks the thermal performance of an existing heat exchanger). The steps for the LMTD method are as follows:

1 Calculate the heat capacity flow rate ratio R from eqn (5.17).
2 Assume the outlet temperature of one fluid stream and then calculate heat load Q as well as the unknown outlet temperature of the second fluid stream from eqns (5.1) and (5.2).
3 Calculate the temperature effectiveness P from eqn (5.16).
4 Determine correction factor F from the available $F - P$ curve for the given flow arrangement.
5 Calculate h_{h} and h_{c} (Chapters 7 and 8), estimate any required fouling factors, and calculate the overall heat transfer coefficient U by eqn (5.4).

6 Calculate ΔT_{LMTD} as per eqn (5.13) and then Q from eqn (5.14).
7 Calculate the outlet temperatures and compare these values with the assumed and calculated values in step 2.
8 Repeat steps 2–7 until the desired convergence is achieved.

In the case of ε–NTU method, the procedure has the following steps:

1 Calculate R from eqn (5.24).
2 Calculate h_{h} and h_{c} (Chapters 7 and 8), fouling factors, and U through eqn (5.4).
3 Determine the heat exchanger effectiveness ε from the ε–NTU–R chart or correlation for the specified flow arrangement.
4 Calculate the total heat load Q from eqn (5.21).
5 Finally, determine the unknown outlet fluid temperatures from the energy balance, eqns (5.1) and (5.2).

A comparison of the steps outlined above clearly shows the relative simplicity and explicitness of the ε–NTU method, which lends to a very straightforward design and thermal analysis.

5.4 Hydrodynamic design methods

A pressure drop is inevitably incurred in each fluid stream as it passes through the heat exchanger, and its estimation is critical to the overall design and selection of a heat exchanger. In both the rating and sizing problems, limits are generally imposed either by economic (pumping cost) considerations or by process limitations (or indeed both). In both cases, the available or affordable or maximum allowable pressure drop is given in the design specification, and an optimized utilization of Δp is often sought as a means of improving the thermal performance (which, in turn, minimizes the heat exchanger size).

The total pressure drop ΔP_{t} in a PHE consists of several friction and head loss elements and can be expressed as

$$\Delta P_{\mathrm{t}} = \left[\Delta P_{\mathrm{f}} + \Delta P_{\mathrm{g}} + \Delta P_{\mathrm{a}} + \sum \Delta P_{\mathrm{Ni}}\right] \tag{5.41}$$

Here ΔP_{f} is the frictional pressure drop or shear loss, ΔP_{g} is the pressure drop due to gravity or static head, ΔP_{a} is the flow acceleration pressure drop, and $\sum \Delta P_{\mathrm{Ni}}$ is the sum of all other pressure losses due to inlet/outlet flow distribution and includes the pressure drop in ports and manifolds.

The frictional pressure drop for single-phase flow applications [12] is usually calculated by

$$\Delta P_{\mathrm{f}} = [2f(L/D_{\mathrm{e}})(G^2/\rho)] \tag{5.42}$$

where L is the plate flow length between ports, G is the mass velocity or flux, ρ is the fluid density, and f is the Fanning friction factor. The Fanning friction factor is a function of the plate surface corrugation pattern, flow Reynolds number, and fluid properties, which is discussed in detail in Chapter 7. For two-phase (boiling

Table 5.1: Values of C in the Lockhart–Martinelli model.

Liquid–vapour flow condition		C
Turbulent–turbulent	T–T	20
Viscous–turbulent	V–T	12
Turbulent–viscous	T–V	10
Viscous–viscous	V–V	5

or condensing) flow applications, the estimation of ΔP_{f} becomes complicated and there is no unified method in the literature. The most often used method is based on the Lockhart–Martinelli model [13, 14] for isothermal air–water flow in horizontal tubes. The Lockhart–Martinelli parameter X and the two-phase friction multiplier ϕ_{l} for the liquid are expressed as

$$X = \sqrt{\Delta P_{\mathrm{l}}/\Delta P_{\mathrm{v}}} \tag{5.43}$$

$$\phi_{\mathrm{l}} = \sqrt{\Delta P_{\mathrm{f}}/\Delta P_{\mathrm{l}}} \tag{5.44}$$

where ΔP_{l} is the liquid-phase frictional pressure drop, ΔP_{v} is the vapour-phase pressure drop, and ΔP_{f} is the two-phase pressure drop. Both ΔP_{l} and ΔP_{v} are calculated from single-phase equations for the prediction of pressure drop in PHEs. The relationship between ΔP_{l} and ΔP_{v} has been established for different applications [14], and for horizontal tubes, the following correlation is recommended:

$$\phi_{\mathrm{l}}^2 = [1 + (C/X) + (1/X^2)] \tag{5.45}$$

The value of constant C in this expression for tube flows is listed in Table 5.1.

For PHEs, Thonon [15] recommends $C = 8$, whereas Wang [16] suggests a value of 16. Clearly there is no unanimity in the literature on this account, and more data is certainly required for developing more precise two-phase flow frictional pressure drop correlations for PHEs.

The gravitational pressure drop or static head loss for single-phase flows in a vertical channel can be calculated from

$$\Delta P_{\mathrm{g}} = \pm \rho \boldsymbol{g} \boldsymbol{L} \tag{5.46}$$

and for two-phase flows from

$$\Delta P_{\mathrm{g}} = \pm \int_0^L [\alpha \rho_{\mathrm{v}} + (1 - \alpha)\rho_l] g \, \mathrm{d}z \tag{5.47}$$

Here the '+' sign holds for vertical up flow and the '−' sign for vertical down flow. Also, g is the gravitational acceleration and α the two-phase void fraction, which is defined as the fraction of the channel flow cross-sectional area that is occupied by the vapour phase. In a vertical channel, the void fraction can be estimated by [17]

$$\alpha = \left[1 + \frac{1-x}{x}(\rho_v/\rho_l)^{2/3}\right]^{-1} \tag{5.48}$$

where x is the vapour quality (fraction of the total mass flow which is composed of vapour).

The pressure drop due to flow acceleration ΔP_a is usually negligible for single-phase flows, but for two-phase flows it can be estimated from [13]

$$\Delta P_a = G^2 \left\{ \left[\frac{(1-x_o)^2}{\rho_l(1-\alpha_o)} + \frac{x_o^2}{\alpha_o \rho_v} \right] - \left[\frac{(1-x_i)^2}{\rho_l(1-\alpha_i)} + \frac{x_i^2}{\alpha_i \rho_v} \right] \right\} \tag{5.49}$$

where α_i and α_o are the inlet and outlet void fractions, respectively, and x_i and x_o the inlet and outlet vapour qualities.

Finally, the additional flow distribution pressure drop $\sum \Delta P_{Ni}$ in a PHE mainly includes: (1) losses in the inlet/outlet connections or manifold $\Delta P_{con,i}$ and $\Delta P_{con,o}$, and (2) losses in the inlet and outlet ports $\Delta P_{port,i}$ and $\Delta P_{port,o}$. In practice, empirical correlations are often used to estimate these additional pressure drops. One method is based on the dynamic pressure VP that is given by [18]

$$\mathrm{VP} = (\rho U^2/2) \tag{5.50}$$

where U is the mean fluid velocity. Also, in two-phase flows through the inlet and outlet ports and connections, the homogenous flow model can be used to calculate the two-phase density [14] as

$$\rho = [(x/\rho_v) + \{(1-x)/\rho_l\}]^{-1} \tag{5.51}$$

Thus, the pressure drops in the inlet and outlet connections are calculated from

$$\Delta P_{con,i} = X_i(c)_i \cdot \mathrm{VP}_i \tag{5.52}$$

$$\Delta P_{con,o} = X_i(c)_o \cdot \mathrm{VP}_o \tag{5.53}$$

where $X_i(c)$ is a coefficient that accounts for the pressure drop due to either contraction or expansion in the connection, and its value depends on the difference between the connection and port diameters. The pressure drops in the inlet and outlet ports are calculated from [16]

$$\Delta P_{port,i} = -\mathrm{VP}_i[0.64 - 0.8(f4L_{port}/3D)] \tag{5.54}$$

$$\Delta P_{port,i} = -\mathrm{VP}_o[2.16 - 0.8(f4L_{port}/3D)] \tag{5.55}$$

where D is the port diameter, f the Fanning friction factor, and L_{port} the port length. *Alternatively*, the additional pressure drop can be estimated from the empirical correlation that approximates it as follows [19]:

$$\sum \Delta P_{Ni} = 1.5(\mathrm{VP}_i)N_{pass} \tag{5.56}$$

The total pressure drop and its distribution in different segments of a PHE are important for its effective operation, as this drives the maximum flow rate sustainable in the exchanger. With low viscosity liquids, pressure drop through the port is normally the limiting factor for achieving high flow capacities. For viscous liquids and gases, however, the channel pressure drop may be more critical. As a general guideline, port pressure drop should be kept below 50% of the total pressure drop; if it exceeds 66% of the total, then design alterations should be effected [20]. Also, after the thermally sizing of the heat exchanger, the pressure drop must be checked to ensure it is less than the maximum allowable value.

5.5 Variable overall heat transfer coefficient

One of the assumptions in the design method outlined in previous sections is the constant heat transfer coefficient in PHEs. Although this is generally acceptable in most practical applications involving single-phase flows, it is rarely true in applications with phase change where U varies along the plate channel. In the latter case, U (or more precisely, h for the phase change fluid) is a function of mass velocity, fluid temperature, and vapour quality, among other factors. The vapour quality can vary significantly with the phase change process along the plate, and so also the fluid temperature (which is a function of fluid pressure) thereby requiring a simultaneous calculation of pressure drop along with the thermal performance. Obviously, the previous design methods (LMTD and ε–NTU) based on constant overall heat transfer coefficient cannot fulfil this task. A stepwise calculation must be performed instead so as to account for this non-linearity [21, 22], and the case of steam condensation by cooling water flow in a PHE is considered as an example [23] in the ensuing discussion.

To facilitate the stepwise calculation, the plate surface is first divided into a number of small segments along the flow length, where, in each zone, heat transfer and pressure drop are calculated simultaneously. Though the number of segments is determined by the required accuracy, calculations suggest that 100 zones are adequate to obtain outlet temperature convergence to an accuracy of 0.001°C. The next step is to assume a temperature on the cooling water side, and thereafter determine the heat flux and vapour quality from the heat balance. The steam-side pressure drop can then be calculated using the Lockhart–Martinelli model [12], e.g. noting that the steam pressure and its saturation temperature are known. Next, from their respective heat transfer correlations, water- and steam-side h, the corresponding wall temperatures, the overall U, and the heat flux can be determined by an iterative procedure. These iterations may require adjusting the water temperature until the heat balance for the sub-zone is accurate, and the overall algorithm is given in Fig. 5.12. This procedure in essence is for the rating problem and, though computationally somewhat tedious, it can be executed for the sizing problem as well.

In addition, this stepwise analysis is also effective in experimental work, where it can be applied to develop correlations for heat transfer coefficients and pressure drops in PHE flows with phase change. This in general requires iterative 'fitting' of a model to experimental data, so as to match the latter closely. Such routines have

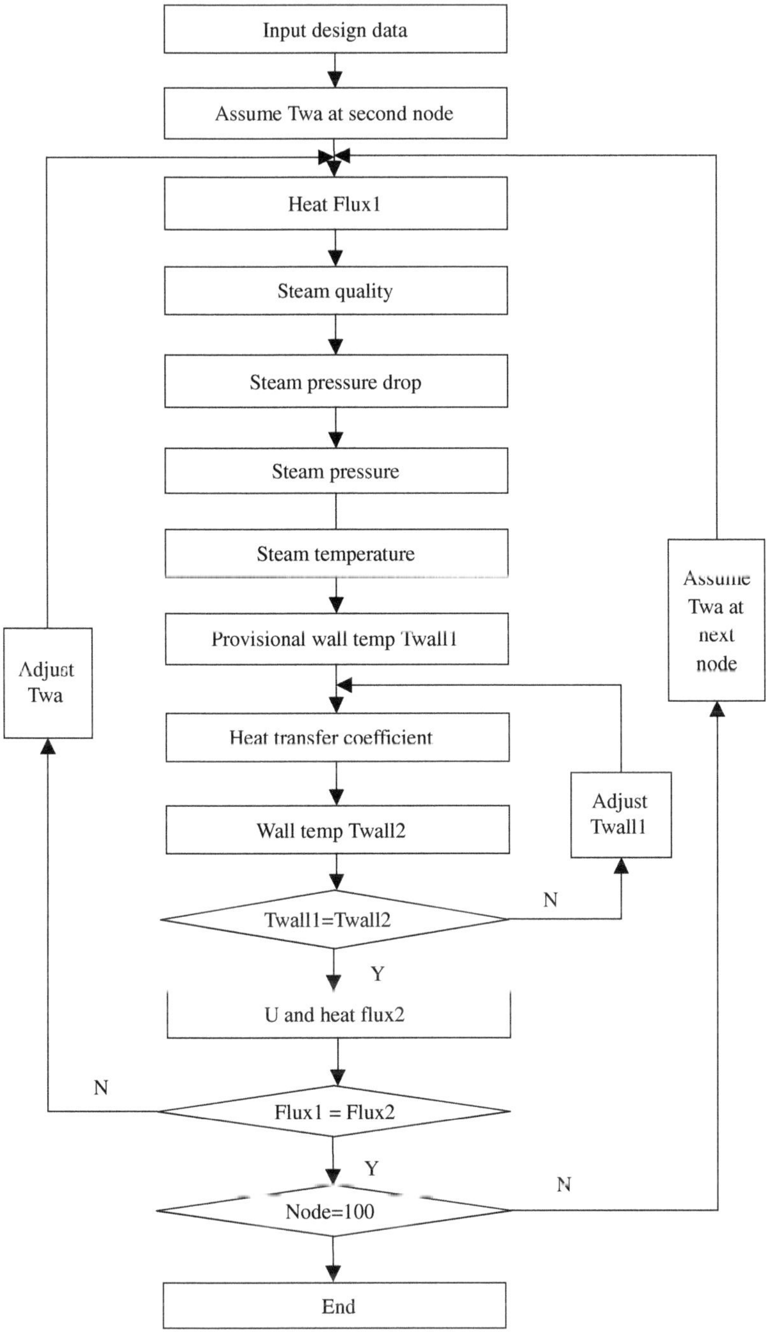

Figure 5.12: Calculation procedure for applications with phase change.

indeed been successfully implemented in the literature [23] for the evaluation of experimental data.

5.6 Thermal mixing

In PHE, design practice, the terms 'long or short plate', 'high theta or low theta plate', and 'hard or soft plate', are often used to characterize their thermal performance. These synonymous terms refer to the NTU that the given plate is capable of sustaining in a PHE. Conventionally, θ represents the plate performance by its definition as the ratio of the temperature change of one fluid stream to the logarithmic mean temperature difference. This parameter can be calculated for both the hot and cold fluid sides as follows:

$$\theta_h = [(T_{h,i} - T_{h,o})/\Delta T_{LMTD}] \tag{5.57}$$

$$\theta_c = [(T_{c,o} - T_{c,i})/\Delta T_{LMTD}] \tag{5.58}$$

For a typical counter-flow arrangement in a PHE, these equations can be restated by applying the basic energy balance expressed in eqns (5.1–5.3) as

$$\theta_h = (\boldsymbol{UA}/C_h) \tag{5.59}$$

$$\theta_c = (\boldsymbol{UA}/C_c) \tag{5.60}$$

Thus, a plate with a high θ value is considered to be a 'thermally long' or 'hard' plate as it lends to comparatively larger heat transfer and pressure drop. This is typically achieved in practice with plates of high chevron angles. Conversely, plates with a low θ value are 'thermally short' or 'soft', as is the case with low chevron angle plates and their comparatively low heat transfer and pressure drop. A PHE with a high θ value is consequently physically long and thin, and that with a low θ value is physically short and fat.

Two types of plates, as shown in Figs 5.13 and 5.14, are generally used in practice so as to obtain three types of flow-channel arrangements. Symmetric arrangement of high θ plate with obtuse-angled chevrons yields the H- or hard-channel arrangement, while the low θ plate with acute-angled chevrons provides the L- or soft-channel arrangement. An intermediate arrangement of M- or mixed channel is obtained with a high and a low θ plate alternating in the PHE plate stack. Also, high and low θ plates can be combined in various other arrangements, e.g. with some parts of the pack consisting of hard channels and other parts having soft channels. The objective in designing with such different arrangements is to achieve a desired thermal mixing.

One problem associated with the PHE design is the precise matching of both the thermal and hydrodynamic loads. It is difficult to accommodate the required thermal duty and at the same time fully utilize the available pressure drop, as the minimum flow cross-section area and the surface area in a PHE are inter-dependent, unlike in other types of compact heat exchangers [19]. This makes the exchanger

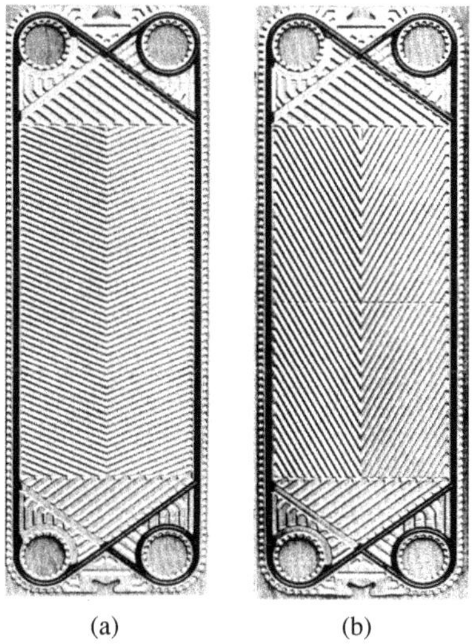

Figure 5.13: Typical (a) high and (b) low θ, or hard and soft, chevron plates.

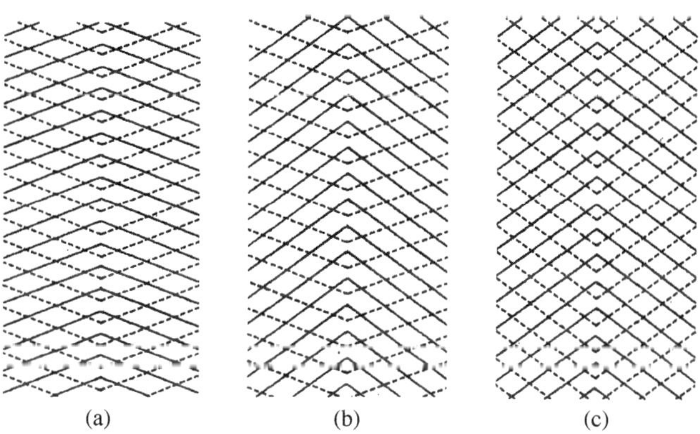

Figure 5.14: Different channel arrangements formed by stacks of identical and mixed high and low θ plates: (a) H Channel, (b) M Channel, and (c) L Channel.

design to be either pressure drop or heat transfer limited. In the pressure drop limited design, the free-flow or channel cross-section area is determined to satisfy the pressure drop limit; the corresponding surface area, however, then tends to be higher than that required to meet the heat duty. In the heat transfer limited design, while the computed surface area meets the heat duty requirement, the corresponding free-flow area tends to be larger than needed for the available pressure drop.

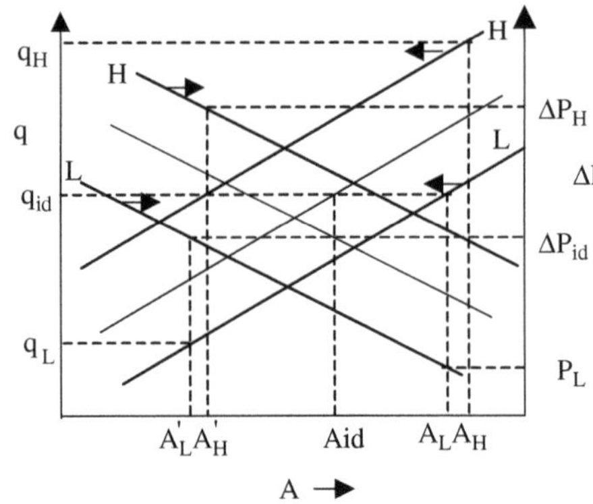

Figure 5.15: Illustration of heat transfer and pressure drop limited design [10].

The thermal mixing design feature can be explained further through Fig. 5.15, and considering two different plates (designated as high and low θ plates) as possible candidates to meet the specified heat duty Q_{id} and pressure drop ΔP_{id} for the fluid stream with the more severe hydrodynamic constraint. If high θ plates are used, the design A_H matches the pressure drop requirement, but utilizes much more surface area which can sustain a higher heat load Q_H. If the heat transfer were to be matched, the design surface area would be A'_H, but the resulting pressure drop ΔP_H would be significantly higher than the available pressure drop ΔP_{id}. The exchanger design with high θ plates is thus seen to be limited by the specified ΔP_{id} constraint and is designated as the pressure drop limited design. If low θ plates are used, then the design surface area A_L matches the heat load but does not completely utilize the available pressure drop ($\Delta P_L < \Delta P_{id}$). When all of the available pressure drop ΔP_{id} is utilized, the design surface area A'_L is significantly lower than that required for the given heat duty ($Q_L < Q_{id}$). In this case the heat exchanger design with low θ theta plates is limited by the surface area A_L required to transfer the heat load Q_{id} and is designated as a heat transfer limited design.

In more closely reviewing Fig. 5.15, it is possible to find an ideal design that can exactly match both the specified heat load Q_{id} and pressure drop limit ΔP_{id} simultaneously. This ideal design would essentially utilize heat transfer area A_{id}, which is smaller than both A_L and A_H and require plates with surface corrugations that are between those of high and low θ plates. Unfortunately, it is not always possible to produce such duty-dependent plates as if entails maintaining costly inventories of a very large number of plate sizes and patterns. One way to overcome this design challenge is to use thermal mixing offered by the non-identical or mixed-channel design [24, 25]. The basic idea behind the non-identical or mixed-channel design is to use parallel arrangements of two different channel types in a manner that accommodates the required overall thermal and hydrodynamic performance. The theoretical basis for the determination of the numbers of each channel type required to achieve this has been outlined by Marriott [25], and the calculations are summarized here. The numbers of channels using low and high θ plates are obtained from

$$N_{\mathrm{L}} = (M/m_{\mathrm{L}})[(\xi - \xi_{\mathrm{H}})/(\xi_{\mathrm{L}} - \xi_{\mathrm{H}})] \tag{5.61}$$

$$N_{\mathrm{H}} = [(M - N_{\mathrm{L}} m_{\mathrm{L}})/m_{\mathrm{H}}] \tag{5.62}$$

In these expressions, M is the total fluid mass flow rate, m_{L} and m_{H} are the mass flow rates per channel in low and high θ channels, respectively, and ξ is a performance parameter that is defined as

$$\xi = [(e^{(1-R)\theta} - 1)/(e^{(1-R)\theta} - R)] \tag{5.63}$$

Here θ is the thermal length as defined earlier, and R is the ratio of heat capacity rate of the process fluid stream under consideration to that of its counterpart fluid stream in the exchanger. Note that the ξ term without a suffix refers to the required overall value whereas that with a suffix refers to the value achieved by the channel type involved.

By using non-identical or mixed channels either the pressure drop or heat transfer limit can be effectively overcome. This enables savings in heat transfer area of up to 25% compared to a conventional PHE of intrinsically equal efficiency. The following example [24] illustrates the utility of designing with such a scheme:

Duty	To cool 400 m^3/hr drier scrubber water		
Temperature change	57–43°C		
Coolant temperature	30°C		
Allowable ΔP	0.15 MPa		
	High θ	Medium θ	Non-identical[a]
No. of plates	211	181	147
ΔP	0.15 MPa	0.08 MPa	0.15 MPa
Remarks	Too much surface area	ΔP not fully utilized	Q met and ΔP fully utilized

Notes:
[a]With 70% high θ and 30% medium θ plates.

Nomenclature

A	heat transfer area, m^2
dA	differential area element, m^2
dQ	differential heat load, W
C	heat capacity rate, W/K
C	constant in Chisholm correlation, eqn (5.45)
D	port diameter, m

D_e	equivalent diameter, m
F	LMTD correction factor
g	gravitational acceleration, m/s^2
G	mass velocity, kg/(m$^2 \cdot$ s)
h	convective heat transfer coefficient, W/(m$^2 \cdot$ K)
k	thermal conductivity, W/(m $\cdot$ K)
L	length, m
m	mass flow rate per channel, kg/s
M	fluid mass flow rate, kg/s;
N	number of channels or passes
NTU	number of heat transfer units
P	temperature effectiveness
Q	heat load, W
R	heat capacity rate ratio
R_f	fouling resistance, m$^2 \cdot$ K/W
U	overall heat transfer coefficient, W/(m$^2 \cdot$ K)
v	velocity, m/s
VP	local dynamic pressure, Pa
x	vapour quality
ΔP	pressure drop, Pa
ΔT	temperature difference, K
ΔT_1	temperature difference at one end, K
ΔT_2	temperature difference at the other end, K
$\sum \Delta P_{Ni}$	distribution pressure drop, Pa

Greek symbols

α	void fraction
δ	plate thickness, m
ε	heat exchanger effectiveness
ϕ	two-phase multiplier
ρ	density, kg/m^3
ξ	performance parameter defined in eqn (5.63)

Subscripts

1	fluid stream 1
2	fluid stream 2
a	acceleration
c	cold fluid stream
con	connection
f	friction
g	gravity
h	hot fluid stream

H	high theta
i	inlet
id	ideal
l	liquid
L	low theta
LMTD	logarithmic mean temperature difference
m	average value
max	maximum
min	minimum
o	outlet
p	plate
pass	pass
port	port
t	total
v	vapour

References

[1] Shah, R.K., Heat Exchanger Design Methodology, *in Heat Transfer Equipment Design*, ed. R.K. Shah, E.C. Subbarao and R.A. Mashelkar, Hemisphere, Washington, pp. 17–22, 1988.

[2] Kakac, S. and Liu, H., *Heat Exchangers: Selection, Rating and Thermal Design*, CRC Press, Boca Raton, FL, 1998.

[3] Bell, K.J., Overall Design Methodology for Shell-and-tube Exchangers, *in Heat Transfer Equipment Design*, ed. R.K. Shah, Subbarao E.C., and Mashelkar R.A., Hemisphere, Washington, pp. 131–144, 1988.

[4] Shah, R.K., Heat Exchanger Basic Design Methods, *in Low Reynolds Number Flow Heat Exchangers*, ed. S. Kakac, R.K. Shah, and A.E. Bergles, Hemisphere, Washington, 1982.

[5] Shah, R.K. and Mueller, A.C., Heat Exchanger Basic Thermal Design Methods, in *Handbook of Heat Transfer Applications*, 2nd edn, ed. W.M. Rohsenow, J.P. Hartnett, and E.N. Ganic, McGraw-Hill, New York, chapter 4, 1985.

[6] Marriott, J., Where and How to Use Plate Heat Exchangers, *Chemical Engineering*, pp. 121–134, April, 1971.

[7] Bassiouny, M.K. and Martin, H., Temperature Distribution in a Four Channel Plate Heat Exchanger, *Heat Transfer Engineering*, vol. 6, no. 2, pp. 58–72, 1985.

[8] Kandlikar, S.G. and Shah, R.K., Asymptotic Effectiveness-NTU Formulas for Multipass Plate Heat Exchangers, *ASME Journal of Heat Transfer*, vol. 111, pp. 314–321, May 1989.

[9] Kandlikar, S.G. and Shah, R.K., Multipass Plate Heat Exchangers-Effectiveness-NTU Results and Guidelines for Selecting Pass Arrangements, *ASME Journal of Heat Transfer*, vol. 111, pp. 300-313, May 1989.

[10] Yang, Q. and Wang, L., Thermal Performance and Maldistribution of Multi-pass Plate Heat Exchangers, in *Heat Transfer Science and Technology*, ed. B. Wang, Higher Education Press, China, pp. 621–626, 1996.

[11] Shah, R.K. and Sekulic, D.P., Heat Exchangers, in *Handbook of Heat Transfer*, 3rd edn, ed. W. Rohsenow, J. Hartnett, and Y. Cho, McGraw-Hill, New York, Chapter 17, 1998.

[12] White, F.M., *Fluid Mechanics*, 4th edn, McGraw-Hill, New York, 1999.

[13] Lockhart, R.W. and Martinelli, R.C., Proposed Correlation of Data for Isothermal Two-phase, Two-component Flow in Pipes, *Chemical Engineering Progress*, vol. 45, no. 1, pp. 39–48, 1949.

[14] Collier, J.G. and Thome, J.R., *Convective Boiling and Condensation*, 3rd edn, Oxford University Press, Oxford, 1994.

[15] Thonon, B., Vidil, R., and Marvillet, C., Recent Research and Developments in Plate Heat Exchangers, *Journal of Enhanced Heat Transfer*, vol. 2, nos. 1–2, pp. 149–155, 1995.

[16] Wang, L., An Experimental Investigation of Steam Condensation Performance in Plate Heat Exchangers, Publ. 99/2009, *Thesis for Licentiate of Engineering Degree*, Lund Institute of Technology, Lund, Sweden, 1999.

[17] Zivi, S.M., Estimation of Steady-state Steam Void-fraction by Means of the Principle of Minimum Entropy Production, *Journal of Heat Transfer*, vol. 12, pp. 247–252, 1964.

[18] Kay, J.M., *An Introduction to Fluid Mechanics and Heat Transfer*, Cambridge University Press, New York, 1963.

[19] Shah, R.K. and Focke, W.W., Plate Heat Exchangers and Their Design Theory, in *Heat Transfer Equipment Design*, ed. R.K. Shah, E.C. Subbarao, and R.A. Mashelkar, Hemisphere, Washington, pp. 227–254, 1988.

[20] Raju, K.S.N. and Bansal, J.C., Design of Plate Heat Exchangers, in *Low Reynolds Number Flow Heat Exchangers*, ed. S. Kakac, R.K. Shah, and A.E. Bergles, Hemisphere, Washington, pp. 913–932, 1983.

[21] Westwater, J.W., Compact Heat Exchangers with Phase Change, *Proceedings of the 8th International Heat Transfer Conference*, San Francisco, pp. 269–278, 1986.

[22] Cooper, A., Condensation of Steam in Plate Heat Exchangers, *AIChE Symposium*, vol. 70, no. 138, pp. 172–177, 1987.

[23] Wang, L., Christensen, R., and Sundén, B., An Experimental Investigation of Steam Condensation in Plate Heat Exchangers, *International Journal of Heat Exchangers*, vol. 1, no. 2, pp. 125–150, 2000.

[24] Clark, D.F., Plate Heat Exchanger Design and Recent Development, *Chemical Engineer*, no. 285, pp. 275–279, 1974.

[25] Marriott, J., Performance of an Alfaflex Plate Heat Exchanger, *Chemical Engineering Progress*, pp. 73–78, February, 1977.

CHAPTER 6

Single- and multi-pass flow arrangement

6.1 Flow arrangement and distribution

An advantage of PHEs is the flexibility to obtain different flow arrangements. This is achieved by appropriate gasket arrangements in PHEs. The flow arrangements are influenced by various factors, such as flow direction, number of passes, number of paths per pass, distributions along the port and inside the channel, end plate, etc.

6.1.1 Flow direction

When the fluid streams flow in the same direction on both sides of the dividing wall, the flow is called parallel flow, and when they flow in opposite directions the flow is called counter flow. This is illustrated in Fig. 6.1. Parallel flow and counter flow are two basic flow arrangements, and the flows in any two adjacent channels are of these two types. As mentioned previously, counter-flow arrangement is usually preferred due to its high thermal efficiency. Parallel-flow arrangement is used only in limited situations, where some special requirements prevent the use of counter-flow arrangement. However, pure counter-flow is difficult to obtain in reality, due to the end plate effect and the multi-pass arrangement.

6.1.2 Pass

A single-pass arrangement is when each fluid flows in just one direction in the channels formed by the adjacent pair of plates in a PHE. A multi-pass arrangement on the other hand is when a plate with less than four port holes is used to make the fluids flow in the opposite direction in a part of the heat exchanger (see Fig. 6.2). Inserting one plate with blind holes results in a two-pass arrangement, two such plates give a three-pass arrangement, and so on. Multi-pass arrangement makes the fluid undergo the heat transfer process during a longer time and a longer distance inside the heat exchanger. The two fluid streams in a PHE can have equal or unequal passes. In general, multi-pass arrangements are used when NTU values greater than

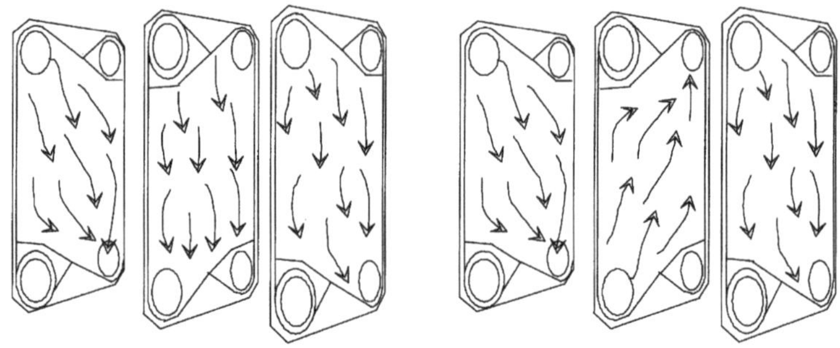

(a) (b)

Figure 6.1: Two basic flow arrangements: (a) parallel and (b) counter-flow.

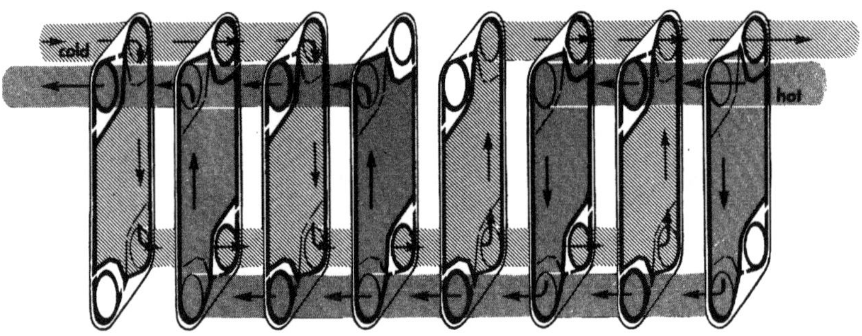

Figure 6.2: Multi-pass arrangement is achieved by using plates with less than four port holes (from reference [1]).

those given by a single plate are required [1]. When flow ratios are high or there is some other reason for minimizing the pressure drop on one side, unequal passes can be used with fewer passes on the low-pressure drop side. There are two deficiencies of multi-pass arrangements compared to single-pass arrangement. One is higher pressure drop due to longer flow distance inside the heat exchangers. The other is low efficiency because of the presence of parallel flows for at least some adjacent channels, which is unavoidable in multi-pass arrangements.

6.1.3 Paths per pass

The number of channels, where the fluid stream flows during one pass, is called the number of paths per pass. In Fig. 6.2, the number of paths per pass for both fluid streams is two. The selection of this value is critical to the flow velocity inside the plate channels. As stated previously, too high velocity can cause flow instability, and too low velocity will have low heat transfer and possibly lead to high fouling effect. Therefore, the selection of path number per pass is related to the flow velocity in addition to the heat transfer surface area.

6.1.4 Distribution along port manifolds

There are two types of distributions, namely U and Z arrangements (see Fig. 6.3). In the U arrangement, the inlet and outlet connections are made to the frame plate of the PHE only, so that the system is restricted to single-pass arrangements. In the Z arrangement, the inlet and outlet connections are on the head and follower, respectively. On small plate packs, the effect of pressure drop along the port manifold can be ignored and calculations can be based on the assumption of equal flow rates down each plate in a pass. On large plate packs, distribution effects can become significant, depending on the relative magnitudes of the plate and manifold pressure drop. This will reduce the effective logarithmic-mean temperature difference. The two factors that affect the port pressure drop are (1) the frictional loss along the manifold and (2) the momentum change due to the variation in liquid velocity. More detailed discussion is in Chapter 10.

However, if uniform mass flow rates in paths are assumed, there is no difference between U and Z arrangements in terms of thermal performance.

6.1.5 Distribution inside channel

Because the diameter of the entry port at either end of the plate is small compared with the width of the main heat transfer area, transverse distribution of flow across the plate can be adversely affected. The most important factors are port arrangement, aspect ratio, and plate entry losses. Ports can be arranged either in diagonal flow (entry and exit at opposite side of the plate) or in vertical flow (entry or exit on the same side of the plate) as shown in Fig. 6.4a and b, respectively [1, 2]. The resulting difference in flow distribution has some effect on the transverse temperature profile, as shown in Fig. 6.4. The effectiveness of air removal is important, particularly for downward flowing plates where the liquid velocity must be high enough to ensure that the plate channels run full [3]. In practice, this can be normally achieved by ensuring that the pressure loss down the plate is not significantly less than the static head corresponding to the vertical distance between the ports.

The effects of poor distribution can be accentuated at low aspect ratio (i.e. mean heat transfer length/flow width). No comprehensive recommendations appear to be

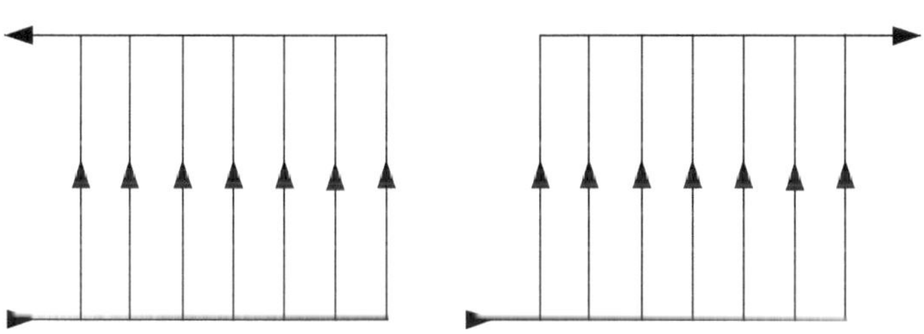

Figure 6.3: Flow distribution along port manifolds: (a) U and (b) Z arrangement.

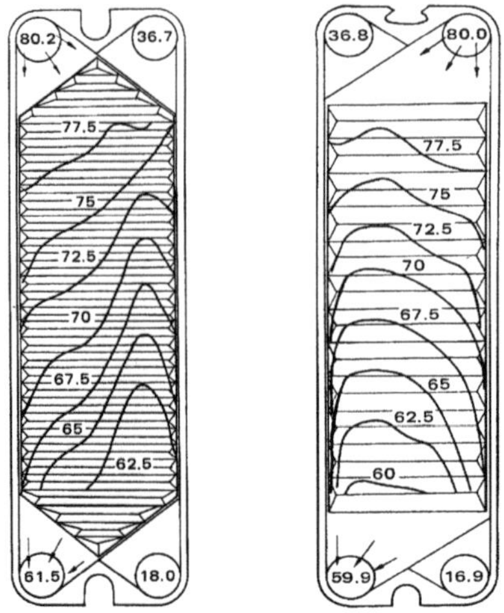

Figure 6.4: Temperature distribution (°C) on (a) vertical and (b) diagonal flow plates (from references [1, 2]).

available for this fact, but on most commercial plates the aspect ratio does not fall much below a value of 1.8 [1].

Extraneous pressure loss must be minimized at the plate entry area by avoiding any unsuitable restriction. Because the entrance region of the plate is the most mechanically weak area of the entire plate, the mutual support between adjacent plates adds to the restriction and therefore requires very careful design. The lower the thermal and pressure drop performance of the plate, the greater will be the proportional effect of these extraneous losses.

6.1.6 End plate

In the U arrangement, the first plate usually has gaskets around all four holes to prevent the fluids from leaking out between the first plate and the frame as the fluid stream enters the PHE via the connections in the frame. The last plate usually has no holes to prevent the fluids from leaking out between the last plate and the frame. Therefore, these two first plates do not transfer any heat and are called end plates. The remaining plates are called thermal plates, which transfer heat. In the Z arrangement, the outer plates are slightly different in terms of gaskets and holes. However, they do not transfer heat as in the U arrangement. Therefore, the channels at both ends of the PHE are heated from one side only, which reduces the thermal efficiency of the exchanger. This influence must be considered in the design of PHEs.

All the above influential factors have impact on the performance of single- and multi-pass arrangements. The thermal efficiency is different from the pure counter-flow arrangement, which needs to be accounted for in the design of the exchanger. In order to simplify the analysis, the flow distributions along the port manifold and inside the channel are assumed uniform in the following analysis. However, this does not diminish the importance of the flow distribution in these two parts.

6.2 Pass arrangement classification

Based on the arrangement of gaskets, many combinations of passes are possible on each fluid side in a PHE. The basic patterns include series flow, looped flow, and complex flow, as shown in Fig. 6.5.

1 Series flow is where the total fluid passes over each plate and changes direction over each subsequent plate. This is feasible only with small flow rates and is rarely used.

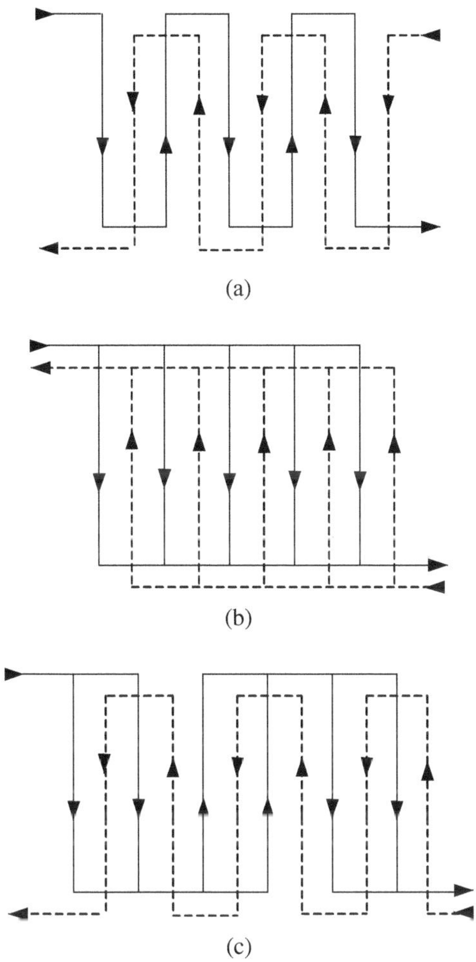

Figure 6.5: Basic flow arrangements: (a) basic, (b) looped, and (c) complex flow.

2 Looped flow is where the fluids are divided into sub-streams that mix before exiting from the PHE. This is done for high flow rate.
3 Complex flow is made up of a combination of the series and looped flows.

One common way to classify the above two-fluid PHEs is on the basis of the number of passes for each stream. Possible arrangements are 1–1 single pass, and 2–1, 2–2, etc., multi-pass arrangements. Here 2 in 2–1 refers to 2 passes on the fluid stream 1 and 1 refers to 1 pass on the fluid stream 2. However, more features of flow arrangements must be accounted for in the classification, such as flow directions of each stream, number of total channels for the two streams, end plate effect, etc. Kandlikar and Shah [4] have presented an effective classification method which can take all these effects into account. This method is adopted here and it results in four configurations each for 1–1 (nos 111–114 in Fig. 6.6), 2–1 (nos 211–214 in Fig. 6.7), 2–2 (nos 221–224 in Fig. 6.8), 3–2 (nos 321–324 in Fig. 6.9), 3–3 (nos 331–334 in Fig. 6.10), 4–1 (nos 411–414 in Fig. 6.11), and six configurations for 3–1 (nos 311–316 in Fig. 6.12). Similar classification results for other arrangements can also be obtained, but are not covered in this book.

These arrangements can be basically classified as overall counter-flow arrangement (nos 111, 112, 211, 212, 213, 214, 221, 222, 311, 312, 313, 321, 322, 331, 332, 411, 412, 413, and 414) and overall parallel-flow arrangement (nos 113, 114, 314, 315, 316, 323, 324, 333, and 334). It must be noted that the overall counter-flow arrangement does not always mean individual counter-flow arrangement. For instance, no. 222 is overall counter-flow arrangement, but most of individual channels are in parallel-flow arrangement. In addition, the use of more than one pass leads to some channels with different flow arrangement compared to the

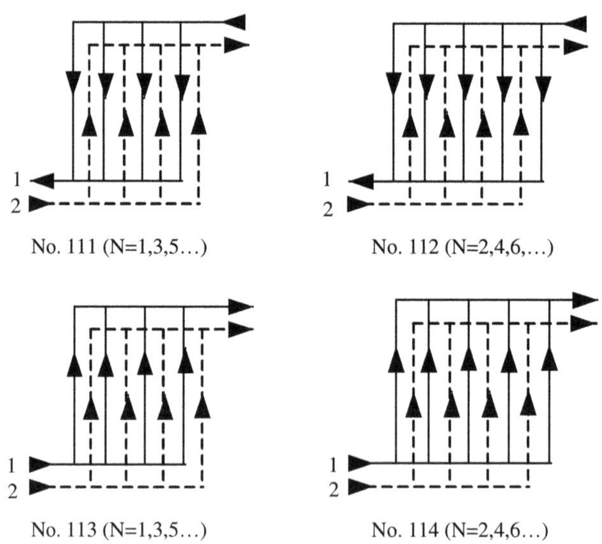

Figure 6.6: Four possible 1–1 arrangements.

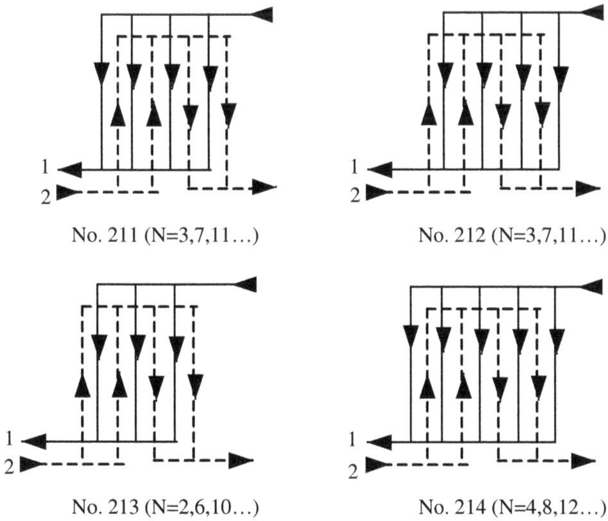

Figure 6.7: Four possible 2–1 arrangements.

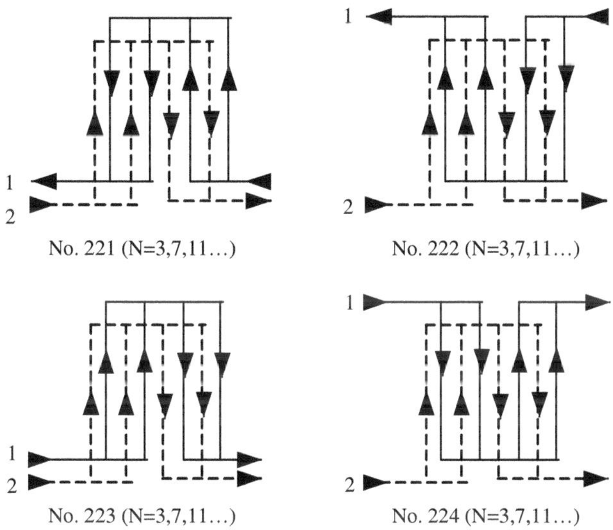

Figure 6.8: Four possible 2–2 arrangements.

other channels. This means that all channels with uniform counter- or parallel-flow arrangement cannot be achieved, and there is at least one pair of channels having different flow arrangements from the rest. This is highlighted in Fig. 6.13, where the parallel flow pointed out by the elliptic symbol cannot be avoided. It should be pointed out that in all these arrangements, the number of channels in each pass of a given fluid side is assumed to be the same.

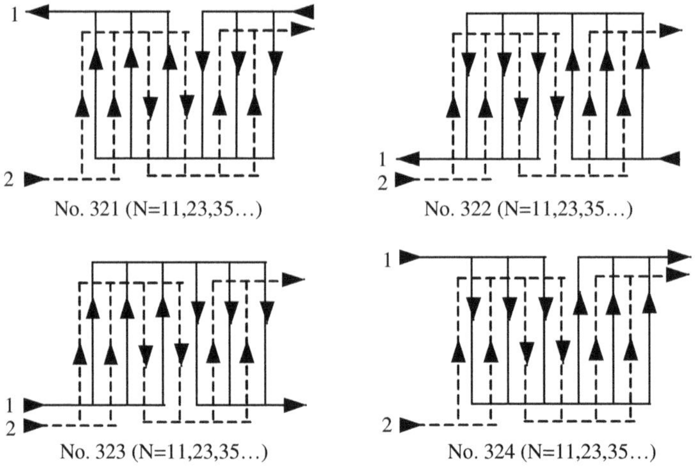

Figure 6.9: Four possible 3–2 arrangements.

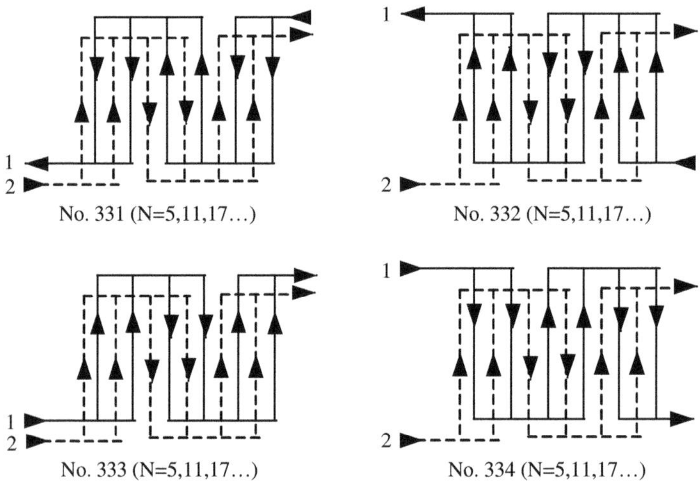

Figure 6.10: Four possible 3–3 arrangements.

In order to demonstrate that all the possible flow arrangements have been included, the flow reversibility principle shown by Pignotti [5] must be introduced first. It states that when two fluid streams in any heat exchanger are reversed simultaneously, the ε−NTU relationship remains the same. Consider the case of 2–1 arrangements as an example. Figure 6.14 shows four additional arrangements compared to four basic arrangements in Fig. 6.7. Reversing the flow directions of both streams in no. 211a will not change ε−NTU relationship from that of the original no. 211a. In the reserved no. 211a (not shown), streams 1 and 2 would enter from the right side and the flow directions in all channels would be reserved. A mirror

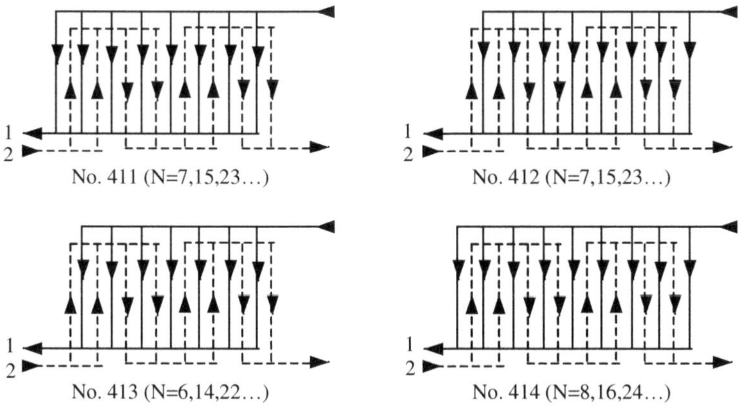

Figure 6.11: Four possible 4–1 arrangements.

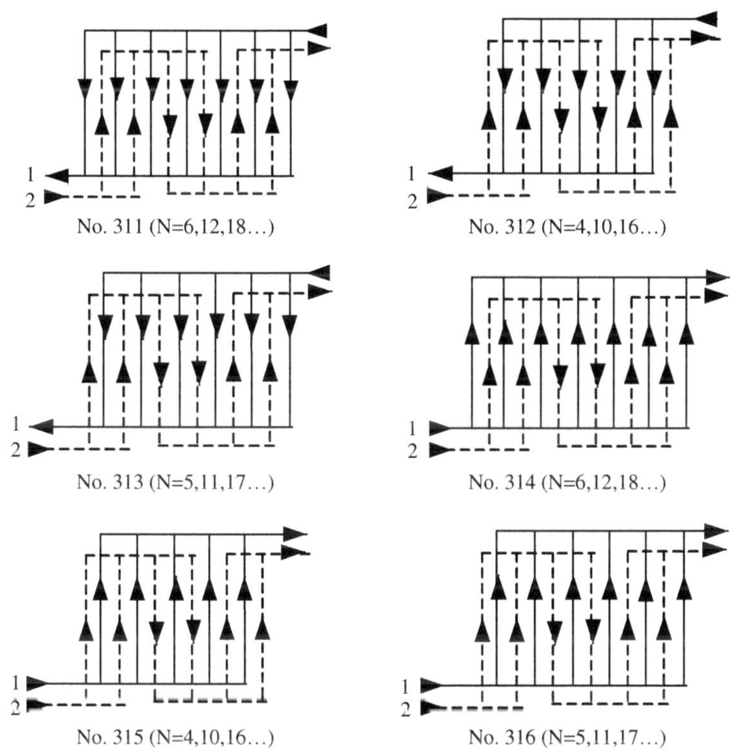

Figure 6.12: Six possible 3–1 arrangements.

image of this reserved no. 211a is the no. 211 arrangement. The same conclusion can be obtained for nos 212 and 212a, 213 and 213a, 214 and 214a. Therefore, it can be concluded that all the possible flow arrangements have been included in this classification.

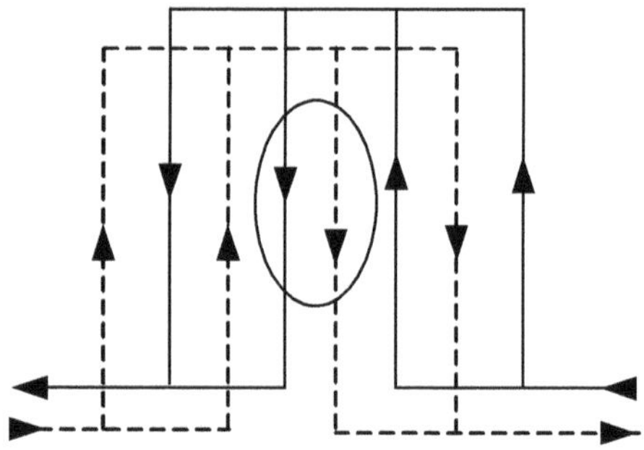

Figure 6.13: Parallel-flow in 2–2 counter-flow arrangement.

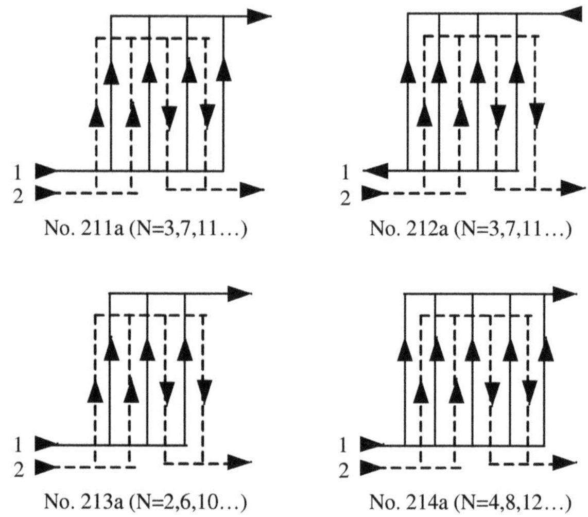

Figure 6.14: Additional equivalent 2–1 arrangements.

6.3 General thermal model

In order to derive the general thermal model for PHEs, some assumptions must be made. They are

- the distribution in different paths of one pass is even
- all physical properties are constant
- steady-state operation
- heat loss to the surroundings is neglected
- heat conduction occurs only perpendicular to the fluid flow in the channels
- single-phase application.

Under these assumptions, the general thermal model can be generated by considering the heat balance of the small control volume ($\mathrm{d}X$) in Fig. 6.15. This means that the heat transferred to the fluid from the adjacent channels is equal to the enthalpy change of the fluid.

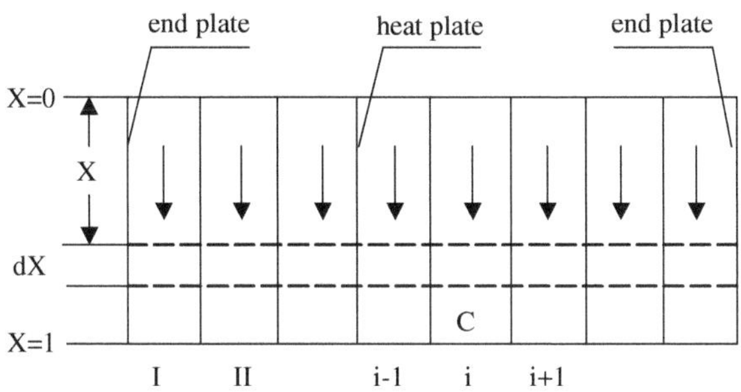

Figure 6.15: Schematic diagram of a PHE.

The differential equations of the temperature variation for the first, intermediate, and last channel can be written as:

$$\pm\frac{\mathrm{d}T_{\mathrm{I}}}{\mathrm{d}x} = \frac{\boldsymbol{UW}}{C_{\mathrm{I}}}(T_{\mathrm{I}} - T_{\mathrm{II}}) \tag{6.1}$$

$$\pm\frac{\mathrm{d}T_{\mathrm{i}}}{\mathrm{d}x} = \frac{\boldsymbol{UW}}{C_{\mathrm{i}}}(2T_{\mathrm{i}} - T_{\mathrm{i}-1} - T_{\mathrm{i}+1}) \tag{6.2}$$

$$\pm\frac{\mathrm{d}T_{\mathrm{n}}}{\mathrm{d}x} = \frac{\boldsymbol{UW}}{C_{\mathrm{n}}}(T_{\mathrm{n}} - T_{\mathrm{n}-1}) \tag{6.3}$$

where the positive and negative signs are applied to the upward and downward flows, respectively. In the above equations, U is the overall heat transfer coefficient, W/(m$^2 \cdot$ K), and W is the plate width, m. In order to make these equations more general, the following different dimensionless parameters are defined.

$$\theta = \frac{T_{\mathrm{i}} - T_{\mathrm{c,in}}}{T_{\mathrm{h,in}} - T_{\mathrm{c,in}}} \tag{6.4}$$

$$X = x/L \tag{6.5}$$

$$\mathrm{NTU}_1 = \frac{\boldsymbol{UA}}{C_1} \tag{6.6}$$

$$R_1 = \frac{C_1}{C_2} \tag{6.7}$$

$$A = N_{\mathrm{p}}A_{\mathrm{p}} \tag{6.8}$$

where N_{p} is the number of heat transfer plates and A_{p} is the heat transfer area of the plate. By employing the dimensionless parameters defined above, the governing eqns (6.1–6.3) can be finally transferred into

$$\pm\frac{\mathrm{d}\theta_{\mathrm{I}}}{\mathrm{d}X} = \frac{\boldsymbol{UA}_{\mathrm{p}}}{C_{1}}(\theta_{\mathrm{I}} - \theta_{\mathrm{II}}) \tag{6.9}$$

$$\pm\frac{\mathrm{d}\theta_{\mathrm{i}}}{\mathrm{d}X} = \frac{\boldsymbol{UA}_{\mathrm{p}}}{C_{\mathrm{i}}}(2\theta_{\mathrm{i}} - \theta_{\mathrm{i-1}} - \theta_{\mathrm{i+1}}) \tag{6.10}$$

$$\pm\frac{\mathrm{d}\theta_{\mathrm{n}}}{\mathrm{d}X} = \frac{\boldsymbol{UA}_{\mathrm{p}}}{C_{\mathrm{n}}}(\theta_{\mathrm{n}} - \theta_{\mathrm{n-1}}) \tag{6.11}$$

For the channels occupied by stream 1,

$$\frac{\boldsymbol{UA}_{\mathrm{p}}}{C_{\mathrm{i}}} = \frac{\mathrm{NTU}_{1} \cdot m_{1}}{N_{\mathrm{p}}} \tag{6.12}$$

For the channels occupied by stream 2,

$$\frac{\boldsymbol{UA}_{\mathrm{p}}}{C_{\mathrm{i}}} = \frac{\mathrm{NTU}_{1} \cdot R_{1} \cdot m_{2}}{N_{\mathrm{p}}} \tag{6.13}$$

where m_1 and m_2 are the number of paths per pass for streams 1 and 2, respectively.

There are several methods to solve the above equations [4, 6, 7]. Among these, the finite difference method is an efficient and easy way to solve the above governing equation. This method requires less computational time and results in temperature distribution in all channels. The temperature change is expressed in terms of the first-order backward difference scheme for the first step, and the second-order backward difference scheme for all subsequent steps [8]. For all the steps in all channels, these create a set of simultaneous linear equations with boundary conditions. These linear equations are solved by the Gauss–Seidel iterative method. The number of steps in the programme is varied, and it was found that 100 steps are sufficient to obtain the outlet non-dimensional temperature within an accuracy of 0.0001. The solutions were found to be insensitive to the initial guesses supplied. Hence, the initial temperature variation was assumed to be linear along the total fluid path with a temperature effectiveness of 80% for each fluid. The iteration number is dependent on the number of total channels, the number of steps, and the resolution. For the single-pass counter-flow arrangement of two-stream PHEs with 80 channels, the iteration number is about 50 and it takes less than 1 s on a personal computer with a Pentium II processor with a frequency of 333 MHz, if the resolution of the outlet temperature is taken as 0.0001. Therefore, the number of steps and the resolution of the outlet temperature in all calculations are taken as 100 and 0.0001, respectively.

Because there is no restriction on the flow configuration, this thermal model is general and valid for all different configurations including counter- and parallel-flow arrangements, multi-pass arrangement, two-stream PHEs as well as multi-stream PHEs. Once the flow configuration is specified, the thermal performance can be easily determined.

6.4 Performance comparison

The thermal performance of PHEs is influenced by end plate, pass arrangement and flow direction, number of transfer units, and heat capacity flow rate ratio. All these factors are analysed in detail in the following sections.

6.4.1 End-plate effect

In PHEs, the two outer plates serve as end plates and ideally do not transfer any heat. This means that the channels at both ends of the PHE are heated or cooled from one side only, and the fluid experiences a smaller temperature change than in the other channels. Therefore, the end plates have a significant influence on the overall thermal performance when the number of total channels is not sufficiently large. There exist two limiting cases in this situation. If there is only one transfer plate in a PHE (corresponding to two flow channels, one for the hot fluid stream and the other for the cold fluid stream), the two streams exchange heat in true counter-flow or parallel-flow style, which means no end-plate effect exists. The other limiting case is when the number of transfer plates is larger than a certain critical value, in which case the end-plate effect becomes negligible. In this case, the flow can be regarded as true counter flow or parallel flow for 1–1 arrangement. In situations deviating from these two limiting cases, the end-plate effect must be taken into account for both the correction factor and the ε–NTU relationship. The determination of the critical number of transfer plates is dependent on the required accuracy for the performance. For instance, 19 is recommended by Heggs and Scheidat [7] for the inaccuracy within 2.5%. However, this number has to increase if higher accuracy is required. For arrangement no. 111a, the end-plate effect can be revealed from Fig. 6.16.

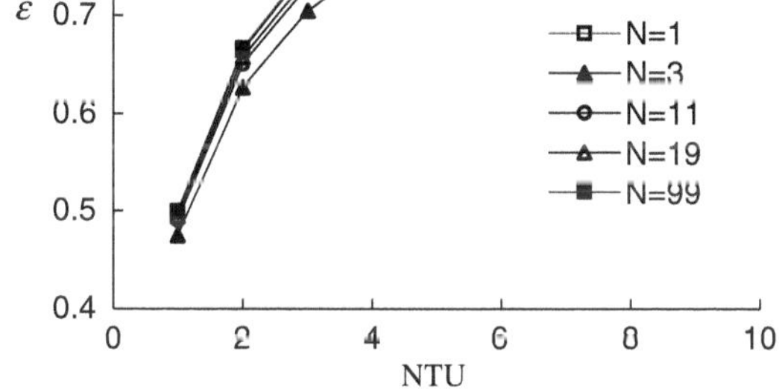

Figure 6.16: End-plate effect for arrangement no.111a under $R = 1$.

As can be seen, the performance of PHEs with three transfer plates is significantly different from the true counter-flow arrangement, corresponding to the case with one transfer plate. For the same values of NTU and R, the difference in the effectiveness is up to 0.07. This is very significant and must be considered in the design. When the number of transfer plates is larger than 99, the performance is almost identical to that of the true counter-flow arrangement (the line with $N = 1$ corresponds to the true counter-flow arrangement), hence no end-plate effect needs to be considered. Once again, this number is determined based on the requirement on accuracy in the thermal performance. Here, a value of 39 is recommended for all different arrangements, which is to achieve an accuracy of 1%. When the number of transfer plates is larger than this value, the end effect can be ignored. Otherwise, this effect must be considered.

In addition, the difference between arrangement nos 111 and 112 is the unequal number of passages on the two sides. The resulting performance is thus different due to the end-plate effect. When the number of transfer plates is larger than 99, the difference in effectiveness between them is smaller than 1%, and it can be ignored. In this case, the performance of the two arrangements becomes identical.

6.4.2 Passage arrangement and flow direction

The performance between counter-flow arrangement and parallel-flow arrangement is very different owing to the effective temperature difference. This has been explained in the previous chapter. Because there is no true counter-flow arrangement existing for multi-pass PHEs, correction factors must be applied to the calculation of average mean temperature difference.

Consider the 2–2 arrangement as an example. Arrangement no. 221 is the overall counter-flow arrangement, and its performance is shown in Fig. 6.17. As can been seen, the performance for $N = 11$ is quite different from the pure counter-flow arrangement. This is caused by two factors: end plate and parallel flow highlighted in Fig. 6.13. These two factors make the performance deviate from the true counter-flow arrangement. However, when the number of total channels exceeds a certain value (here 99), the performance is almost identical to the pure counter-flow arrangement. In this case, the end effect and non-counter-flow effect can be ignored.

Arrangement no. 223 is the overall parallel-flow arrangement, and its performance is shown in Fig. 6.18. Because of the end-plate effect and because non-parallel-flow exists, the performance for $N = 11$ is quite different from the pure parallel-flow arrangement. Similar with arrangement no. 221, the performance is almost identical to the true parallel-flow arrangement when the number of total channels exceeds a certain value (here 99). However, one can find that the performance of arrangement no. 223 is much worse than arrangement no. 221. Therefore, arrangement no. 221 is generally preferred instead of arrangement no. 223.

Arrangement no. 222 is the overall counter-flow arrangement with individual parallel-flow arrangement. There is also non-parallel flow existing within the plate package. In addition, the end-plate effect also exists. Its performance is shown in Fig. 6.19. One can easily find that the performance of arrangement no. 222

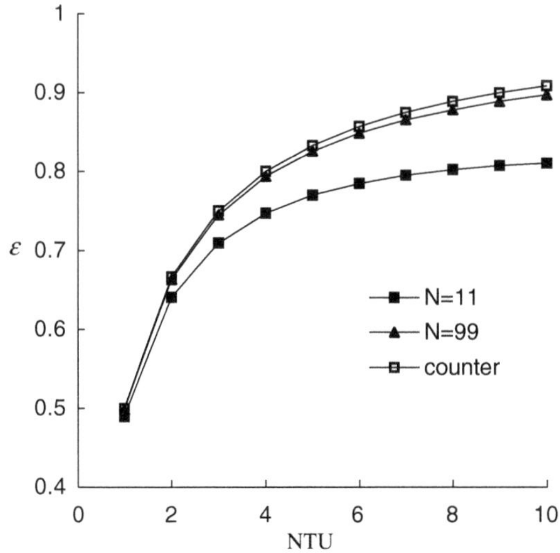

Figure 6.17: ε—NTU relationship for no. 221 under $R = 1$.

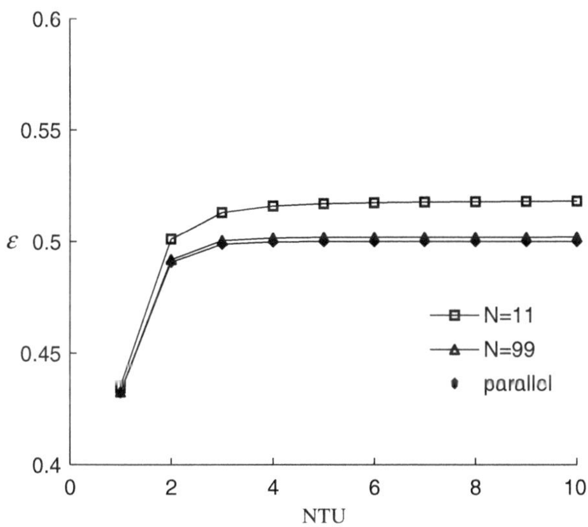

Figure 6.18: ε—NTU relationship for no. 223 under $R = 1$.

lies between true counter-flow arrangement and true parallel-flow arrangement. Regardless of the number of plates, the performance cannot approach either counter- or parallel-flow arrangement. Because this performance is also lower than that of arrangement no. 221, it is generally not preferred in practice.

Arrangement no. 224 is the overall parallel-flow arrangement with individual counter-flow arrangement. There is also non-counter flow existing within the plate

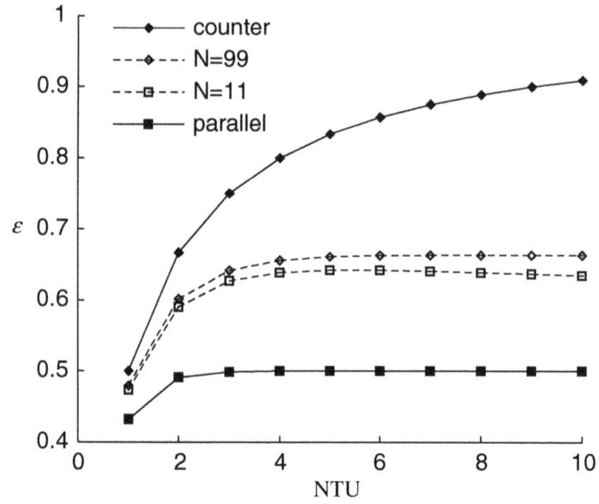

Figure 6.19: ε–NTU relationship for no. 222 under $R = 1$.

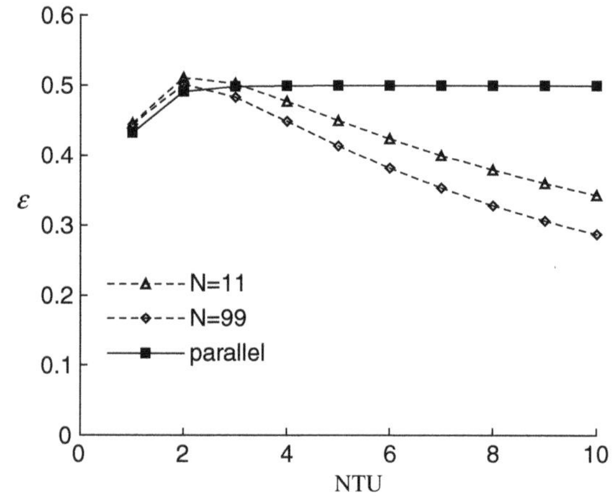

Figure 6.20: ε–NTU relationship for no. 224 under $R = 1$.

package, in addition to the end-plate effect. Its performance is shown in Fig. 6.20. It might appear as a surprise that the performance is even worse than pure parallel-flow arrangement. The reason is quite simple: inverse heat transfer has occurred due to the flow arrangement. The cold fluid stream is first heated by the hot fluid stream. After a certain flow length, the temperature of the cold fluid stream is higher than the hot fluid stream. Under this condition, heat is transferred back to the hot fluid stream (currently colder) from the cold fluid stream (currently hotter). This inverse heat flow makes the overall thermal efficiency even smaller than that in the parallel-flow arrangement. Such inverse flow should be avoided, and arrangement no. 224 should never be used under any practical condition.

The general conclusion from the above analysis is that the configuration, which has overall counter-flow arrangement and individual counter-flow arrangement, has the highest effectiveness among their respective arrangements. Under this condition, the best configurations among the listed configurations in this chapter are nos 111, 211, 221, 311, 321, 331, and 411. This will be further verified in the next section.

6.4.3 Number of transfer units

The influence of the number of transfer units (NTU) has already been illustrated in Figs 6.17–6.20. Except for arrangement no. 224, the performance is always enhanced with the increase of NTU. However, the performance increases very sharply when NTU is small. As NTU has reached a certain value, the performance approaches an asymptotic value. For arrangement no. 221 with infinite transfer plates (true counter flow), the effectiveness increases very sharply before NTU reaches 4, but increases very slowly after that, and finally approaches an asymptotic value of 1.0. For arrangement no. 222 with infinite transfer plates (true parallel flow), the effectiveness increases very sharply before NTU reaches 3, but beyond this value, the performance approaches an asymptotic value of 0.5. Therefore, using large NTU to enhance performance is only valid when NTU is small. The effect will become less significant as NTU reaches a certain value.

6.4.4 Heat capacity flow rate ratio

This influence is illustrated in Fig. 6.21 for arrangement no. 221 with infinite number of transfer plates. For the same value of NTU, the performance is better for a smaller value of *R*. However, in the design of PHEs, the heat capacity flow rate ratio is usually given, and it can not be changed during the design. The value of *R* mainly has effect on the selection of pass arrangement because the flow rates in

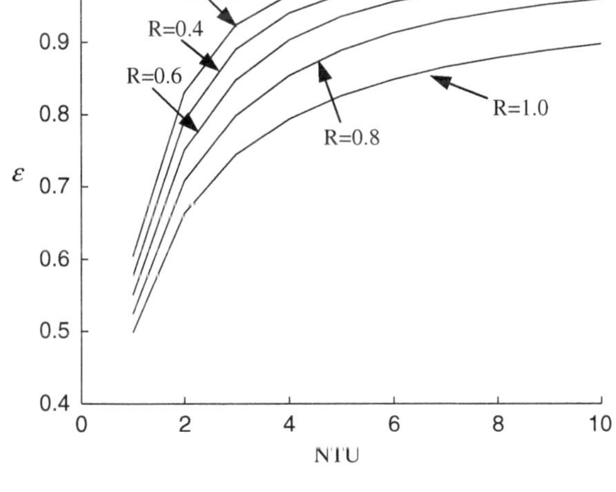

Figure 6.21: No. 221 with infinite transfer plates.

individual channels are critical to the overall performance. This is illustrated in the following section.

6.5 Guidelines of pass selection

In practice, the counter-flow arrangement is used exclusively because it yields higher effectiveness and heat transfer performance than that obtained by the parallel-flow arrangement, for single-pass arrangements. In multi-pass arrangements, both the counter-flow and parallel-flow arrangements exist, at least one pair of channels different from others. Therefore, one must find the configuration that is closest to the true counter-flow arrangement in such cases. As stated in the previous chapter, the performance of PHEs can also be expressed by the correction factor for logarithmic-mean temperature difference in addition to ε−NTU relationship. Thus, the correction factor will be employed in thermal performance comparison in this section because the comparison with true counter-flow arrangement can be easily viewed. The correction factor F is equal to 1 for true counter-flow arrangement. Therefore, the configuration with the closest value to 1 for correction factor is most closely to true counter-flow arrangement and is attempted in practice. This provides the basis for selecting the best pass arrangement.

6.5.1 Best arrangement within individual arrangement category

In order to know the best configuration within the individual arrangement category, correction factors for different configurations have to be constructed. Figures 6.22–6.28 are for 1–1, 2–1, 2–2, 3–1, 3–2, 3–3, and 4–1 arrangements, respectively. From these figures, one can easily distinguish that arrangements

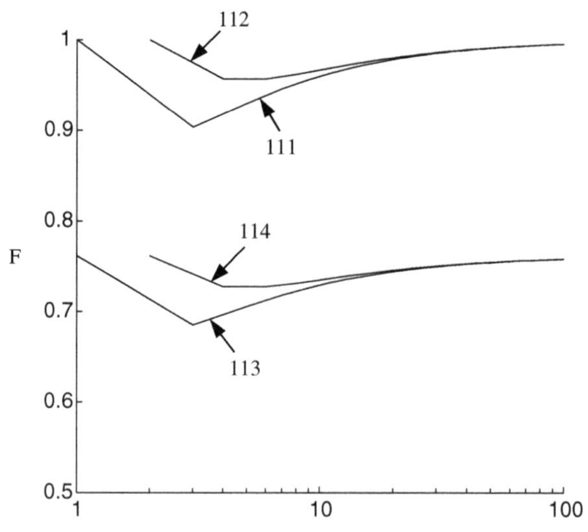

Figure 6.22: Correction factor for 1–1 arrangement at R = 1.0 and NTU = 1.0.

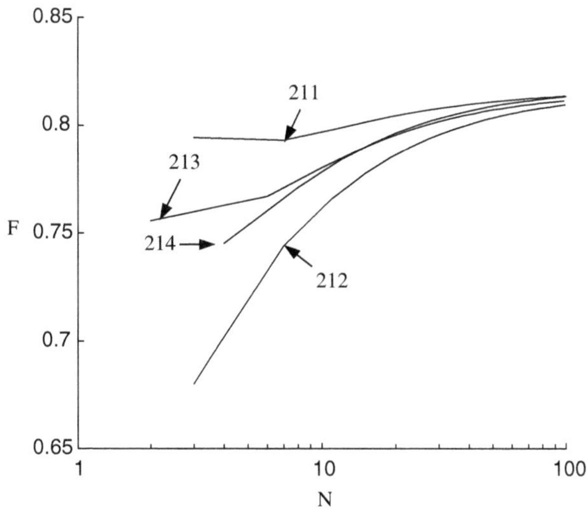

Figure 6.23: Correction factor for 2–1 arrangement at $R = 0.5$ and NTU = 2.0.

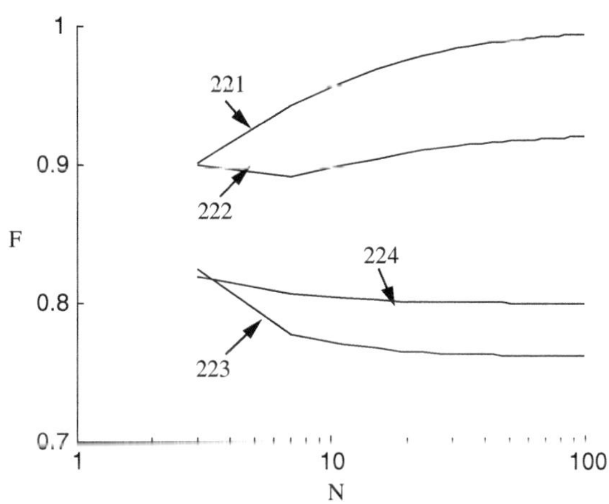

Figure 6.24: Correction factor for 2–2 arrangement at $R = 0.5$ and NTU = 1.0.

nos 112, 211, 221, 311, 321, 331, and 411 are the best arrangements in their respective categories. It should be pointed out that the same conclusion can be obtained for the situations with different values of R and NTU. Among these best configurations, nos 112, 221, and 331 have correction factors approaching 1 if the number of transfer plates is sufficiently large. This implies that these three configurations are the closest configuration to true counter-flow arrangement. For the rest of the configurations, the correction factor will not approach 1 under any condition. This implies that these configurations are a combination of counter-flow and parallel-flow arrangement.

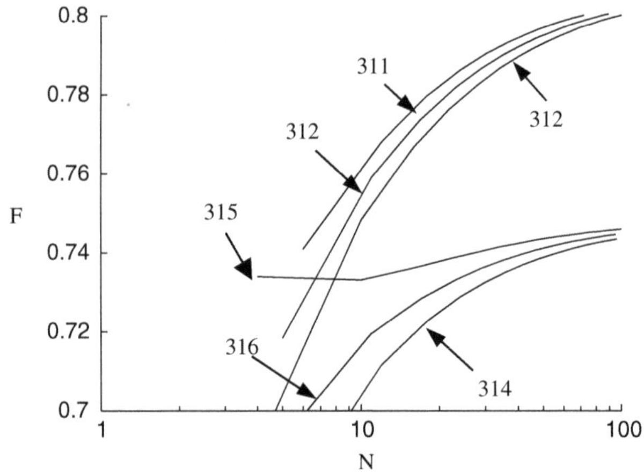

Figure 6.25: Correction factor for 3–1 arrangement at $R = \frac{1}{3}$ and NTU $= 3.0$.

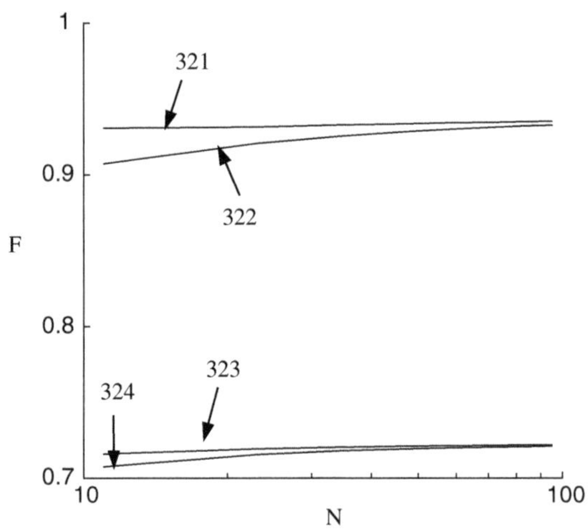

Figure 6.26: Correction factor for 3–2 arrangement at $R = \frac{2}{3}$ and NTU $= 1.5$.

It should be pointed out that arrangement no. 112 is only better than no. 111, when the number of transfer plates is less than 10. Beyond this, both no. 111 and no. 112 have almost identical performance. The main difference between them is that arrangement no. 111 has equal number of flow channels for the two fluid streams, and no. 112 has unequal number. In practice, it is sometimes preferred to have two outer channels occupied by cold fluid streams, in order to prevent the frame from becoming too hot due to heat release to environment. From this point of view, arrangement no. 112 is better than arrangement no. 111. Otherwise, both configurations have high performance and can be employed in practical applications.

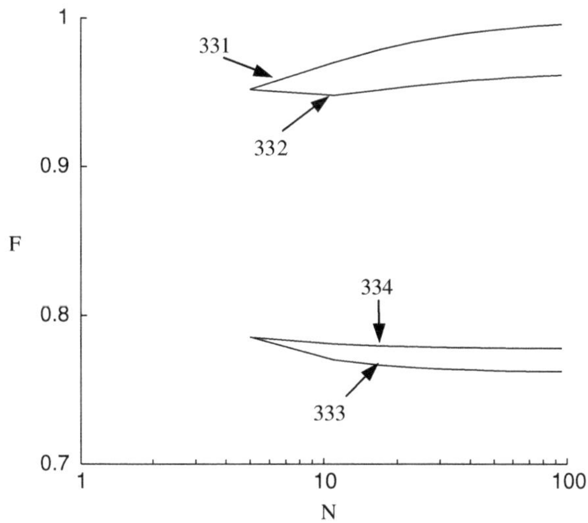

Figure 6.27: Correction factor for 3–3 arrangement at $R = 1.0$ and NTU $= 1.0$.

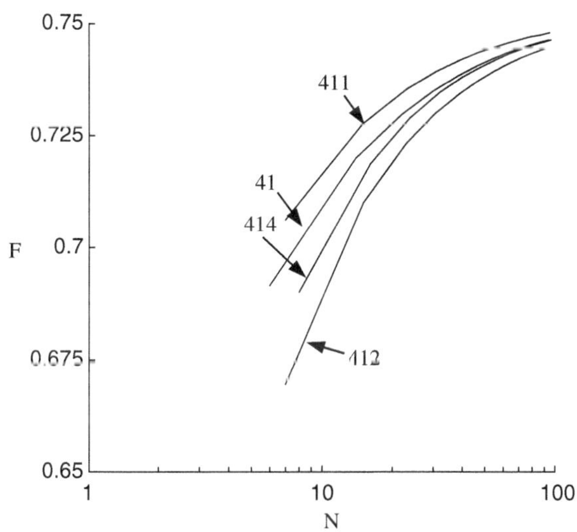

Figure 6.28: Correction factor for 4–1 arrangement at $R = \frac{1}{4}$ and NTU $= 1.0$.

6.5.2 Selecting best arrangement in all categories

It is a very complex task to select the best arrangement for a particular design situation. Pressure drop is another criterion in addition to thermal performance. If the flow rates (or heat capacity flow rates) of the two fluid streams are approximately equal, arrangement no. 112 or 111 is the first choice. This is because arrangement nos 112 and 111 are true counter-flow arrangement excluding the end-plate effect, which yields the highest thermal performance. However, if the number of transfer plates is too large, low heat transfer coefficient (due to low velocity) and large maldistribution may occur. This certainly decreases the thermal efficiency. In order to remedy this, arrangement no. 221 or no. 331 can be used for the same transfer plates. Due to the smaller number of total channels per pass, the maldistribution is suppressed and the heat transfer coefficient is enhanced. Hence a higher value of NTU is achieved, which leads to higher thermal efficiency. However, the pressure drop is increased dramatically in an exponential manner. This must be considered in selecting the pass arrangement.

When there is significant imbalance in the flow rate of the two streams, the fluid having the lower flow rate is generally channelled through multiple passes and the other fluid through either a lower number of passes or one pass, such as 2–1, 3–1, 3–2, 4–1 arrangements, etc. This results in approximately the same flow rate through individual flow passages. Such an arrangement would yield a higher heat transfer coefficient for the fluid having multiple passes and minimize the flow maldistribution. Compared with 1–1 counter-flow arrangement no. 112, there are two mechanisms in determining the thermal efficiency for these multi-pass arrangements. On one hand, multi-pass arrangement nos 211, 311, 321, and 411 introduce at least one pass of parallel flow, which decreases the thermal efficiency. On the other hand, the heat transfer coefficient is enhanced due to high velocity in multi-pass arrangements. This results in higher value of NTU, which can increase the thermal efficiency. In addition, the end-plate effect also has influence on the overall thermal efficiency. Therefore, the general conclusion on the selection of specific arrangement is difficult to reach, and a detailed design calculation has to be carried out for the individual situation.

6.6 Correction factors and effectiveness

LMTD and ε–NTU methods are the general design methods for PHEs. As described in the previous chapter, the correction factor F is function of the temperature effectiveness P and the heat capacity flow rate ratio R. Thermal effectiveness ε is the function of number of transfer units NTU and heat capacity flow rate ratio R. These relationships are required in these two methods, respectively. In order to assist design in a general purpose, these relationships are presented in this section for the best arrangements – nos 111, 211, 221, 311, 321, 331, and 411. These figures are based on the results from the general thermal model described above. Due to the complexity introduced by the end-plate effect, the following results are only for the case with infinite number of transfer plates. Those results with finite number of transfer plates are presented in Appendix A.1. If one wants to design PHEs with other configurations, those relationships can be obtained through solving the general thermal model by the suggested method.

Arrangement nos 111, 221, and 331 When the number of transfer plates is infinite, the flow in these arrangements can be regarded as true counter flow. Therefore, the correction factors are equal to 1, and the ε−NTU relationship is given by eqn (6.27).

Arrangement no. 211 The correction factors and the ε−NTU relationship for arrangement no. 211 are presented in Figs 6.29 and 6.30, respectively.

Arrangement no. 311 The correction factors and the ε−NTU relationship for arrangement no. 311 are presented in Figs 6.31 and 6.32, respectively.

Arrangement no. 321 The correction factors and the ε−NTU relationship for arrangement no. 321 are presented in Figs 6.33 and 6.34, respectively.

Arrangement no. 411 The correction factors and the ε−NTU relationship for arrangement no. 411 are presented in Figs 6.35 and 6.36, respectively.

For the PHE with infinite number of transfer plates, the temperature effectiveness P_1 (definition in eqn (5.31)) can be computed using closed-form formulas for each arrangement [9]. The temperature effectiveness P_1 is equal to the effectiveness ε if the heat capacity flow rate of stream 1 is smaller than that of stream 2. These formulas are presented in Table 6.1, which is reproduced from reference [9] without any modification. Therefore, all variables except P_1, R_1, NTU_1, and F are local or dummy variables which are not necessarily related to those defined in the nomenclature.

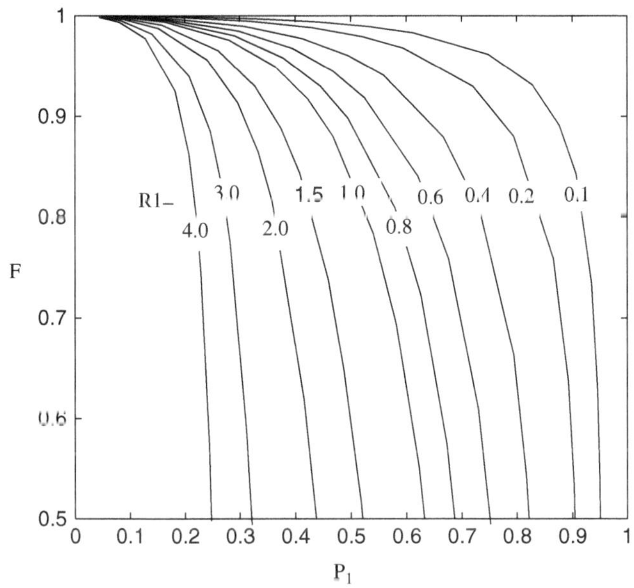

Figure 6.29: Correction factors for arrangement no. 211 with infinite number of transfer plates.

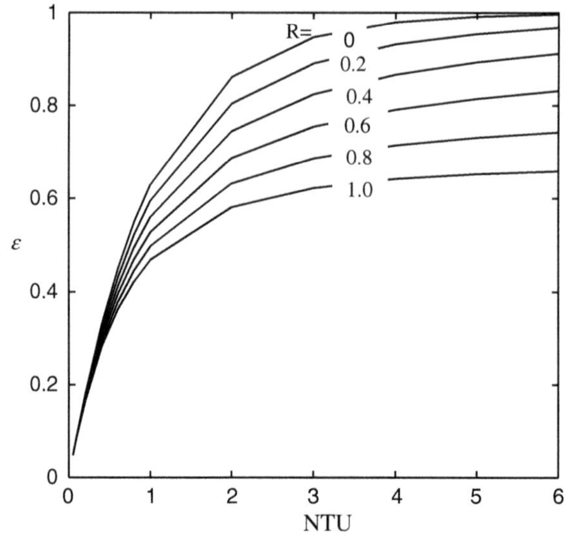

Figure 6.30: ε–NTU relationship for arrangement no. 211 with infinite number of transfer plates.

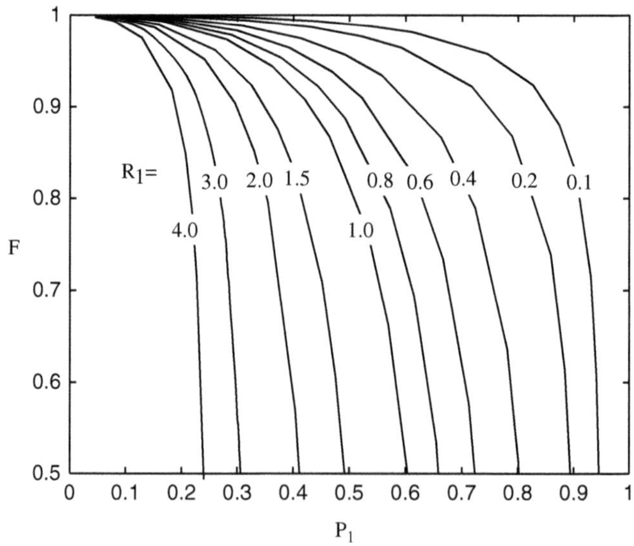

Figure 6.31: Correction factors for arrangement no. 311 with infinite number of transfer plates.

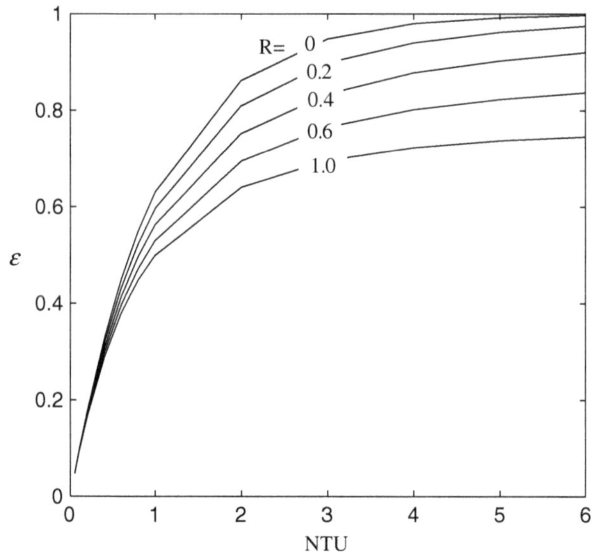

Figure 6.32: ε–NTU relationship for arrangement no. 311 with infinite number of transfer plates.

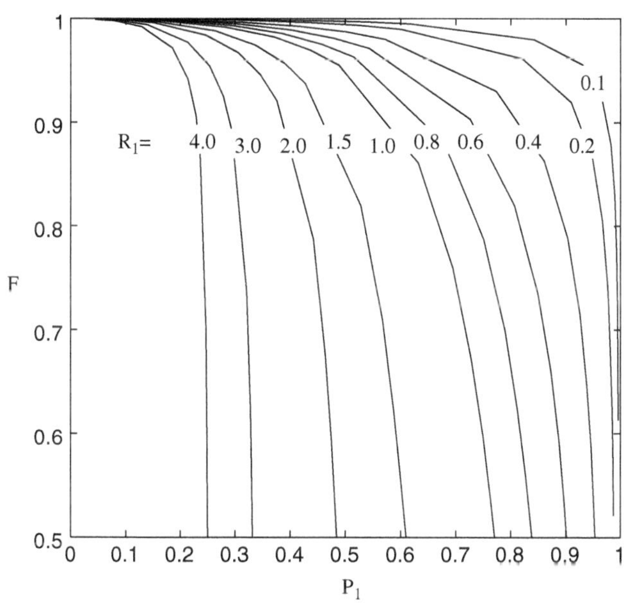

Figure 6.33: Correction factors for arrangement no. 321 with infinite number of transfer plates.

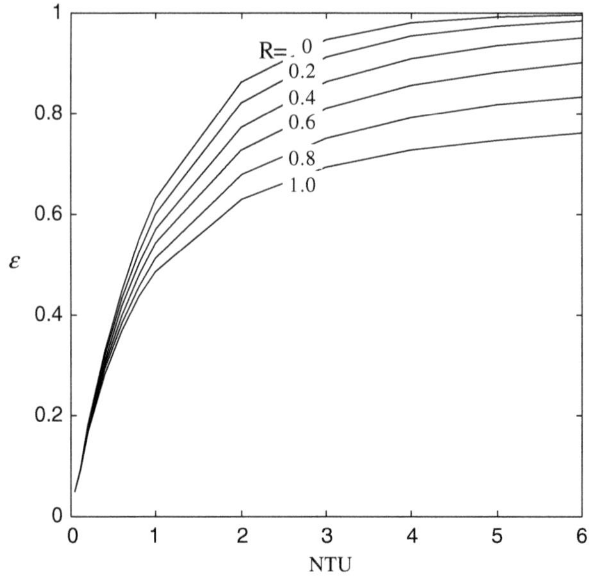

Figure 6.34: ε−NTU relationship for arrangement no. 321 with infinite number of transfer plates.

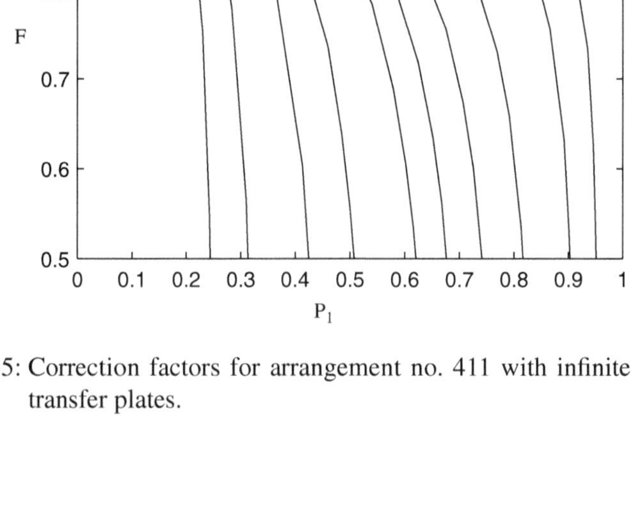

Figure 6.35: Correction factors for arrangement no. 411 with infinite number of transfer plates.

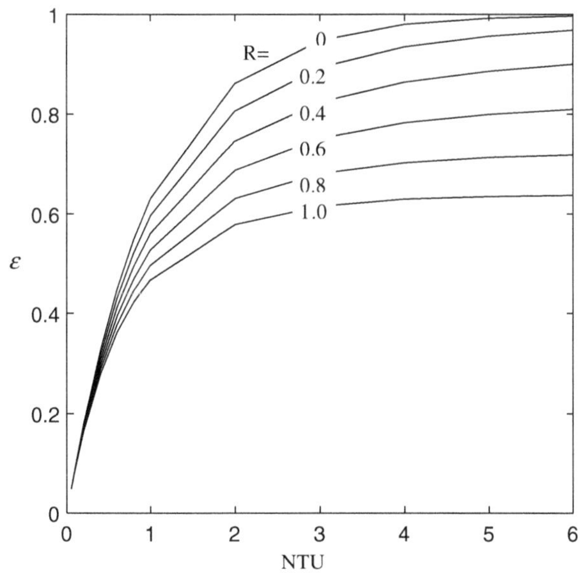

Figure 6.36: ε−NTU relationship for arrangement no. 411 with infinite number of transfer plates.

Nomenclature

A	heat transfer area, m^2
C	heat capacity flow rate, W/K
F	correction factor for LMTD
L	plate length, m
LMTD	logarithmic-mean temperature difference, K
m	number of paths per pass
N	number of plates
NTU	number of transfer units
P	temperature effectiveness
R	heat capacity flow rate ratio
T	temperature, K
U	overall heat transfer coefficient, $W/(m^2\ K)$
W	plate width, m
x	plate length, m

Greek symbols

θ	dimensionless temperature
ε	heat exchanger effectiveness

Table 6.1: P_1–NTU_1 formulas and limiting values P_1 and $R_1 = 1$ and $\mathrm{NTU}_1 = \infty$ for multi-pass PHEs.

Flow arrangement	Eqn no.	General formula	Value for $R_1 = 1$ unless specified differently	Value for $\mathrm{NTU}_1 \to \infty$
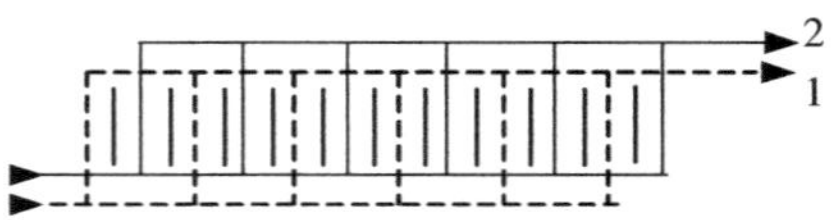 1 pass – 1 pass parallel-flow plate exchange stream symmetric	V. 1	$P_1 = A$ $A = P_\mathrm{p}(\mathrm{NTU}_1, R_1)$	$P_1 = \frac{1-\exp(-2\mathrm{NTU}_1)}{2}$	$P_1 = \frac{1}{1+R_1}$
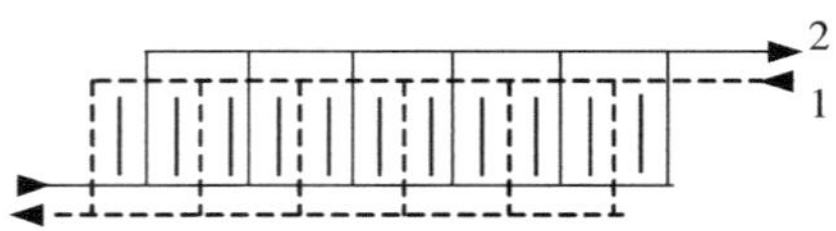 1 pass – 1 pass counter-flow plate exchanger, stream symmetric	V. 2	$P_1 = B$ $B = P_\mathrm{c}(\mathrm{NTU}_1, R_1)$	$P_1 = \frac{\mathrm{NTU}_1}{1+\mathrm{NTU}_1}$	$P_1 = 1$ for $R_1 \leq 1$ $P_1 = 1/R_1$ for $R_1 > 1$
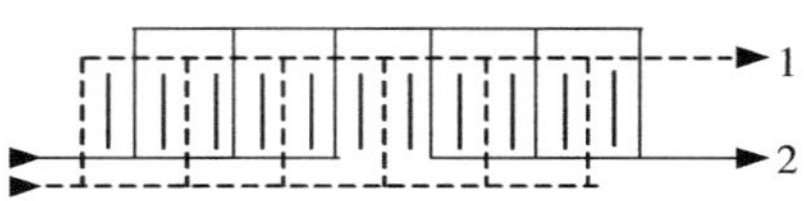 1 pass – 2 pass plate exchanger	V. 3	$P_1 = \frac{1}{2}(A + B - \frac{1}{2}ABR_1)$ $A = P_\mathrm{p}(\mathrm{NTU}_1, R_1/2)$ $B = P_\mathrm{c}(\mathrm{NTU}_1, R_1/2)$	Same as eqn (V.3) with $B = \frac{\mathrm{NTU}_1}{1+\mathrm{NTU}_1}$ For $R_1 = 2$	$P_1 = \frac{2}{2+R_1}$ for $R_1 \leq 2$ $P_1 = 1/R_1$ for $R_1 > 2$

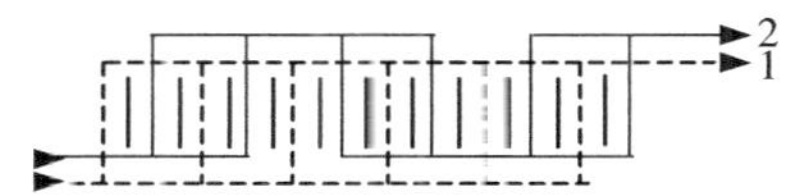 1 pass – 3 pass plate exchanger with two end plates in parallel flow	V. 4	$P_1 = \frac{1}{3}[B + A(1 - R_1B/3)(2 - R_1A/3)]$ $A = P_p(\mathrm{NTU}_1, R_1/3)$ $B = P_c(\mathrm{NTU}_1, R_1/3)$	Same as eqn (V.4) with $B = \frac{\mathrm{NTU}_1}{1+\mathrm{NTU}_1}$ For $R_1 = 3$	$P_1 = \frac{9+R_1}{(3+R_1)^2}$ for $R_1 \leq 3$ $P_1 = 1/R_1$ for $R_1 > 3$
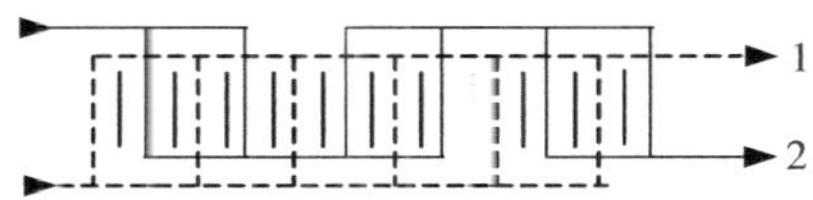 1 pass – 3 pass plate exchanger with two end plates in counter flow	V. 5	$P_1 = \frac{1}{3}[A + A(1 - R_1A/3)(2 - R_1B/3)]$ $A = P_p(\mathrm{NTU}_1, R_1/3)$ $B = P_c(\mathrm{NTU}_1, R_1/3)$	Same as eqn (V.5) with $B = \frac{\mathrm{NTU}_1}{1+\mathrm{NTU}_1}$ For $R_1 = 3$	$P_1 = \frac{9-R_1}{3+R_1}$ for $R_1 \leq 3$ $P_1 = 1/R_1$ for $R_1 > 3$
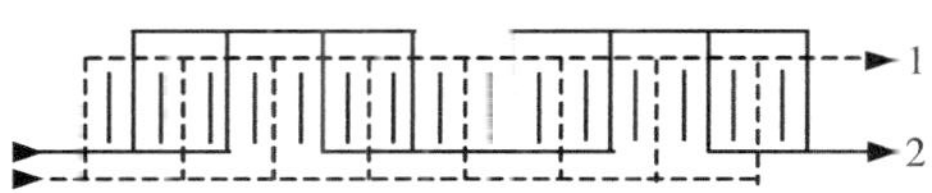 1 pass – 4 pass plate exchanger	V. 6	$P_1 = (1 - Q)/R_1$ $Q = (1 - AR_1/4)^2 (1 - \boldsymbol{BR}_1/4)^2$ $A = P_p(\mathrm{NTU}_1, R_1/4)$ $B = P_c(\mathrm{NTU}_1, R_1/4)$	Same as eqn (V.6) with $B = \frac{\mathrm{NTU}_1}{1+\mathrm{NTU}_1}$ For $R_1 = 4$	$P_1 = \frac{16}{(4+R_1)^2}$ for $R_1 \leq 4$ $P_1 = 1/R_1$ for $R_1 > 4$

(*Continued*)

Table 6.1: (*Continued*).

Flow arrangement	Eqn no.	General formula	Value for $R_1 = 1$ unless specified differently	Value for $\mathrm{NTU}_1 \to \infty$
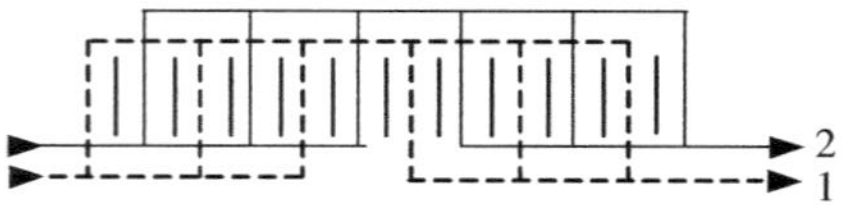 2 pass – 2 pass plate exchanger with overall parallel flow and individual passes in parallel flow, stream symmetric	V. 7	P_1 same as eqn (V.1)	Same as eqn (V.1)	Same as eqn (V.1)
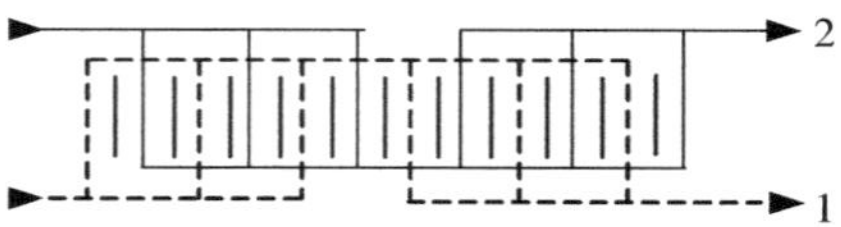 2 pass – 2 pass plate exchanger with overall counter flow and individual passes in parallel flow, stream symmetric	V. 8	$P_1 = B[2 - B(1 + R_1)]$ $B = P_c(\mathrm{NTU}_1/2, R_1)$	Same as eqn (V.8) with $B = \frac{\mathrm{NTU}_1}{2+\mathrm{NTU}_1}$ for $R_1 = 1$	$P_1 = 1 - R_1$ for $R_1 \leq 1$ $P_1 = (R_1 - 1)/R_1^2$ for $R_1 > 1$
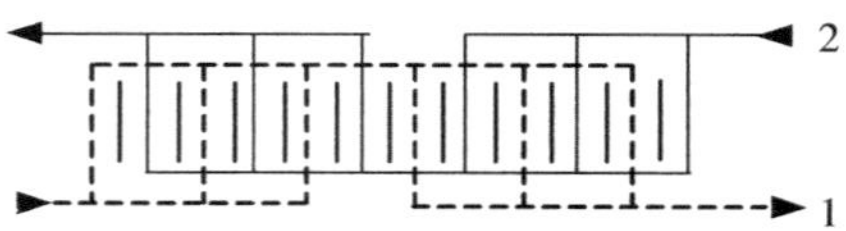 2 pass – 2 pass plate exchanger with overall counter flow and individual passes in parallel flow, stream symmetric	V. 9	$P_1 = \frac{2A - A^2(1+R_1)}{1-R_1A^2}$ $A = P_p(\mathrm{NTU}_1/2, R_1)$	Same as eqn (V.9)	$P_1 = \frac{1+R_1}{1+R_1+R_1^2}$

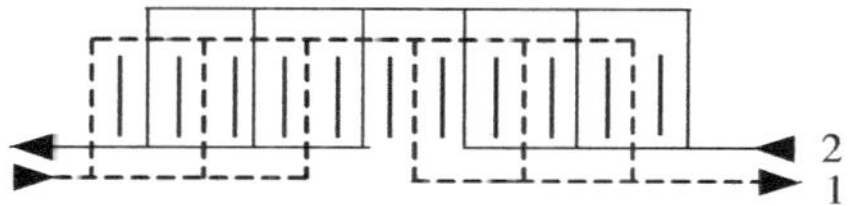2 pass – 2 pass plate exchanger with overall counter flow and individual passes in counter flow, stream symmetric	V. 10	P_1 same as eqn (V.2)	Same as eqn (V.2)	Same as eqn (V.2)
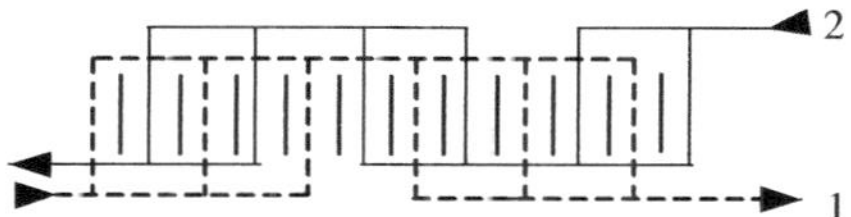2 pass – 3 pass plate exchanger with overall counter flow	V. 12	$P_1 = (A + 0.5B + 0.5C + D)/R_1$ $A = \frac{2R_1 EF^2 - 2EF + F - F^2}{2R_1 E^2 F - E^2 - F^2 - 2EF + E + F}$ $B = A(E-1)/F$ $C = (1-A)/E$ $D = R_1 E^2 C - R_1 E + R_1 - C/2$ $E = 1/(2R_1 G/3)$ $F = 1/(2R_1 H/3)$ $G = P_c(\mathrm{NTU}_1/2, 2R_1/3)$ $H = P_p(\mathrm{NTU}_1/2, 2R_1/3)$	Same as eqn (V.12) with $G = \frac{\mathrm{NTU}_1}{2+\mathrm{NTU}_1}$ for $R_1 = \frac{3}{2}$	$P_1 = \frac{27+12R_1-4R_1^2}{27+12R_1+4R_1^2}$ for $R_1 \leq \frac{3}{2}$ $P_1 = \frac{1}{R_1}$ for $R_1 > \frac{3}{2}$
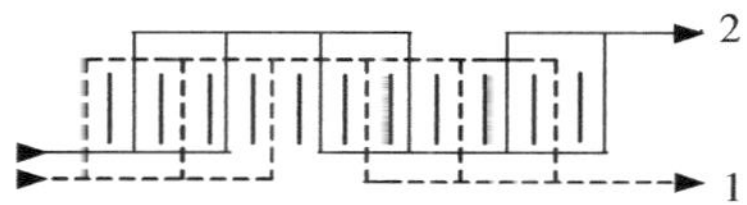2 pass – 3 pass plate exchanger with overall parallel flow	V. 11	$P_1 = A + B - (2/9 + D/3)(A^2 + B^2) - (5/9 + 4D/3)AB + D(1 + D)AB(A + B)/3 - D^2A^2B^2/9$ $A = P_p(\mathrm{NTU}_1/2, D)$ $B = P_c(\mathrm{NTU}_1/2, D)$ $D = 2R_1/3$	Same as eqn (V.11) with $B = \frac{\mathrm{NTU}_1}{2+\mathrm{NTU}_1}$ for $R_1 = \frac{3}{2}$	$P_1 = \frac{9-2R_1}{9+6R_1}$ for $R_1 \leq \frac{3}{2}$ $P_1 = \frac{4R_1^2+2R_1-3}{2R_1^2(3+2R_1)}$ for $R_1 > \frac{3}{2}$

(Continued)

Table 6.1: (*Continued*).

Flow arrangement	Eqn no.	General formula	Value for $R_1 = 1$ unless specified differently	Value for $\mathrm{NTU}_1 \to \infty$
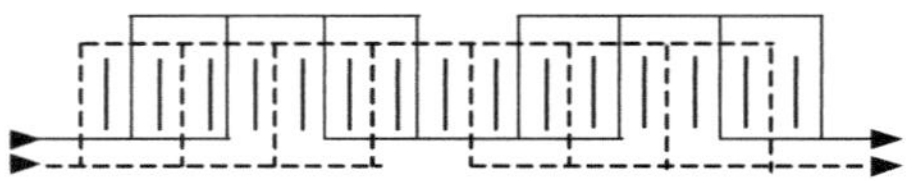 2 pass – 4 pass plate exchanger with overall parallel flow	V. 13	$P_1 = 2D - (1+R_1)D^2$ $D =$ $(A + B - \boldsymbol{ABR}_1/2)/2$ $A = P_\mathrm{p}(\mathrm{NTU}_1/2, D)$ $B = P_c(\mathrm{NTU}_1/2, D)$	Same as eqn (V.13) with $B = \frac{\mathrm{NTU}_1}{2+\mathrm{NTU}_1}$ for $R_1 = 2$	$P_1 = \frac{4}{(2+R_1)^2}$ for $R_1 \le 2$ $P_1 = \frac{R_1-1}{R_1^2}$ for $R_1 > 2$
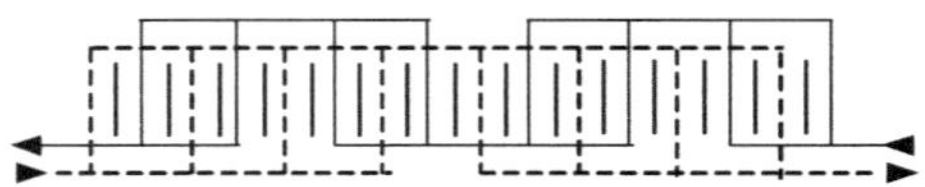 2 pass – 4 pass plate exchanger with overall counter flow	V. 14	$P_1 = \frac{2D-(1+R_1)D^2}{1-D^2R_1}$ $D =$ $(A + B - \boldsymbol{ABR}_1/2)/2$ $A = P_\mathrm{p}(\mathrm{NTU}_1/2, D)$ $B = P_\mathrm{c}(\mathrm{NTU}_1/2, D)$	Same as eqn (V.14) with $B = \frac{\mathrm{NTU}_1}{2+\mathrm{NTU}_1}$ for $R_1 = 2$	$P_1 = \frac{4}{4+R_1^2}$ for $R_1 \le 2$ $P_1 = \frac{1}{R_1}$ for $R_1 > 2$

Note: In all the formulas of PHEs with the number of thermal plates $N \to \infty$ (equation numbers starting with V), the single-pass parallel-flow and counter-flow temperature effectivenesses are presented in implicit forms. Their explicit forms are as follows, with x and y representing the appropriate values of the number of transfer units and heat capacity rate ratios, respectively.

Single-pass parallel flow

$$P_\mathrm{p}(x,y) = \frac{1 - \exp[-x(1+y)]}{1+y}$$

$$P_\mathrm{p}(x,1) = [1 - \exp(-2x)]/2$$

$$P_\mathrm{p}(\infty,y) = 1/(1+y)$$

Single-pass counter flow

$$P_\mathrm{c}(x,y) = \frac{1 - \exp[-x(1-y)]}{1 - y\exp[-x(1-y)]}$$

$$P_\mathrm{c}(x,1) = x/(1+x)$$

$$P_\mathrm{c}(\infty,y) = 1 \text{ for } y \le 1$$

$$P_\mathrm{c}(\infty,y) = 1/y \text{ for } y > 1$$

Subscripts

1	fluid stream 1
2	fluid stream 2
c	cold fluid stream
h	hot fluid stream
i	channel i
I	channel 1
II	channel 2
in	inlet
n	channel n
p	transfer plates

References

[1] Cooper, A. and Usher, J.D., Plate Heat Exchangers, in *Heat Exchanger Design Handbook*, ed. E.U. Schlunder, vol. 3, Hemisphere, Washington, section 3.7, 1983.

[2] Okada, K., et al., Temperature Distribution of Fluids in Plate Type Heat Exchangers, *Kagaku Kogaku*, vol. 34, no. 1, pp. 93–95, 1970.

[3] Watson, E.L., et al., Plate Heat Exchangers, Flow Characteristics, *Industrial Engineering Chemistry*, vol. 52, no. 9, pp. 733–740, 1960.

[4] Kandlikar, S.G. and Shah, R.K., Multipass Plate Heat Exchangers – Effectiveness-NTU Results and Guidelines for Selecting Pass Arrangements, *ASME Journal of Heat Transfer*, vol. 111, pp. 300–313, 1989.

[5] Pignotti, A., Flow Reversibility of Heat Exchangers, *ASME Journal of Heat Transfer*, vol. 106, pp. 361–368, 1984.

[6] Yang, Q. and Wang, L., Thermal Performance and Maldistribution of Multi-pass Plate Heat Exchangers, in *Heat Transfer Science and Technology*, ed. B. Wang, Higher Education Press, China, pp. 621–626, 1996.

[7] Heggs, P.J. and Scheidat, H.J., Thermal Performance of Plate Heat Exchangers with Flow Maldistribution, in *Compact Heat Exchangers for Power and Process Industries*, ed. R.K. Shah, T.M. Rudy, J.M. Robertson, and K.M. Hostetler, HTD, vol. 201, ASME, New York, pp. 87–93, 1992.

[8] Anderson, D.A., Tannehill, J.C., and Pletcher, R.H., *Computational Fluid Mechanics and Heat Transfer*, Hemisphere, Washington, 1984.

[9] Shah, R.K. and Sekulic, D.P., Heat Exchangers, in *Handbook of Heat Transfer*, 3rd edn, ed. W. Rohsenow, J. Hartnett, and Y. Cho, McGraw-Hill, New York, chapter 17, 1998.

CHAPTER 7

Thermal-hydraulic performance in single-phase flows

7.1 Introduction

The manufacture of end products in the chemical, pulp and paper, food, personal care and hygiene, pharmaceutical, and biochemical industries, to name a few, invariably involves heating or cooling of viscous process media. Single-phase fluid flow conditions often exist in such processing applications that generally form the controlling large thermal resistance in a heat exchanger. As such, the use of high-performance (or enhanced) exchangers can significantly improve the process efficiency, generate substantial energy, materials, and cost savings, as well as help in the abatement of environmental degradation. Furthermore, besides the maintenance of hygienic conditions, many applications also require close thermal quality control of the end products, for which it is necessary to operate with smaller temperature differences and thus mitigate thermal degradation of the process media. This is particularly critical to most foods, pharmaceutical, and personal care product processing. To meet these demands, PHEs have increasingly become the heat exchanger of choice because they are compact, have the flexibility of altering the thermal size, easy to clean, and have attractive enhanced heat transfer characteristics [1–4].

To recapitulate, PHEs consist of thin, rectangular, pressed sheet metal plates that are sandwiched between full peripheral gaskets and clamped together in a frame, such that the hot and cold fluid streams alternate through the inter-plate passages. The plates are stamped with corrugated patterns that not only provide larger effective heat transfer surface area but also modify the flow field, thereby promoting a significantly enhanced thermal-hydraulic performance [1–5]. Typical structural design and single-pass flow arrangement features of a PHE are described in Chapters 1 and 2. It may be noted again that their modular construction provides the flexibility for increasing or decreasing the thermal size for accommodating varying heat load capacities, by simply adding or removing thermal plates without incurring additional changes in either the process piping or space requirements. This also allows easy disassembly for periodic cleaning and fouling prevention. Several earlier reviews, texts, and technical reports [1–8] provide some guidelines on the different aspects of their design, development, and applications in single-phase flows.

Although plates with a variety of surface-corrugation patterns are commercially available (primarily as a result of parallel, competitive commercial development among different PHE manufacturers, as described in Chapters 1 and 2), the chevron corrugation geometry and its variation as a herringbone or zig-zag pattern are perhaps the most widely used [1, 2, 4, 5]. Typical chevron plates of low and high corrugation inclination angle β and their stack arrangements are shown in Fig. 7.1a. Again, the surface waviness can be of different profile shapes (triangular with rounded edges, trapezoidal, sinusoidal, etc.) of which the sinusoidal profile is more commonly used [1, 5, 9–11]. Low and high β plates can be stacked together in either a symmetric or mixed arrangement; in a mixed arrangement, plates of two different β values are employed alternately. The multiple contact points between adjacent plates lend to their increased rigidity and mechanical support of the stack. The effective surface area, section *a–b–c–d–e–f–a* in Fig. 7.1a, which primarily accounts for the exchanger's compactness and enhanced heat transfer performance, includes the main chevron corrugation area as well as the triangular fluid distribution and collection areas.

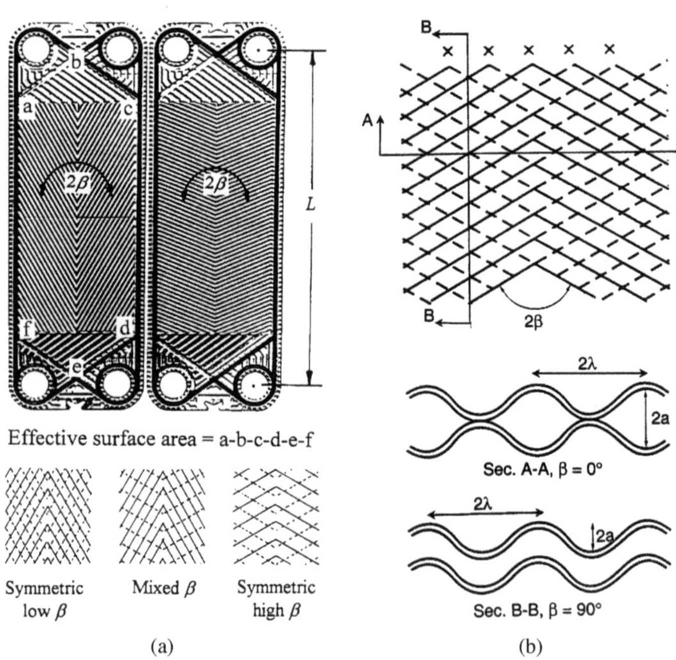

Figure 7.1: Corrugation features of chevron plates: (a) low and high chevron-angle β plates and their different arrangements and (b) geometrical description of sinusoidal surface corrugates.

The importance of PHEs to the highly competitive operating environment of the chemical and process industry is amply reflected in the extensive available literature. However, the effective usage of PHEs in diverse single-phase flow applications requires reliable design data and/or predictive tools. Over- or under-sizing often results in large local temperature gradients and flow maldistribution, which greatly affect process efficiency and cause thermal degradation of the process media. Also, the associated uncertainty with the pressure drop adversely impacts plant scale-up operations. In order to address the concomitant design problems, an extended review of the performance of chevron plates, along with a critical assessment of the relative influence of the geometrical features of chevron corrugations and the associated heat transfer enhancement, available thermal-hydraulic design correlations, and performance optimization strategies are presented in the ensuing sections.

7.2 Chevron-plate performance literature

The single-phase flow thermal-hydraulic performance of PHEs with chevron plates is greatly influenced by the geometrical attributes of plate-surface corrugations. These are primarily described by the chevron corrugation inclination angle β, its wavelength λ, and amplitude a, as schematically illustrated in Fig. 7.1b. The corrugation geometry and the consequent complex inter-plate flow channels directly promote enhanced heat transfer by increasing the effective heat transfer area, the disruption and reattachment of boundary layers, and generation of swirl or vortex or helical secondary flows [12–15]. The surface waviness can be essentially represented by two dimensionless parameters, namely, the surface enlargement factor ϕ, which is given by

$$\phi = \frac{\text{Effective area } a\text{–}b\text{–}c\text{–}d\text{–}e\text{–}f\text{–}a}{\text{Projected area } a\text{–}b\text{–}c\text{–}d\text{–}e\text{–}f\text{–}a} \tag{7.1}$$

and the corrugation aspect ratio γ, which is defined as

$$\gamma = (4a/\lambda) \tag{7.2}$$

It may be noted that the increased effective heat transfer area (Fig. 7.1b) includes the main chevron area (a–c–d–f–a) as well as the triangular fluid distribution and collection areas (a–b–c–a and e–f–d–e). Both the heat transfer coefficients and flow friction losses generally increase with β, ϕ, and γ. Furthermore, the inter-plate spacing that confines the flow channel is twice the corrugation depth, or $4a$, in order to assure metal-to-metal contact at the peak-points of the surface waviness. The length scale or the equivalent diameter for the inter-plate channel then becomes

$$d_e = 8a \tag{7.3}$$

Inherent in this definition is the assumption that the width of the plate is much greater than the inter plate separation ($W \gg 4a$), and is essentially the same length scale as that in flat, parallel plate channels. In some literature, however, a modified length scale in the form of an effective hydraulic diameter has been used that includes the surface area enlargement and is given by

$$d_h = (d_e/\phi) = (8a/\phi) \tag{7.4}$$

In this context, the discussion and definitions given in Chapter 2, eqns (2.1) through (2.3), may also be referred. The definition given by eqn (7.3) is used in the ensuing presentation as it has the advantage of facilitating a more direct comparison of the wavy plate performance with that in a flat-plate channel.

Emerson [11] was perhaps the first to report experimental results for heat transfer and friction factor for water flow in the laminar and transition regimes in chevron-plate PHEs. Subsequently, one of the earliest studies to systematically investigate the effects of chevron angle β was reported by Savostin and Tikhonov [16], with experimental Nu and f data for water-cooled air flows in chevron-plate passages with $0° \leq \beta \leq 72.2°$. Similarly, experimenting with turbulent water flows, Okada et al. [17] have reported Nu and Δp results for chevron plates with $30° \leq \beta \leq 75°$. Rosenblad and Kullendorff [18] and Focke et al. [19] have used mass transfer techniques with scaled-up plates to evaluate the effects of β. In all of these studies, only symmetric β arrangements have been considered, and the heat transfer coefficients have been found to increase with β; of course, there is also a larger associated pressure drop penalty. More recently, Muley et al. [13], and Muley and Manglik [20] have reported Nu and f data for Newtonian vegetable oil and water flows in laminar and turbulent regimes, with symmetric- and mixed-plate arrangements where $\beta = 30°$ and $60°$.

The thermal-hydraulic performance of mixed chevron-plate arrangements has been investigated by Marriott [21], Chisholm and Wanniarachchi [22], and Heavner et al. [23], among a few others [1, 2, 24]. Marriott21 has reported results for $\beta = 32.5°/65°$ that suggest a rather moderate increase in Δp relative to symmetric arrangements, but with a much greater increase in the corresponding heat transfer coefficients. The experimental data of Heavner et al. [23] are for turbulent water flows ($400 < \text{Re} < 10^4$) with both symmetric and mixed chevron-plate arrangements where $\beta = 23°, 45°$, and $90°$.

In many of these studies, correlations have been devised for predicting the heat transfer coefficients and frictional loss in PHEs with chevron plates. Although the majority of equations are based on rather sparse data, or the investigators' own data, it is nevertheless instructive to discuss and contrast some of them. In each case in the ensuing presentation, the hydraulic-diameter length scale given in eqn (7.3) has been used for consistency. One of the earliest effort [1] was made in 1970 by Savostin and Tikhonov [16], who were able to describe their air flow results for symmetric plates with $0° \leq \beta \leq 33°$ by

$$f = 6.25(1 + 0.95\psi^{1.72})\phi^{1.84}\text{Re}^{-0.84} \tag{7.5a}$$

$$\text{Nu} = 1.26[0.62 + 0.38\cos(2.3\psi)]\phi^{1-a1}\text{Pr}^{1/3}\text{Re}^{a1};$$

$$a1 = 0.22(1 + 1.1\psi^{1.5}) \tag{7.5b}$$

for $200 \leq (\mathrm{Re}/\phi) \leq 600$, and

$$f = 0.925[0.62 + 0.38\cos(2.6\psi)]\phi^{1+a2}\mathrm{Re}^{-a2};$$
$$a2 = 0.53[0.58 + 0.42\cos(1.87\psi)] \tag{7.5c}$$
$$\mathrm{Nu} = 0.072\phi^{0.33}\mathrm{Pr}^{1/3}\mathrm{Re}^{0.67}\exp(0.5\psi + 0.17\psi^2) \tag{7.5d}$$

for $600 \leq (\mathrm{Re}/\phi) \leq 4000$; here $\psi = 2\beta$ radians. Okada et al. [17] devised separate curve-fit equations for the Nusselt number to fit their water data for $700 < \mathrm{Re} < 25{,}000$ with different β symmetric-plate arrangements as follows:

$$\mathrm{Nu} = \begin{cases} 0.157\mathrm{Pr}^{0.4}\mathrm{Re}^{0.66} & \beta = 30^\circ \\ 0.249\mathrm{Pr}^{0.4}\mathrm{Re}^{0.64} & \beta = 45^\circ \\ 0.327\mathrm{Pr}^{0.4}\mathrm{Re}^{0.65} & \beta = 60^\circ \\ 0.478\mathrm{Pr}^{0.4}\mathrm{Re}^{0.62} & \beta = 75^\circ \end{cases} \tag{7.6}$$

No equations or data for friction factors has been given.

Almost a decade later in 1980, Tovazhnyanski et al. [25] devised the following correlations for turbulent flows ($2000 < \mathrm{Re} < 25{,}000$) in PHEs with three different symmetric-plate arrangements of $\beta = 30^\circ, 45^\circ$, and 60°:

$$f = 0.085\exp[1.52\tan\beta]\mathrm{Re}^{-(0.25-0.06\tan\beta)} \tag{7.7a}$$
$$\mathrm{Nu} = 0.051e^{[0.64\tan\beta]}\mathrm{Re}^{0.73}\mathrm{Pr}^{0.43}(\mathrm{Pr}/\mathrm{Pr_w})^{0.25} \tag{7.7b}$$

For the same β set of plates and based on their mass transfer measurements with scaled-up plates, Focke et al. [19] have devised multiple sets of equations for each arrangement and flow regime as follows:

Chevron plates with $\beta = 60^\circ$

$$f = \begin{cases} 1.2575 + 188.75/\mathrm{Re} & 90 < \mathrm{Re} < 400 \\ 6.7\mathrm{Re}^{-0.209} & 400 < \mathrm{Re} < 16{,}000 \end{cases} \tag{7.8a}$$

$$\mathrm{Nu} = \begin{cases} 1.89\mathrm{Re}^{0.46}\mathrm{Pr}^{1/2} & 20 < \mathrm{Re} < 150 \\ 0.57\mathrm{Re}^{0.7}\mathrm{Pr}^{1/2} & 150 < \mathrm{Re} < 600 \\ 1.12\mathrm{Re}^{0.6}\mathrm{Pr}^{1/2} & 600 < \mathrm{Re} < 16{,}000 \end{cases} \tag{7.8b}$$

Chevron plates with $\beta = 45^\circ$

$$f = \begin{cases} 0.3025 + 91.75/\mathrm{Re} & 150 < \mathrm{Re} < 1800 \\ 1.46\mathrm{Re}^{-0.177} & 1800 < \mathrm{Re} < 30{,}000 \end{cases} \tag{7.8c}$$

$$\mathrm{Nu} = \begin{cases} 1.67\mathrm{Re}^{0.44}\mathrm{Pr}^{1/2} & 45 < \mathrm{Re} < 300 \\ 0.405\mathrm{Re}^{0.7}\mathrm{Pr}^{1/2} & 300 < \mathrm{Re} < 2000 \\ 0.84\mathrm{Re}^{0.6}\mathrm{Pr}^{1/2} & 2000 < \mathrm{Re} < 20{,}000 \end{cases} \tag{7.8d}$$

Chevron plates with $\beta = 30°$

$$f = \begin{cases} 0.0925 + 57.5/\text{Re} & 260 < \text{Re} < 3000 \\ 0.8975\text{Re}^{-0.263} & 3000 < \text{Re} < 50{,}000 \end{cases} \tag{7.8e}$$

$$\text{Nu} = \begin{cases} 0.77\text{Re}^{0.54}\text{Pr}^{1/2} & 120 < \text{Re} < 1000 \\ 0.44\text{Re}^{0.64}\text{Pr}^{1/2} & 1000 < \text{Re} < 42{,}000 \end{cases} \tag{7.8f}$$

Chisholm and Wanniarachchi [22] represented their data with both symmetric and mixed plates, where $30° \leq \beta \leq 80°$, by

$$f = 0.8\text{Re}^{-0.25}\phi^{1.25}(\beta/30)^{3.6} \quad 10^3 < \text{Re} < 4 \times 10^3 \tag{7.9a}$$

$$\text{Nu} = 0.72\text{Re}^{0.59}\text{Pr}^{0.4}\phi^{0.41}(\beta/30)^{0.66} \quad 10^3 < \text{Re} < 4 \times 10^3 \tag{7.9b}$$

These were later modified into a much more complex form by Wanniarachchi et al. [26] to represent the same set of results.

For one mixed-plate arrangement, perhaps the first predictive equations for Δp and h, in dimensional form, were proposed in 1977 by Marriott [21]. More than a decade and a half later in 1993 and with a rather extended set of mixed plates and flow rates ($400 < \text{Re}/\phi < 10{,}000$), Heavner et al. [23] devised the following set of expressions for each plate arrangement:

$$f = \begin{cases} 1.715\phi^{1.0838}\text{Re}^{-0.0838} & \beta = 45°/90° \\ 1.645\phi^{1.1353}\text{Re}^{-0.1353} & \beta = 23°/90° \\ 0.810\phi^{1.1405}\text{Re}^{-0.1405} & \beta = 45°/45° \\ 0.649\phi^{1.1555}\text{Re}^{-0.1555} & \beta = 23°/45° \\ 0.571\phi^{1.1814}\text{Re}^{-0.1814} & \beta = 23°/23° \end{cases} \tag{7.10a}$$

$$\text{Nu} = \text{Pr}^{1/3}\left(\frac{\mu_b}{\mu_w}\right)^{0.17} \times \begin{cases} 0.278\phi^{0.317}\text{Re}^{0.683} & \beta = 45°/90° \\ 0.308\phi^{0.333}\text{Re}^{0.667} & \beta = 23°/90° \\ 0.195\phi^{0.308}\text{Re}^{0.692} & \beta = 45°/45° \\ 0.118\phi^{0.280}\text{Re}^{0.720} & \beta = 23°/45° \\ 0.089\phi^{0.282}\text{Re}^{0.718} & \beta = 23°/23° \end{cases} \tag{7.10b}$$

Muley and Manglik [24] considered a single mixed-plate arrangement with chevron angles of $\beta = 30°$ and $60°$ and proposed the following for their Newtonian vegetable oil and water data:

$$f = \begin{cases} [(40.32/\text{Re})^5 + (8.12\text{Re}^{-0.5})^5]^{0.2} & 2 \leq \text{Re} \leq 200 \\ 1.274\text{Re}^{-0.15} & \text{Re} \geq 1000 \end{cases} \tag{7.11a}$$

$$\text{Nu} = \begin{cases} 0.471\text{Re}^{0.5}\text{Pr}^{1/3}(\mu/\mu_w)^{0.14} & 20 \leq \text{Re} \leq 400 \\ 0.10\text{Re}^{0.76}\text{Pr}^{1/3}(\mu/\mu_w)^{0.14} & \text{Re} \geq 1000 \end{cases} \tag{7.11b}$$

In all of these cases, it has been suggested that $\beta_{avg} = [(\beta_1 + \beta_2)/2]$ is representative of the effective chevron angle in mixed-plate arrangements.

From the above it is evident that there are several different correlations given in the literature for f and Nu characteristics of chevron plates, and there appears to be little consensus among them for predicting the performance of plates with identical chevron angle. Almost all equations are curve fits through sparse empirical data, where invariably log-linear power-law-type functions have been used. The expressions generally do not specifically identify the contributions of the different geometrical attributes of the corrugated passages. Also, with the exception of Savostin and Tikhonov [16], Tovazhnyanski et al. [25], and Chisholm and Waniarachchi [22], separate equations for different values of β that are valid over a limited range of operating conditions have been devised. In the three cases where the β-dependency has been incorporated in the f and Nu correlations, their functional forms are quite different.

7.3 Thermal-hydraulic characteristics

The enhanced convection heat transfer in the corrugated inter-plate channels is primarily due to the relatively higher effective heat transfer area, small flow cross-section hydraulic diameter, disruption and reattachment of boundary layers, and swirl- or vortex-flow generation. Generally, both the heat transfer coefficient and friction factor increase with increasing chevron angle β, besides γ or the severity of corrugation waviness [1, 2, 12–14, 19, 20, 27–30]. While PHEs generally use chevron plates with $0° < \beta < 90°$, the limiting effects of β on the thermal-hydraulic behaviour are described by the inter-plate channel flows when $\beta = 0°$ and $90°$, and these cases are also found in some practical applications.

7.3.1 Single-phase convection in $\beta = 0°$ plate channels

With the chevron inclination angle in the range $0° < \beta < 90°$, cross-corrugated channels with multiple metal-to-metal inter-plate point contacts are obtained. The case when $\beta = 0°$ yields a set of parallel double-sine-shaped ducts in the inter-plate passages, as shown schematically in sec. A–A of Fig. 7.1b.

Laminar, constant property, fully developed forced convection in double-sine-shaped straight channels, which represents inter-plate flow passages in $\beta = 0°$ plates, has been considered by Manglik and Ding [31]. The Galerkin integral method was employed to obtain solutions for the flow velocity and temperature fields. These theoretical results reveal some interesting influences of the geometrical features of the flow passages.

Typical distributions for the dimensionless normalized axial velocity (w/w_m) and temperature (T/T_m) in ducts of three different aspect ratios ($\gamma = 0.25$, 1.0, and 4.0) are shown in Fig. 7.2. For the thermal problem, the uniform wall temperature (UWT or **T**) as well as the uniform wall heat flux (UHF or **H1**) boundary condition are considered. The thermal condition of UWT is commonly encountered in two-fluid stream exchangers where the heat capacity rate of the heating/cooling

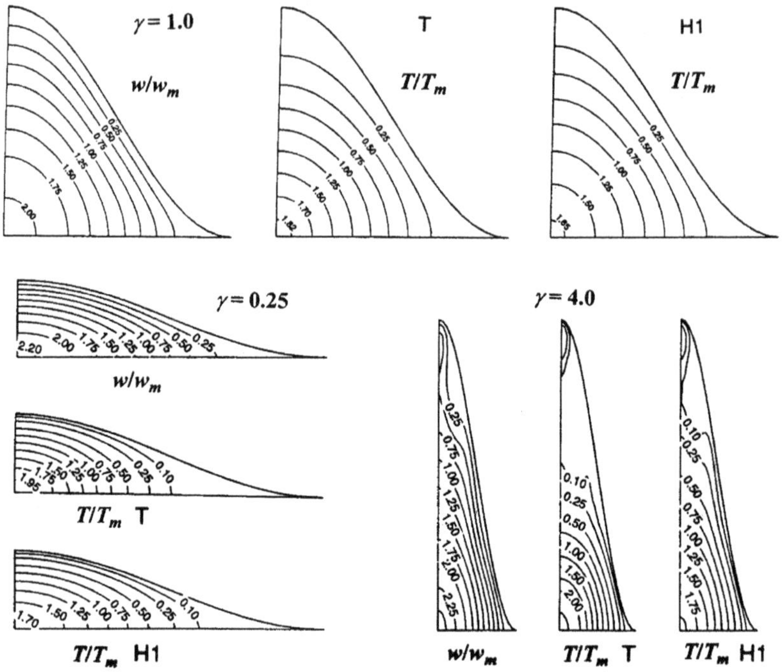

Figure 7.2: Axial velocity (w/w_{m}) and temperature (T/T_{m}) contours for fully developed laminar flows in straight double-sine-shaped ducts of different cross-sectional aspect ratios ($\gamma = 1.0$, 0.25, and 4.0) [31].

fluid is relatively very large, as well as in steam heating and refrigerant evaporative/condensing cooling and heating of a process stream. The second fundamental boundary condition of UHF represents the fluid-to-fluid heating and cooling conditions where the heat capacity rate of the two streams is the same. As seen from the axial isovelocity and isotherm contour plots, when the channel aspect ratio is either very small ($\gamma = 0.25$) or large ($\gamma = 4.0$), the 'squeezing' effect of the channel geometry produces higher peak or centreline velocities. Although this causes greater fluid mobility in the core portion, when compared to that in a $\gamma = 1.0$ channel, the flow in the corner regions tends to stagnate. As a result, considerable thermal maldistribution is obtained where much of the corner fluid flow attains the wall temperature, while the core fluid has a significantly cooler or hotter temperature. Thermal stratification is also seen to be greater with the UWT condition as compared to that with the UHF condition. It may be noted, however, that corrugations with $\gamma = 4.0$ would perhaps be impractical to manufacture (typical limit is $\gamma \sim 2.0$), but the results for this case are instructive in dramatizing the anomalous convection in deep-grooved plate passages.

The effects of flow maldistribution in double-sine ($\beta = 0°$) corrugated channels with increasing aspect ratio γ on the friction factor and Nusselt number for fully developed laminar flows are depicted in Fig. 7.3. While both heat transfer coefficient and pressure drop increase with γ, the thermal performance is seen to level off when

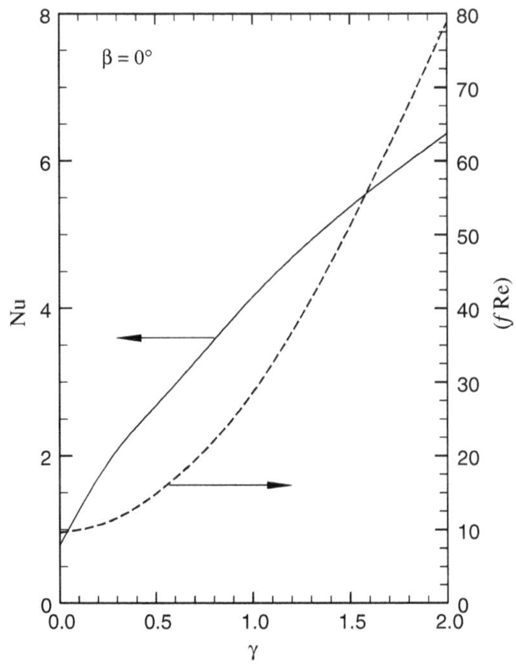

Figure 7.3: Variation in Nusselt number and isothermal Fanning friction factor with cross-sectional aspect ratio γ in fully developed laminar flows in $\beta = 0^\circ$ inter-plate channels [31, 34].

$\gamma > 1$ but the flow friction loss increases exponentially; there is relatively a very small increase in friction factor in channels with $\gamma < 1$. These results have an important implication for PHE usage: the benefits of larger effective plate-surface area (or greater area enlargement ϕ) from deeper or smaller pitch corrugations may be significantly offset by lower Nusselt numbers and larger friction factors. There is, in any case as pointed out earlier, a limit to the depth of corrugations that can be safely pressed and manufactured without compromising the structural integrity of the plates. Furthermore, the flow maldistribution and inhomogeneity in the deeper corrugated channels may also be very detrimental in fouling situations.

Forced convection in turbulent flows through straight double-sine channels does not appear to have been considered in the literature [32, 33]. In the absence of such results, a first-order assessment of the thermal-hydraulic performance can perhaps be made by following the usual convention of employing the straight circular duct correlation for f and Nu, where the length scale used is the actual hydraulic diameter of the duct being considered; eqn (7.4) in the present case.

7.3.2 Single-phase convection in $\beta = 90^\circ$ plate channels

When the chevron inclination angle $\beta = 90^\circ$ in these corrugations, in-phase sinusoidal wavy plate channels are obtained in the inter-plate flow passages. The peaks and valleys in the top and bottom plates in this configuration lie on the same longitudinal plane, as seen in the schematic of sec. B–B of Fig. 7.1b, with plate spacing equal to twice the waviness amplitude. Also, because the plate spacing $2a$ is much smaller than the plate width w, the flow passages can essentially be modelled as two-dimensional wavy, parallel-plate channels. The severity of wall waviness or corrugation is nevertheless described by γ; it may be noted that when $\gamma = 0$, a flat parallel-plate channel is obtained, and $\gamma > 1$ is thus indicative of surface area enlargement ($\phi > 1$) as well. Metwally and Manglik [12] have computationally modelled the low Reynolds number ($\mathrm{Re} \leq 1000$) forced-convective behaviour in such parallel-plate channels that have in-phase sinusoidal wall corrugations, and their results are summarized below.

As seen in Fig. 7.4, variations in the pitch and amplitude of the sinusoidal plate-surface corrugations (or γ) and the flow rate (or Re) have a significant effect on the behaviour of the flow field. The change in flow structure for a fixed flow rate ($\mathrm{Re} = 300$) with varying severity of the corrugated-plate-surface waviness is depicted in Fig. 7.4a, and the strong influence of the duct geometry on the stream function distribution is evident. With increasing corrugation aspect ratio γ (increasing amplitude a or decreasing pitch λ), fluid re-circulation or swirl flows are generated in the corrugation troughs. The intensity and flow area coverage of this counter-rotating lateral vortex grows with γ, which in turn lends to higher momentum transfer. Increasing flow Re also produces a similar effect, as displayed by the stream function distribution in a channel with fixed corrugation geometry ($\gamma = 0.5$) in Fig. 7.4b, where the vortex cell is seen to grow with Re. With both increasing γ and Re, the flow separation point preceding the onset of re-circulation moves upstream towards the peak of the plate-surface corrugation, and the concomitant flow re-attachment point moves further downstream on the oblique face of the wavy wall. At very low Reynolds number ($\mathrm{Re} \sim O[10]$), or with small plate waviness ($\gamma < 0.375$), on the other hand, the surface geometry has no effect on the flow, and the fluid moves undisturbed through the channel with no re-circulation by simply 'adopting' the wavy passage shape.

The re-circulating flow behaviour brings about changes in the local wall shear stress, and the variations in $(\tau_w/\tau_{w,m})$ with Re and γ are graphed in Fig. 7.5. At low flow rates ($\mathrm{Re} = 10$) in a channel with $\gamma = 0.5$, there is very little change in the local shear stress along the length of the duct; the small undulations seen in Fig. 7.5a result from local core velocity variations as the flow adapts to the wavy wall contours. As Re increases, however, the shear–stress distribution clearly reflects the influence of the wavy wall trough-region lateral vortex and its streamwise growth. The profile shows a wall-velocity gradient peak at the part of wall exposed to flow acceleration (i.e. the side opposite the trough and flow recirculation therein). Also, the longitudinal size of the re-circulation bubble can be estimated from the points of separation and reattachment, or zero wall shear stress locations. A similar behaviour is observed in flows with a fixed Re but increasing severity of plate-surface corrugation γ in Fig. 7.5b. The shear stress again increases with γ, has a local maximum value in the flow acceleration region, and the flow separation

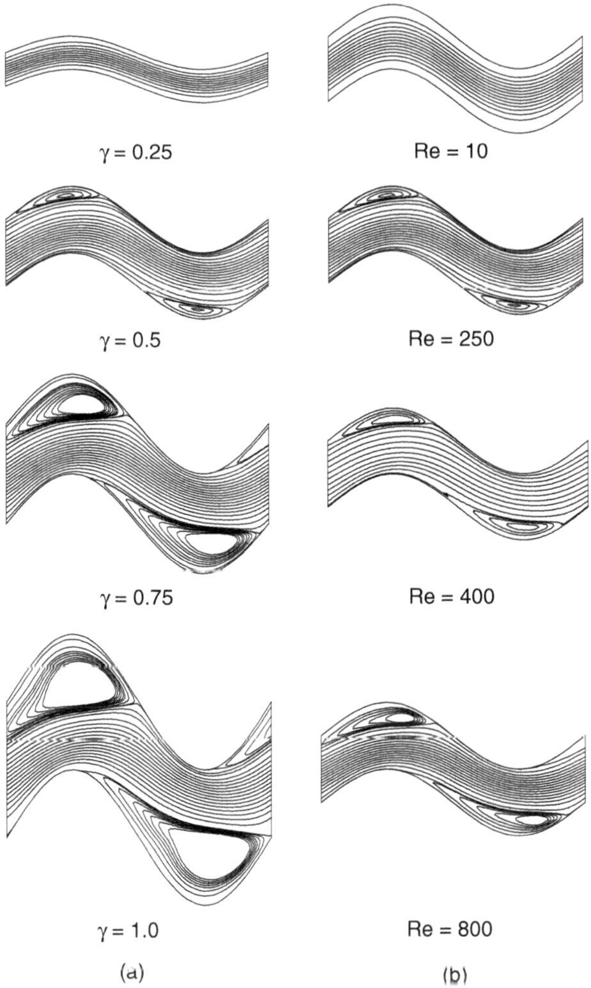

Figure 7.4: Streamline distributions in steady laminar flows in sinusoidal corrugated-plate channels [12]: (a) effect of corrugation severity with $0.25 \leq \gamma \leq 1.0$ and $\mathrm{Re} = 300$, and (b) effect of flow rate with $\gamma = 0.5$ and $10 \leq \mathrm{Re} \leq 800$.

and reattachment points (upstream and downstream ends of the wall trough) that encapsulate the lateral swirl.

Furthermore, the temperature distributions depicted in Fig. 7.6 are atypical of the enhanced laminar flow thermal fields due to the wall corrugations and the swirl flows they induce. The periodically developed temperature fields under constant wall temperature (UWT or **T**) conditions for a viscous liquid with $\mathrm{Pr} = 35$ are graphed. (It may be noted that though the UHF or **H1** condition has not been considered by

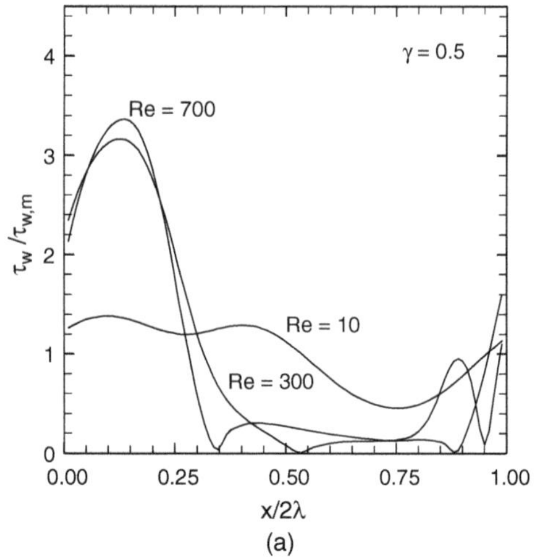

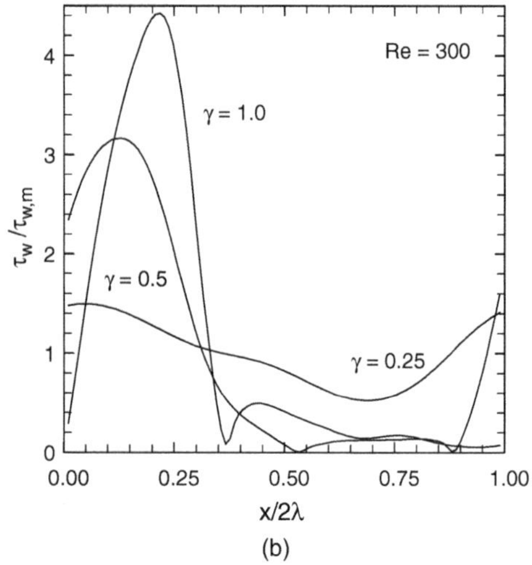

Figure 7.5: Local wall shear stress distribution in steady laminar flows in sinusoidal wavy plate channels [12] and the effect of (a) Re and (b) γ.

Metwally and Manglik [12], the thermal-hydraulic behaviour is qualitatively similar to that with the UWT or **T** condition.) With the growth of transverse vortices in the wall troughs, which encompass much of the bulk flow field with increasing γ, the convective transport is enhanced substantially, as seen from the isotherm maps of Fig. 7.6a. There is considerable thinning of the thermal boundary layer with higher wall temperature gradients as $\gamma = 0.25 \rightarrow 1.0$. This behaviour is also evident in Fig. 7.6b, where the local wall temperature gradients are seen to increase with flow rate ($\text{Re} = 10 \rightarrow 800$) in the channel with $\gamma = 0.5$. Once again, the onset

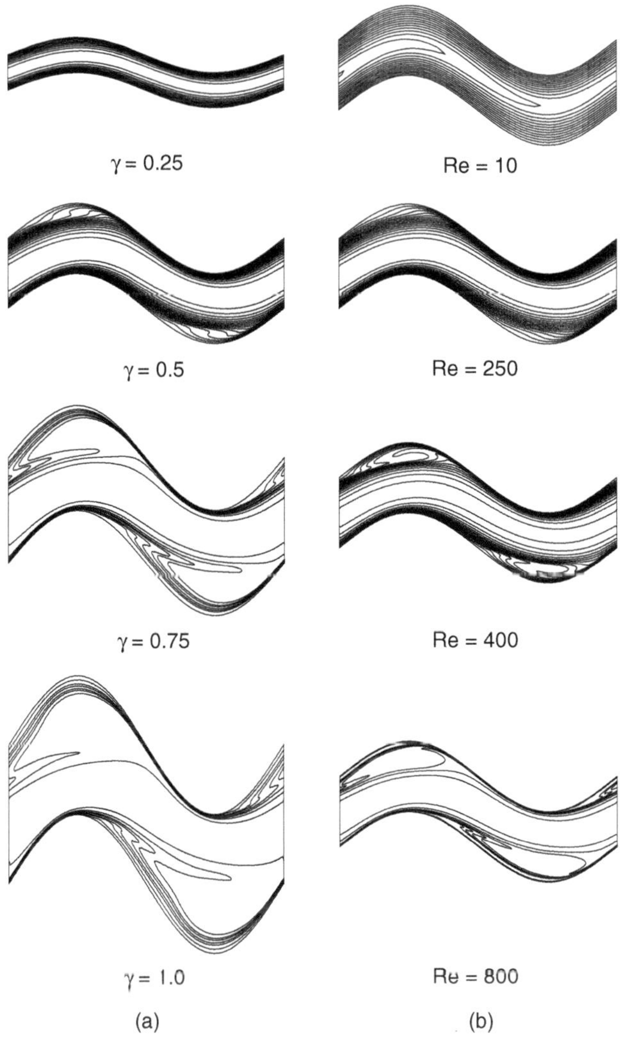

Figure 7.6: Temperature distributions in steady periodically developed laminar viscous liquid (Pr = 35) flows in sinusoidal corrugated-plate channels with UWT condition [12]: (a) effects of corrugation severity with $0.25 \leq \gamma \leq 1.0$ and Re = 300 and (b) effects of flow rate with $\gamma = 0.5$ and $10 \leq \mathrm{Re} \leq 800$.

and growth of swirl flows promotes a more uniform temperature field with sharper wall gradients thereby enhancing the convective heat transfer.

The influence of swirl on heat transfer is further seen in Fig. 7.7, where the variation in the local wall heat flux, normalized by its mean value over the one-period duct length ($q''_w/q''_{w,m}$), virtually mimics the normalized local wall shear stress behaviour. As the swirl develops and grows in the wall troughs of a channel with

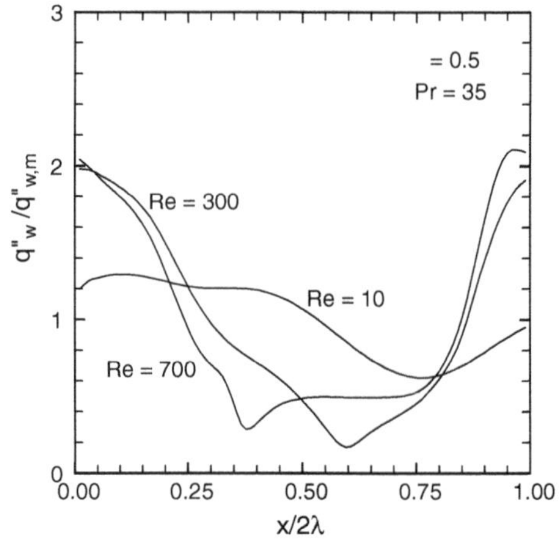

(a)

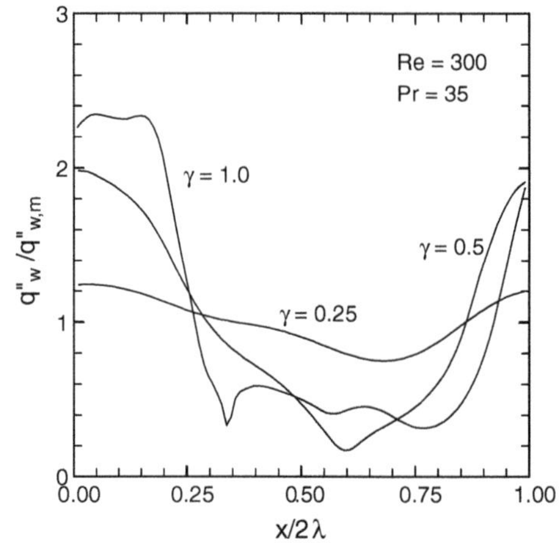

(b)

Figure 7.7: Local wall heat flux distribution in steady laminar flows in sinusoidal wavy plate channels [12] and the effects of (a) Re and (b) γ.

fixed $\gamma(=0.5)$ and increasing Re, large variations in $(q''_{\rm w}/q''_{\rm w,m})$ are seen in Fig. 7.7a, with peak values at the wall exposed to flow acceleration on the side opposite the re-circulation zone. A relatively greater impact on the local heat transfer enhancement is perhaps due to increasing duct corrugation aspect ratio γ, as seen in Fig. 7.7b, where considerably higher local peak heat fluxes are obtained as $\gamma \rightarrow 1$; the level of heat transfer deterioration in the re-circulation region decreases somewhat and is more than offset by the larger peak heat flux values. On the other hand with low Re flows or less severe corrugation (low γ) channels, the local heat flux remains virtually constant with small changes due to local fluid acceleration caused by wall undulations.

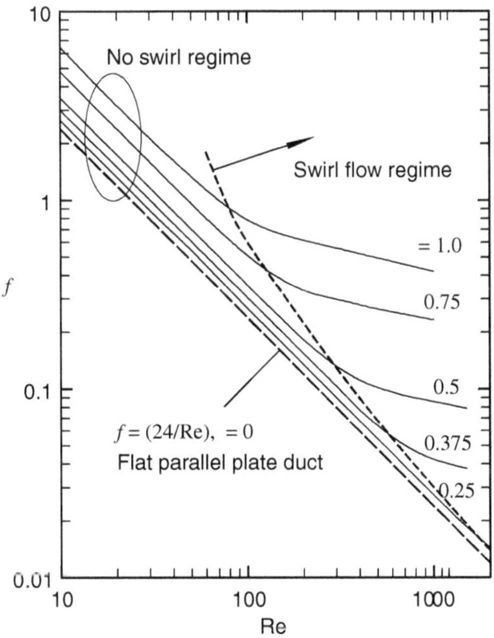

Figure 7.8: Periodically developed laminar flow isothermal Fanning friction factors in sinusoidally corrugated chevron-plate channels [12] with $\beta = 90°$.

The effects of plate-wall corrugations and Reynolds number on the isothermal Fanning friction factor in the laminar regime are seen in Fig. 7.8, where results for the changes in f with Re are graphed for $0.25 \leq \gamma \leq 1.0$. Also included is the fully developed laminar flow result for a flat parallel-plate duct or when $\gamma = 0$. The higher frictional losses that are incurred with increasing severity of the plate corrugation ($\gamma > 0 \rightarrow 1$) are clearly evident. Also seen is the influence of the onset of swirl flow in the troughs of the wall corrugation on f, demarcated by the deviation from the log-linear behaviour of (f – Re) at higher Reynolds number. The increased friction loss in this regime is predominantly due to the onset and growth of lateral vortices in the channel troughs, which increase the fluid momentum transfer and thus the wall shear stress. In the low Re regime, however, where (fRe) is constant, the increase in the frictional loss with γ is primarily due to the increased effective flow length or surface area. Furthermore, the inter-play of plate-wall corrugations and flow distribution is seen to produce two distinct regimes: (1) an undisturbed streamline flow regime and (2) a steady swirl flow regime that is characterized by the presence of self-sustained transverse vortices in the wavy wall troughs. This categorization is graphically shown in the f–Re plots of Fig. 7.8. A general critical Reynolds number, however, does not seem to uniquely represent the transition from one regime to the other, and it is seen to change with different corrugation aspect ratio γ.

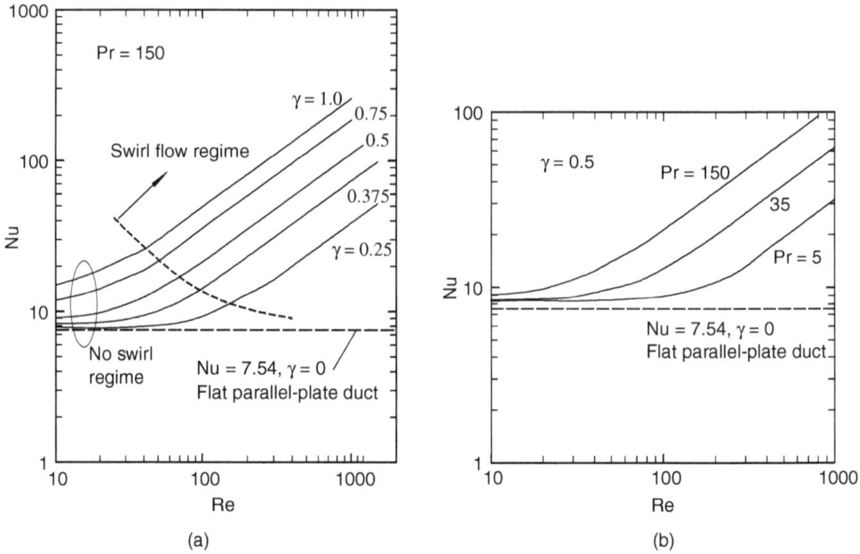

Figure 7.9: Periodically developed laminar flow Nusselt numbers in sinusoidally corrugated chevron-plate channels [12] with $\beta = 90°$: (a) effect of γ in viscous liquid (Pr = 150) flows and (b) effect of Pr when $\gamma = 0.5$.

Furthermore, the Nusselt number results for a highly viscous liquid (Pr = 150) graphed in Fig. 7.9a further demonstrate the effects of transverse vortices that are induced in the troughs of the plate-surface corrugations in promoting enhanced convection in the flow cross-section. Relative to the performance of the parallel-plate channel in fully developed laminar flow with constant wall temperature conditions (Nu = 7.541; UWT or **T** condition), and depending upon γ and Re, the heat transfer coefficient is seen to increase several folds. The largest enhancement in the period-averaged Nusselt number is obtained with higher γ and Re, or in the swirl flow regime. With $\gamma = 1.0$ and Re = 1000, for example, Nu is 34 times higher than that in a parallel-plate channel. Furthermore, similar to the $(f - \text{Re})$ behaviour depicted in Fig. 7.8, the effect of swirl on Nu also diminishes as Re and/or γ decreases. The higher heat transfer coefficients in the no-swirl regime are once again due to the larger surface area of corrugated channels. Moreover, the effect of Prandtl number (Pr = 5, 35, and 150, which are representative of most viscous liquids encountered in thermal processing applications) is also seen in Fig. 7.9b for a typical case where $\gamma = 0.5$. The improved convective behaviour with increasing Pr, when higher Nusselt numbers are obtained, is clearly evident. With a typical $Re = 700$ flow rate, for example, which corresponds to the periodically developed swirl flow regime for $\gamma = 0.5$, Nu (Pr = 35) is 99% higher than Nu (Pr = 5). The corresponding enhancement in Nu for Pr = 150 over that for Pr = 35 and the same flow rate is 77%. Another notable feature of the results in Figs 7.8 and 7.9 is that the surface area enlargement in the non-swirl regime has a relatively greater impact on enhancing heat transfer, compared to its effect on flow frictional loss.

As in the case of $\beta = 0°$ inter-plate channels, very little attention has been given to turbulent flows in wavy parallel-plate or rectangular channels ($\beta = 90°$) in the literature. In one of the earliest work, Goldstein and Sparrow [35] measured mass transfer rates in the laminar-transition ($400 < \mathrm{Re} < 4000$) regime in corrugated-plate channels that had sharp-cornered triangular profiles. This study was further extended [36, 37] to characterize the turbulent flow behaviour ($2000 < \mathrm{Re} < 40{,}000$) and the associated thermal-hydraulic performance, and to investigate the effect of rounding triangular apex profiles. In the latter case at higher flow rates ($\mathrm{Re} > 20{,}000$), the edge rounding produced 5–18% reduction in Nu; for lower Re flows the differences were minimal. Based on a limited data set, Okada et al. [17] have presented the following curve-fit equation for their own heat transfer results with round-edged triangular-profiled plates:

$$\mathrm{Nu} = \begin{cases} 0.358\mathrm{Re}^{0.63}\mathrm{Pr}^{0.4} & (4a/d_{\mathrm{h}}) = 1.612 \\ 0.321\mathrm{Re}^{0.66}\mathrm{Pr}^{0.4} & (4a/d_{\mathrm{h}}) = 1.659 \\ 0.345\mathrm{Re}^{0.67}\mathrm{Pr}^{0.4} & (4a/d_{\mathrm{h}}) = 1.695 \end{cases} \tag{7.12}$$

For sinusoidal corrugations with $\gamma = 1.0$, Focke et al. [19] used diffusion-limited current technique to measure forced convective mass transfer. They correlated their pressure drop and mass transfer coefficient data for $3000 < \mathrm{Re} < 16{,}000$ by the following curve fits:

$$f = (15.95/\mathrm{Re}^{0.289}) \tag{7.13a}$$

$$\mathrm{Nu} = 0.98\mathrm{Re}^{0.63}\mathrm{Pr}^{1/2} \tag{7.13b}$$

Nishimura et al. [38] have also considered turbulent flows in sinusoidal wavy plate channels, but with inter-plate spacing $>2a$, and as such their results are not applicable to PHEs. A relatively more detailed study by Motamed Ektesabi et al. [39] reports experimental friction factor data for flow rates in the range $1000 \leq \mathrm{Re} \leq 40{,}000$, and the following curve-fit equations:

$$f = C\mathrm{Re}^{n}, \quad \text{where} \begin{cases} C = 17, n = 0.49 \text{ for } \gamma = 0.5 \\ C = 16, n = 0.34 \text{ for } \gamma = 0.75 \\ C = 9.5, n = 0.16 \text{ for } \gamma = 1.0 \end{cases} \tag{7.14}$$

Flow visualization experiments were also conducted that demonstrated typical path-lines traced by injected polystyrene particles in the turbulent flow field.

Computationally, Tochon and Mercier [40] have modelled unsteady flows using large eddy simulation (LES) techniques for two flow rates ($\mathrm{Re} = 5000$ and 10,000), and have also reported a single case of experimental flow visualization. In an earlier work Amano [41] had employed the k–ε model for numerically simulating turbulent flows and heat transfer in corrugated-wall channels that had a sharp right-angled triangular profile. More recently, Gradeck and Lebouché [42] have employed an electrochemical technique to measure local wall shear stresses and turbulence intensity in sinusoidal corrugated-wall parallel-plate channels with flow rates of $2000 < \mathrm{Re} < 30{,}000$.

7.3.3 Single-phase convection in $0° < \beta < 90°$ plate channels

It has generally been recognized in the literature that the heat transfer coefficient as well as the concomitant wall shear stress is strongly dependent upon the geometrical attributes of plate-surface corrugations. As pointed out earlier, these features give rise to several enhancement mechanisms that include disruption and reattachment of boundary layers, swirl or vortex flow generation, effective surface-area enlargement, and small hydraulic diameter flow channels. Also, besides the forced convection conditions (flow Reynolds number Re, fluid Prandtl number Pr, and the heat transfer rate q or driving temperature difference ΔT), the enhanced thermal-hydraulic performance generally increases with increasing chevron inclination angle β.

While there is a fairly large body of literature that has dealt with determining the thermal-hydraulic performance of cross-corrugated-plate or chevron-plate heat exchangers, as documented in Section 7.2, there is little consensus between the different sets of results. This has been more quantitatively highlighted in a recent experimental study [13], where Nu and f data for three different chevron-plate arrangements ($\beta = 30°/30°, 30°/60°$, and $60°/60°$) for cooling of hot vegetable oil ($130 < \mathrm{Pr} < 290, 0.118 < \mu/\mu_w < 0.393$) and a wide laminar flow range ($2 < \mathrm{Re} < 400$) were acquired. These results, along with those of Focke et al. [19] and Thonon et al. [27], are graphed in Figs 7.10 and 7.11. Also included is the performance of an equivalent (same d_e) flat-plate pack predicted by the following [32], in order to evaluate the heat transfer enhancement due to chevron plates:

$$\mathrm{Nu} = 1.849[(d_e/L)\mathrm{RePr}]^{1/3}(\mu/\mu_w)^{0.14}, (L/d_e) = 55.41$$
$$f = (24/\mathrm{Re}) \tag{7.15}$$

The enhanced performance is clearly evident with up to three times higher Nu in chevron plate passages, and, for the same Re, Nu is clearly seen to increase with β. However, there is a correspondingly higher pressure drop penalty as well, and up to 6.6 times higher f values are obtained in comparison to an equivalent flat-plate duct. It should be noted here that conventionally, the mixed-plate arrangement is usually designated by the average chevron angle, i.e. for a $30°/60°$ arrangement, $\beta \cong 45°$. Also, there is fair agreement between the data and those of Thonon et al. [27] for $\beta = 30°$ and $45°$, but their results for $60°$ plates are considerably higher. The results of Focke et al. [19] that are based on mass transfer experiments with simulated scaled-up plate surfaces are significantly higher for all cases. It is rather difficult to explain this, though, among several factors, the differences in the chevron-plate corrugations in these studies might be primarily responsible [13, 20]. Different Pr exponents (0.5 and 1/3) and absence of a viscosity variation correction factor could also account for part of the deviations. In the latter case, for typical viscous liquids ($\mathrm{Pr} \sim 185$), a variation of $0.80 > \mu/\mu_w) > 0.20$ could result in a 3–20% difference in Nu.

The f–Re data in Fig. 7.11 provide some evidence of the inter-play between the flow conditions and plate-surface corrugations and give some idea of the enhancement mechanisms that are thereby effected. In the $15 < \mathrm{Re} < 25$ region,

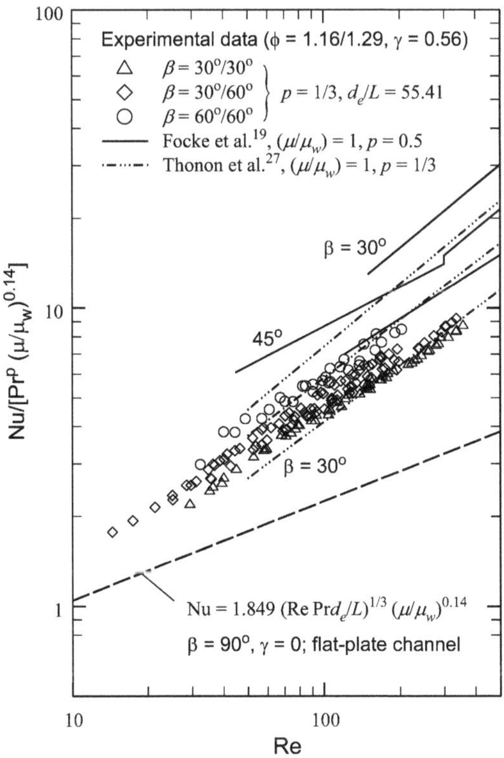

Figure 7.10: Variation of laminar flow Nu with chevron angle β.

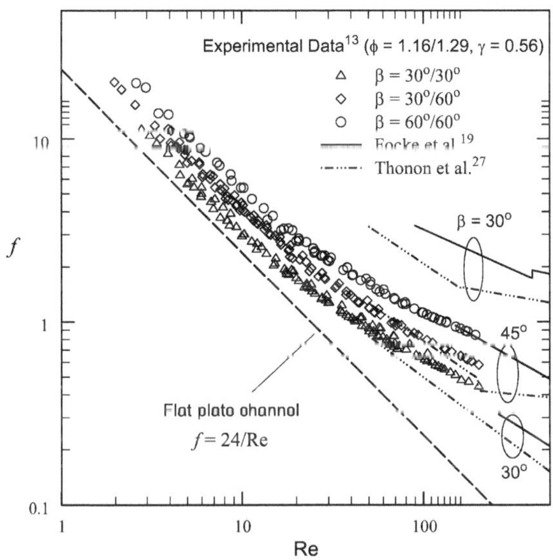

Figure 7.11: Variation of laminar flow isothermal f with chevron angle β.

all the three different β data set show a distinct change in the log-linear slope. This is indicative of the onset and intensification of swirling flows that are induced by the wavy, cross-corrugated flow passages (obtained from the mating of corrugated chevron plates) when Re > 25. With lower flow rates (Re < 25), however, there appears to be no swirl generation, as is evident from the log-linear trend that is parallel to the flat-plate performance. The higher f data in this regime in Fig. 7.11 are primarily due to the larger convoluted flow path, and they essentially exhibit the classical straight duct, fully developed laminar flow behaviour. Given the variations in results, clearly there are rather complex effects of plate-surface corrugations (β and γ or ϕ) on the convective behaviour. A case in point is the evaluation of ϕ in the Muley et al. [13, 20, 24] data with commercial plates, where $\phi = 1.16$ if only the middle corrugations are considered, and $\phi = 1.29$ when the total surface area enlargement (including the deeper grooves in the flow distribution and collection areas, ~20% of total area) is considered. Such issues make the development of reliable predictive tools rather difficult.

Nevertheless, based on their own measurements, Muley et al. [13] have proposed the following equations for predicting single-phase laminar flow Nu and f in chevron-plate exchangers:

$$\mathrm{Nu} = 1.6774(d_e/L)^{1/3}(\beta/30°)^{0.38}\mathrm{Re}^{0.5}\mathrm{Pr}^{1/3}(\mu/\mu_w)^{0.14}$$
$$30° \le \beta \le 60°, \gamma = 0.56, 30 \le \mathrm{Re} \le 400 \qquad (7.16)$$
$$f = [(30.20/\mathrm{Re})^5 + (6.28/\mathrm{Re}^{0.5})^5]^{0.2}(\beta/30°)^{0.83}$$
$$30° \le \beta \le 60°, \gamma = 0.56, 2 \le \mathrm{Re} \le 300 \qquad (7.17)$$

Evidently, additional data over an extended range of plate-surface characteristics (β, γ, andϕ) and flow conditions (Re, Pr, and q or ΔT) would help in developing more generalized Nu and f predictive correlations.

In turbulent flows also, there is little agreement between the different set of data available in the literature [1, 20] and this is evident from Figs 7.12 and 7.13, where experimental data [20] for cooling of hot water (2 < Pr < 6) flows in the 600 < Re < 104 range are compared with the predictions of a few correlations listed in Section 7.2. The data represent two symmetric ($\beta = 30°/30°$ and $60°/60°$) and one mixed-plate ($\beta = 30°/60°$) arrangements, where the latter chevron angle can again be approximated by $\beta_{avg} = 45°$, and predictions for an equivalent flat-plate channel given by eqn (7.15) for the following correlations [32] are also included:

$$\mathrm{Nu} = 0.023\,\mathrm{Re}^{0.8}\mathrm{Pr}^{1/3}(\mu/\mu_w)^{0.14} \quad \mathrm{Re} > 4000 \qquad (7.18a)$$

$$f = 0.1268\,\mathrm{Re}^{-0.3} \quad \mathrm{Re} > 2000 \qquad (7.18b)$$

Higher Nu and f are obtained with increasing β, which essentially reflects the increased intensity of swirl flows generated by the larger β chevron plates. Also, in Fig. 7.12, the Nu data are seen to be in good agreement with the Okada et al. [17] results for all three β arrangements, while the Focke et al. [19] results for both Nu and f show considerable disagreements.

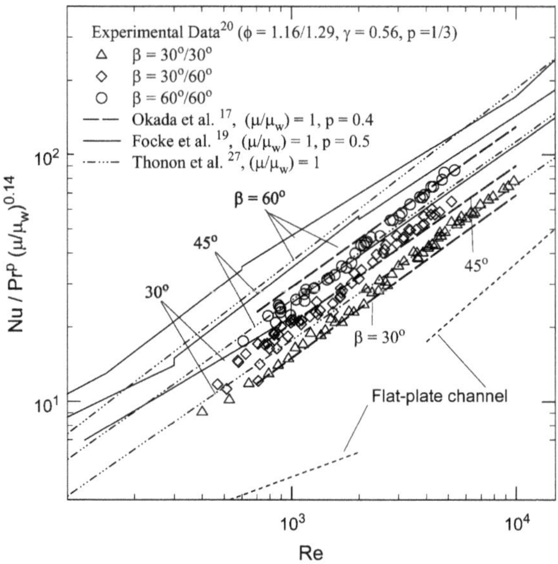

Figure 7.12: Variation of turbulent flow Nu with chevron angle β.

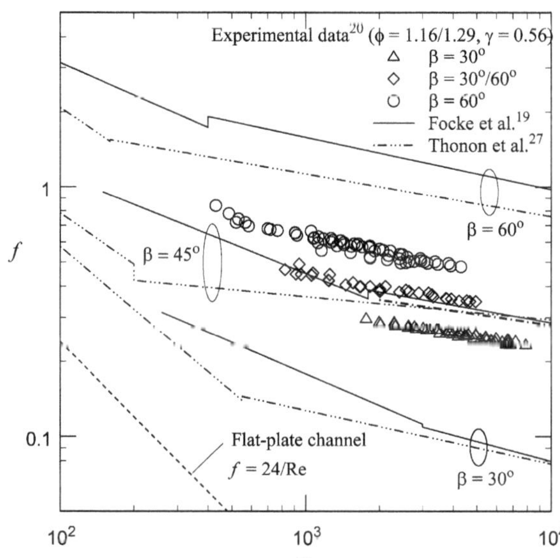

Figure 7.13: Variation of turbulent flow isothermal f with chevron angle β.

These variations can perhaps be attributed to, among some other factors, the geometric differences in the chevron-plate-surface corrugations and flow distributions channel configurations used in these investigations. For example, $\phi = 1.464$ in the scaled-up surfaces employed by Focke et al. [19] in their mass transfer study and such deeper furrows would tend to induce greater swirl mixing and hence have higher Nu, whereas $\phi = 1.29$ in the Muley and Manglik [20] data and 1.294 in the Okada et al. [17] data as such gives good agreement between the two. The plate-surface geometry features have not been reported by Thonon et al. [27] In fact, few studies provide such details that make a meaningful comparison of results rather difficult. Varying Pr exponent (0.5, 0.4, and 0.33), and absence of a correction factor for temperature-dependent viscosity variations in the other studies could also account for some part of the deviations. In the latter case, as shown by Muley and Manglik [24] for typical viscous liquids (Pr $\sim$ 185), a variation of $0.20 < (\mu/\mu_w) < 0.80$ could result in a 3–20% difference in Nu.

In an effort to correlate their own experimental data as well as those of several other studies [17, 19, 23, 25], Muley and Manglik [20] have proposed the following power-law expressions based on regression analyses for the turbulent flow regime characterized by Re $\geq$ 1000:

$$\begin{aligned} \mathrm{Nu} &= C_1(\beta)C_1'(\phi)\mathrm{Re}^{p1(\beta)}\mathrm{Pr}^{1/3}(\mu/\mu_w)^{0.14} \\ f &= C_2(\beta)C_2'(\phi)\mathrm{Re}^{p2(\beta)} \end{aligned} \tag{7.19}$$

In the expression for Nu, the (1/3) Pr exponent and classical [43] viscosity ratio correction factor were in conformance to much of the literature [1, 10]. Whether the complex inter-plate channel geometry would dictate different exponents perhaps requires verification with additional data. Also, the lead coefficients C_1, C_2 and Reynolds exponents $p1$, $p2$ were found to be functions of β and are described by the following second-order polynomials and trigonometric functions:

$$\begin{aligned} C_1 &= (0.2668 - 6.967 \times 10^{-3}\beta + 7.244 \times 10^{-5}\beta^2) \\ C_2 &= (2.917 - 0.1277\beta + 2.016 \times 10^{-3}\beta^2) \end{aligned} \tag{7.20}$$

$$\begin{aligned} p1 &= [0.728 + 0.0543 \sin\{(\pi\beta/45) + 3.7\}] \\ p2 &= [0.2 + 0.0577 \sin\{(\pi\beta/45) + 2.1\}] \end{aligned} \tag{7.21}$$

Besides the corrugation inclination angle β, the area enlargement factor ϕ also has an influence. Higher heat transfer and pressure drop are obtained with increasing ϕ as deeper corrugations increase the effective surface area as well as promote greater swirl mixing. The lead coefficients C_1' and C_2' account for this, and they were correlated to include the Muley and Manglik [20] data for $\phi = 1.29$ as well as those of others [17, 19, 23, 25] for $1 \leq \phi \leq 1.5$ as follows:

$$\begin{aligned} C_1' &= (20.78 - 50.94\phi + 41.16\phi^2 - 10.51\phi^3) \\ C_2' &= (5.474 - 19.02\phi + 18.93\phi^2 - 5.341\phi^3) \end{aligned} \tag{7.22}$$

The correlations in eqns (7.19)–(7.22) were found to represent much of the available literature data very well and are generally valid for Re $\geq 10^3$, $30° \leq \beta \leq 60°$, and $1 \leq \phi \leq 1.5$. With regard to the depth of corrugations and the associated surface enlargement, it may be noted here that manufacturing plates with $\phi \approx 1.5$ may be very difficult because of tooling, fixtures, and material strength limitations, and this value reflects the $\phi = 1.412$ and 1.464 plates, respectively, used by Okada et al. [19] and Focke et al. [19]. Additional data for plates with a broader set of geometric parameters, however, would indeed be desirable to further establish the validity of these correlations or develop alternative expressions.

A few computational efforts have also been made to simulate the forced convective behaviour in chevron-plate passages and predict their thermal-hydraulic performance. Notable among these are the unsteady-flow models reported by Ciofalo et al. [44] and Blomerius et al. [30], and the steady-flow simulations of Sawyers et al. [29] and Sundén and Di Piazza [45]. While their results are useful in providing some insights in the flow structure in the complex inter-plate channels, considerable work is still needed to produce mature computational models that can then be directly used for the thermal-hydraulic design of chevron-plate heat exchangers. Some aspects of numerical modelling and mathematical formulation for turbulent flows and heat transfer in cross-corrugated channels of PHEs are also discussed in Section 10.2.

7.4 Heat transfer enhancement

In order to employ PHEs effectively in different heating/cooling applications, the trade-off between the enhanced heat transfer coefficients and accompanying pressure drop penalties needs to be evaluated. This relative quantification of the enhanced thermal performance of chevron plate PHEs can be made by using different figures of merit that have been proposed in the literature [46–49]. For compact heat exchangers, as proposed by Shah and London [49], the flow *area goodness* and *volume goodness factors* are traditionally used. Additionally, several energy-conservation-based methods have been developed, as outlined in Manglik [46], Bergles [47], and Webb [48], which evaluate the increased heat load or surface area reduction relative to a reference 'plain' or 'non-enhanced' surface for different application constraints.

The *area goodness factor*, which is represented by the ratio j/f, can be expressed in terms of the fluid flow and chevron-plate characteristics as

$$(j/f) = \frac{\mathrm{Nu}\mathrm{Pr}^{-1/3}}{f\mathrm{Re}} = \left(\frac{4L}{A_c d_e^2}\right)^2 \left[\frac{\dot{m}hA\mathrm{Pr}^{2/3}}{\rho \Delta p}\right] \tag{7.23}$$

However, unlike other heat exchangers, in a PHE the heat transfer surface area and free flow or cross-section area are not independent. It is thus not possible to match the specified heat transfer and pressure drop identically by this figure of merit. Nevertheless, it is useful for evaluating the relative influence of the corrugation aspect ratio γ on the heat transfer surface area compactness [34]. This is seen in Fig. 7.14, where the variations in j/f and $(\mathrm{Nu}/f\mathrm{Re})$ with γ for $\beta = 90°$ and $0°$, respectively, are depicted. While the heat transfer is enhanced for all values of γ, the optimum surface compactness is obtained with $0.2 < \gamma < 0.6$. For very low

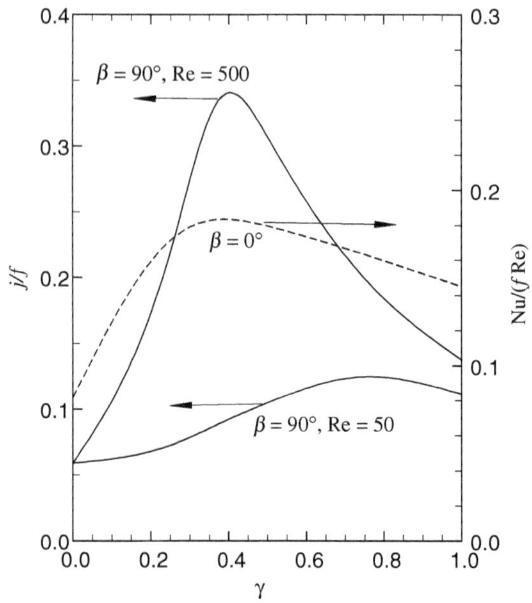

Figure 7.14: Enhanced performance for $\beta = 0°$ and 90° chevron plates in laminar flows as measured by the area goodness factor [34].

flow rates (Re $\leq$ 50), however, much deeper grooves with $0.6 < \gamma < 0.9$ may be more advantageous.

The *volume goodness factor*, on the other hand, represents the balance between the heat transfer coefficient and the friction power per unit surface area. The heat transfer coefficient h can be expressed as

$$h = (k\text{Nu}/d_e) = (\mu c_p/d_e \,\text{Pr}^{2/3})(j\text{Re}) \tag{7.24}$$

and the fluid friction power per unit area W is given as

$$W = (\dot{m}\Delta p/A\rho) = (\mu^3/2\rho d_e^3)(f\text{Re}) \tag{7.25}$$

Thus, the trade-off between increased heat transfer rate and pumping power to overcome frictional loss can be represented by the $(j\text{Re}) - (f\text{Re}^3)$ behaviour. This is depicted in Fig. 7.15, along with a comparison of the flat-plate channel performance. In order to accommodate a more convenient scale, $(f\text{Re}3)^{1/3}$ is taken as the abscissa in this figure, and variable viscosity effects on j are also included in the ordinate for completeness [34]. The superior thermal performance of chevron plates, relative to an equivalent flat-plate stack, for any given friction power is evident from this figure. The heat transfer rates are more than two times higher for the same friction power, except when flow rates are either very small or very large, and the higher β plates provide the largest enhancement benefits. In the turbulent region, however, $\beta = 30°/30°$ plates perform more poorly than flat plates. In any case, the higher

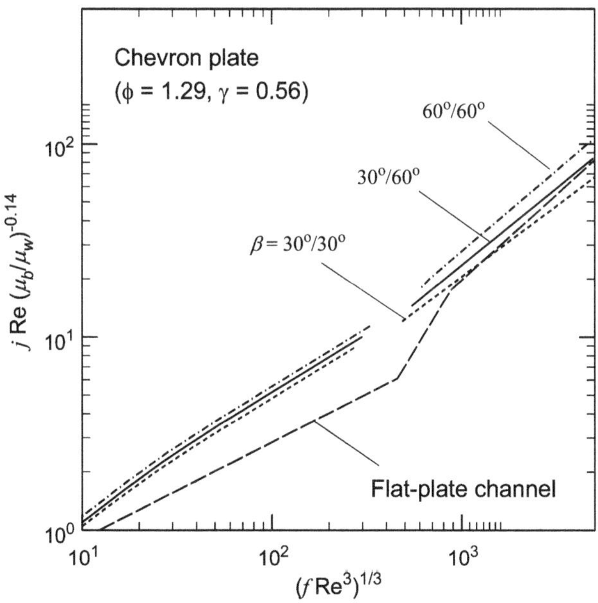

Figure 7.15: Enhanced performance for 30° $= \beta =$ 60° chevron plates in laminar and turbulent flows as measured by the volume goodness factor [34].

position of the corrugated-plate channel performance on this plot, relative to a flat-plate channel, suggests that the PHE would require a smaller core volume, and hence will be more compact for the same heat duty application.

A criterion often used in many chemical and process industry applications is to evaluate the thermal performance of enhanced surfaces for the constraint of fixed geometry (overall size and plate separation) and equal pumping power [46–48]. This criterion can be implemented by expressing the ratio of heat transfer rate for chevron-plate q_{ch} and flat-plate channels q_{fp}, respectively, as

$$(q_{ch}/q_{fp}) = (\mathrm{Nu}_{ch}/\mathrm{Nu}_{fp})_{L,d_e,N,A,\dot{m},\Delta T} \tag{7.26a}$$

and the constraint of equal pumping power as

$$(f\,\mathrm{Re}^3)_{ch} = (f\,\mathrm{Re}^3)_{fp} \tag{7.26b}$$

Here the latter essentially establishes the relationship between Re for chevron and flat-plate channels. Because the chevron-plate geometry has higher friction factors compared to the flat-plate channel, this constraint requires that for a given Re_{ch}, Re_{fp} must be higher. The results according to this criterion are presented in Fig. 7.16 for both laminar and turbulent flows. It may be noted that to obtain these results, the Nu data for chevron plates have been modified to account for the surface enlargement factor ($\phi = 1.29$ for the test plates of the present study) so that the chevron-plate

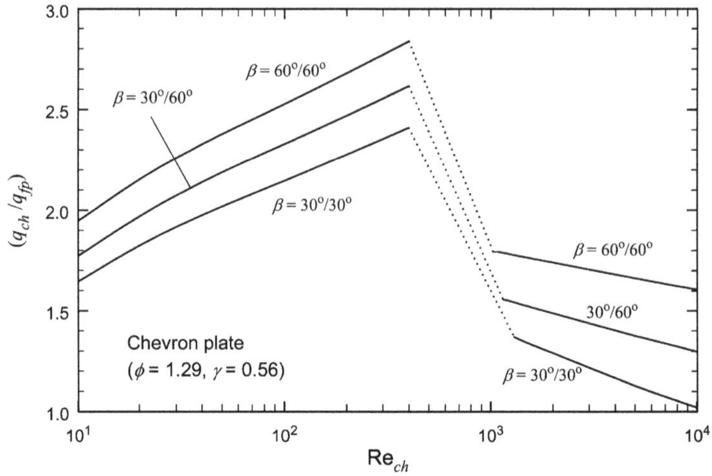

Figure 7.16: Enhanced heat transfer performance of chevron-plate PHEs based on fixed geometry and pumping power criteria.

and flat-plate channels have the same heat transfer area. The flat plates, of course, represent the limiting baseline reference when $\gamma \rightarrow 0$, or absence of corrugations. Figure 7.16 shows that in laminar flows, depending upon β and Re, the heat transfer for chevron plates is enhanced by 1.5–2.9 times than in flat-plate channels of the same geometry at equal pumping power; increases of 1.1–2.8 times are obtained in turbulent flows. Also, while the heat transfer enhancement increases with increasing Re_{ch} in laminar flow, it diminishes in turbulent flow. Obviously, plate corrugation-induced swirl mixing produces a disproportionately larger pressure drop penalty relative to the increase in heat transfer in the already well-mixed turbulent flows. Furthermore, contrary to some previously held views [21], there seems to be no significant advantage in using a mixed-plate arrangement over a symmetric arrangement. Typically with $\text{Re}_{\text{ch}} = 1500$, heat transfer rates are 1.2, 1.4, and 1.6 times higher in $\beta = 30°/30°$, $30°/60°$, and $60°/60°$ chevron plates, respectively, than that in equivalent flat-plate channels. The $30°/60°$ mixed chevron-plate arrangement, of course, provides an intermediate performance and thus obviates the need for specifically manufacturing 45° chevron plates.

The use of enhanced-surface heat exchangers with the constraint of fixed heat duty and pressure drop, likewise, results in the reduction of heat transfer surface area requirement for a variety of applications [46–48]. The decreased need for heat transfer surface area as per this criterion can be expressed as

$$(A_{\text{ch}}/A_{\text{fp}}) = (\text{Nu}_{\text{ch}}/\text{Nu}_{\text{fp}})_{q,d_e,\Delta p,\Delta T} \tag{7.27a}$$

In this case, for a fixed pressure drop, the Reynolds numbers in the chevron-plate channel and that in the flat-plate channel are inter-dependent as per the following relationship:

$$(f\,\mathrm{Re}^2)_{\mathrm{ch}} = (f\,\mathrm{Re}^2)_{\mathrm{fp}} \tag{7.27b}$$

The consequent surface area savings due to the enhanced performance of chevron plates in laminar flows is depicted in Fig 7.17a. Depending upon the flow rate and β, up to 48% reduction in heat transfer surface area is obtainable with chevron plates

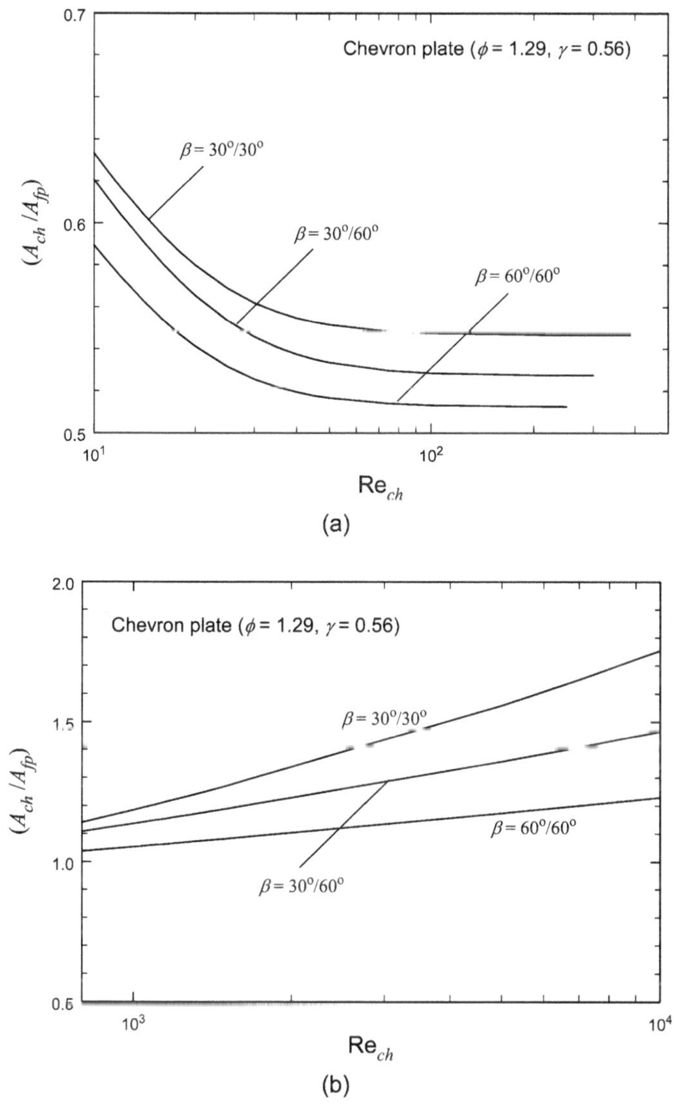

Figure 7.17: Enhanced heat transfer performance of chevron-plate PHEs based on (a) laminar flow and (b) turbulent flow.

in comparison to equivalent flat-plate stacks. In turbulent flows, however, chevron plates have an adverse performance, as is evident from the results in Fig. 7.17b. The most significant enhancement is restricted to low flow rates (low Re); high β plates promote larger reductions in the required surface area. Once again, the mixed chevron-plate (30°/60°) performance is in-between that of the symmetric-plate arrangement performance envelope in both laminar and turbulent flows.

Several other operating and optimization constraints can be considered in a similar fashion to evaluate the enhanced performance of PHEs. Extended listings of the different figures of merit or performance evaluation criteria are given by Manglik [46], Bergles [47], and Webb [48]. Furthermore, perhaps another way to supplement these energy-conservation principle-based optimization criteria is to consider the second law of thermodynamics or the minimization of entropy generation analysis [34]. This facilitates the evaluation of the impact of enhanced thermal-hydraulic performance on the overall system efficiency.

Nomenclature

A_{ch}	chevron plate geometry heat transfer surface area, m^2
A_{fp}	flat plate geometry heat transfer surface area, m^2
A	heat transfer area, m^2
a	thermal diffusivity, m^2/s
d_e	equivalent diameter, m
f	friction factor
h	heat transfer coefficient, $W/(m^2K)$
j	Colburn factor
k	thermal conductivity, W/(mK)
m	rate of mass flow, kg/s
Nu_{ch}	Nusselt number chevron plate geometry
Nu_{fp}	Nusselt number flat plate geometry
Pr	Prandtl number
q_{ch}	chevron plate channel heat transfer rate, W
q_{fp}	flat plate channel heat transfer rate, W
Re	Reynolds number
Re_{ch}	Reynolds number chevron plate geometry
w	plate width, m
W	fluid friction power per unit surface area, W/m^2

Greek symbols

β	chevron angle
γ	corrugation aspect ratio

Δp	pressure drop, Pa
ϕ	surface enlargement factor
λ	corrugation wavelength, m
μ	dynamic viscosity, kg/ms
ρ	density, kg/m^3
τ_w	local wall shear stress, Pa
$\tau_{w,m}$	mean wall shear stress, Pa

References

[1] Manglik, R.M., Plate Heat Exchangers for Process Industry Applications: Enhanced Thermal-Hydraulic Characteristics of Chevron Plates, in *Process, Enhanced, and Multiphase Heat Transfer*, ed. R.M. Manglik, and A.D. Kraus, Begell House, New York, pp. 267–276, 1996.

[2] Manglik, R.M. and Muley, A., *Heat Transfer and Pressure Drop Characteristics of Plate-and-Frame Heat Exchangers: A Literature Review*, Report No. TFL-Int-1, University of Cincinnati, Cincinnati, OH, 1993.

[3] Hewitt, G.F., Shires, G.L. and Bott, T.R., *Process Heat Transfer*, Begell – CRC Press, Boca Raton, FL, 1994.

[4] Kakaç, S. and Liu, H., *Heat Exchangers: Selection, Rating and Thermal Design*, 2nd edn, CRC Press, Boca Raton, FL, 2002.

[5] Shah, R.K. and Focke, W.W., Plate Heat Exchangers and Their Design Theory, in *Heat Transfer Equipment Design*, ed. R.K. Shah, E.C. Subbarao, and R.A. Mashelkar, Hemisphere, New York, pp. 227–254, 1988.

[6] Raju, K.S.N. and Bansal, J.C., Plate Heat Exchangers and Their Performance, in *Low Reynolds Number Flow Heat Exchangers*, ed. S. Kakaç, R.K. Shah, and A.E. Bergles, Hemisphere, Washington, DC, pp. 899–912, 1983.

[7] Gupta, J.P., *Fundamentals of Heat Exchanger and Pressure Vessel Technology*, Hemisphere, Washington, DC, 1986.

[8] Saunders, E.A.D., *Heat Exchangers – Selection, Design and Construction*, Wiley, New York, 1988.

[9] Ding, J. and Manglik, R.M., Analytical Solutions for Laminar Fully Developed Flows in Double-Sine Shaped Ducts, *Heat and Mass Transfer*, vol. 31, pp. 269–277, 1996.

[10] Manglik, R.M. and Bergles, A.E., Numerical Modeling and Analysis of Laminar Flow Heat Transfer in Non-Circular Compact Channels, in *Computer Simulations in Compact Heat Exchangers*, eds. B. Sundén, and M. Faghri, Computational Mechanics, Southampton, UK, chapter 2, 1998.

[11] Emerson, W.H., *The Thermal and Hydrodynamic Performance of a Plate Heat Exchanger: I – Flat Plates, II – A deLaval Exchanger, and III – A Rosenblad Exchanger*, Report Nos. 283, 285, and 286, National Engineering Laboratory, Ministry of Technology, UK.

[12] Metwally, H.M. and Manglik, R.M., Enhanced Heat Transfer Due to Curvature-Induced Lateral Vortices in Laminar Flows in Sinusoidal

Corrugated-Plate Channels, *International Journal of Heat and Mass Transfer*, vol. 47, Nos. 10–11, pp. 2283–2292, 2004.

[13] Muley, A., Manglik, R.M. and Metwally, H.M., Enhanced Heat Transfer Characteristics of Viscous Liquid Low Reynolds Number Flows in a Chevron Plate Heat Exchanger, *Journal of Heat Transfer*, vol. 121, no. 4, pp. 1011–1017, 1999.

[14] Focke, W.W. and Knibbe, P.G., Flow Visualization in Parallel-Plate Ducts with Corrugated Walls, *Journal of Fluid Mechanics*, vol. 165, pp. 73–77, 1986.

[15] Garg, V.K. and Maji, P.K., Flow and Heat Transfer in a Sinusoidally Curved Channel, *International Journal of Engineering Fluid Mechanics*, vol. 1, no. 3, pp. 293–319, 1988.

[16] Savostin, A.F. and Tikhonov, A.M., Investigation of the Characteristics of Plate-Type Heating Surfaces, *Thermal Engineering*, vol. 17, no. 9, pp. 113–117, 1970.

[17] Okada, K., Ono, M., Tominura, T., Okuma, T., Konno, H., and Ohtani, S., Design and Heat Transfer Characteristics of New Plate Heat Exchanger, *Heat Transfer – Japanese Research*, vol. 1, no. 1, pp. 90–95, 1972.

[18] Rosenblad, G. and Kullendorff, A., Estimating Heat Transfer Rates from Mass Transfer Studies on Plate Heat Exchanger Surfaces, *Wärme- und Stoffübertragung*, vol. 8, pp. 187–191, 1975.

[19] Focke, W.W., Zachariades, J., and Olivier, I., The Effect of the Corrugation Inclination Angle on the Thermohydraulic Performance of Plate Heat Exchangers, *International Journal of Heat and Mass Transfer*, vol. 28, no. 8, pp. 1469–1479, 1985.

[20] Muley, A. and Manglik, R.M., Experimental Study of Turbulent Flow Heat Transfer and Pressure Drop in a Plate Heat Exchanger with Chevron Plates, *Journal of Heat Transfer*, vol. 121, no. 1, pp. 110–117, 1999.

[21] Marriott, J., Performance of an Alfaflex Plate Heat Exchanger, *Chemical Engineering Progress*, vol. 73, no. 2, pp. 73–78, 1977.

[22] Chisholm, D. and Wanniarachchi, A.S., Maldistribution in Single-pass Mixed-channel Plate Heat Exchangers, *Compact Heat Exchangers for Power and Process Industries*, HTD, vol. 201, ASME, New York, pp. 95–99, 1992.

[23] Heavner, R.L., Kumar, H. and Wanniarachchi, A.S., Performance of an Industrial Plate Heat Exchanger: Effect of Chevron Angle, *AIChE Symposium Series*, vol. 89, no. 295, AIChE, New York, pp. 262–267, 1993.

[24] Muley, A. and Manglik, R.M., Enhanced Heat Transfer Characteristics of Single-Phase Flows in a Plate Heat Exchanger with Mixed Chevron Plates, *Journal of Enhanced Heat Transfer*, vol. 4, no. 3, pp. 187–201, 1997.

[25] Tovazhnyanski, L.L., Kapustenko, P.A., and Tsibulnik, V.A., Heat Transfer and Hydraulic Resistance in Channels of Plate Heat Exchangers, *Energetika*, vol. 9, pp. 123–125, 1980.

[26] Wanniarachchi, A.S., Ratnam, U., Tilton, B.E. and Dutta-Roy, K., Approximate Correlations for Chevron-type Plate Heat Exchangers, *Proceedings of*

the 30th National Heat Transfer Conference, vol. 12, HTD-vol. 314, ASME, New York, pp. 145–151, 1995.
[27] Thonon, B., Vidil, R. and Marvillet, C., Recent Research and Developments in Plate Heat Exchangers, *Journal of Enhanced Heat Transfer*, vol. 2, pp. 149–155, 1995.
[28] Gaiser, G. and Kottke, V., Effects of Wavelength and Inclination Angle on the Homogeneity of Local Heat Transfer Coefficients in Plate Heat Exchangers, in *Heat Transfer 1998*, vol. 6, ed. J.S. Lee, KSME, Seoul, Korea, pp. 203–208, 1998.
[29] Sawyers, D.R., Sen, M. and Chang, H.C., Heat Transfer Enhancement in Three-dimensional Corrugated Channel Flow, *International Journal of Heat and Mass Transfer*, vol. 41, pp. 3559–3573, 1998.
[30] Blomerius, H., Hölsken, C. and Mitra, N.K., Numerical Investigation of Flow Field and Heat Transfer in Cross-Corrugated Ducts, *Journal of Heat Transfer*, vol. 121, no. 2, pp. 314–321, 1999.
[31] Manglik, R.M. and Ding, J., Laminar Flow Heat Transfer to Viscous Power-law Fluids in Double-sine Ducts, *International Journal of Heat and Mass Transfer*, vol. 40, no. 6, pp. 1379–1390, 1997.
[32] Kakaç, S., Shah, R.K. and Aung, W., *Handbook of Single-Phase Convective Heat Transfer*, Wiley-Interscience, New York, 1987.
[33] Bejan, A. and Kraus, A.D., *Heat Transfer Handbook*, Wiley, Hoboken, NJ, 2003.
[34] Muley, A. and Manglik, R.M., Enhanced Thermal-hydraulic Performance Optimization of Chevron Plate Heat Exchangers, *International Journal of Heat Exchangers*, vol. I, no. 1, pp. 3–18, 2000.
[35] Goldstein, L., Jr. and Sparrow, E.M., Heat/Mass Transfer Characteristics for Flow in a Corrugated Wall Channel, *Journal of Heat Transfer*, vol. 99, pp. 187–195, 1977.
[36] O'Brien, J.E. and Sparrow, E.M., Corrugated-duct Heat Transfer, Pressure Drop, and Flow Visualization, *Journal of Heat Transfer*, vol. 104, pp. 410–416, 1982.
[37] Sparrow, E.M. and Hossfeld, L.M., Effect of Rounding of Protruding Edges on Heat Transfer and Pressure Drop in a Duct, *International Journal of Heat and Mass Transfer*, vol. 27, no. 10, pp. 1715–1723, 1984.
[38] Nishimura, T., Kajimoto, Y. and Kawamura, Y., Mass Transfer Enhancement in Channels with a Wavy Wall, *Journal of Chemical Engineering of Japan*, vol. 19, no. 2, pp. 142–144, 1986.
[39] Motamed Ektesabi, M.R., Sako, M. and Chiba, T., Fluid Flow and Heat Transfer in Wavy Sinusoidal Channels (Part II, Pressure Drop and Flow Pattern of Turbulent Flow Field), *Heat Transfer – Japanese Research*, pp. 32–43, 1989.
[40] Tochon, P. and Mercier, P., Thermal-hydraulic Investigations of Turbulent Flows in Compact Heat Exchangers, in *Compact Heat Exchangers and Enhancement Technology for the Process Industry*, ed., R.K. Shah et al., Begell House, New York, pp. 97–101, 1999.

[41] Amano, R.S., A Numerical Study of Laminar and Turbulent Heat Transfer in a Periodically Corrugated Wall Channel, *Journal of Heat Transfer*, vol. 107, no. 3, pp. 564–569, 1985.
[42] Gradeck, M. and Lebouché, M., Wall Shear Measurements Inside Corrugated Channels using the Electrochemical Technique, *Experiments in Fluids*, vol. 24, pp. 17–26, 1998.
[43] Sieder, E.N. and Tate, G.E., Heat Transfer and Pressure Drop of Liquids in Tubes, *Industrial and Engineering Chemistry*, vol. 28, pp. 1429–1435, 1936.
[44] Ciofalo, M., Stasiek, J., and Collins, M.W., Investigation of Flow and Heat Transfer in Corrugated Passages – II. Numerical Simulations, *International Journal of Heat and Mass Transfer*, vol. 19, no. 1, pp. 165–192, 1996.
[45] Sundén, B. and Di Piazza, I., Numerical Analysis of Fluid Flow and Heat Transfer in Plate-and-frame Heat Exchangers, *ASME HTD vol. 361–3*, ASME, New York, pp. 287–293, 1998.
[46] Manglik, R.M., Heat Transfer Enhancement, in *Heat Transfer Handbook*, ed., A. Bejan and A.D. Kraus, Wiley, Hoboken, NJ, chapter 14, 2003.
[47] Bergles, A.E., Techniques to Enhance Heat Transfer, in *Handbook of Heat Transfer*, ed., W.M. Rohsenow, J.P. Hartnett, and Y.I. Cho, McGraw-Hill, New York, NY, chapter 11, 1998.
[48] Webb, R.L., *Principles of Enhanced Heat Transfer*, Wiley, New York, 1994.
[49] Shah, R.K. and London, A.L., Laminar Flow Forced Convection in Ducts, in *Advances in Heat Transfer*, ed., T.F. Irvine, Jr. and J.P. Hartnett, supplement 1, Academic, New York, 1978.

CHAPTER 8

Thermal-hydraulic performance in condensers and evaporators

A condenser is a two-phase flow heat exchanger in which vapour is converted into liquid by a coolant, and an evaporator is a two-phase flow heat exchanger in which liquid is converted into vapour by the heat supplied by a heating medium. When PHEs are used for condensation and evaporation processes, they are called plate condenser and plate evaporator, respectively. Sometimes, a PHE can act as both condenser and evaporator at the same time, e.g. fluid evaporation by steam condensation. Although PHEs were originally developed for single-phase applications and indeed were used only in single-phase applications for many years after their original appearance, their presence in two-phase applications has emerged and expanded significantly during the last two decades. This development attributes to their construction advances in addition to their intrinsic advantages, which have been described in previous chapters. Today, typical applications of plate condensers and evaporators can be found in process industries (e.g. reboiler and top condenser), energy systems (e.g. steam generator in power generation and steam condenser in district heating networks), refrigeration systems (e.g. refrigeration condenser and evaporator), etc.

Generally, condensation can be classified into two main types: those in which the coolant and condensate steam are separated by a solid surface (heterogeneous condensation), and those in which the coolant and condensing vapour are brought into direct contact (homogeneous condensation). Obviously, a plate condenser always belongs to the heterogeneous condensation type, which can be further classified into filmwise condensation and dropwise condensation, depending on whether the condensed liquid formed on the surface will exist either as a wetted film or in droplets. If the condensate does not wet the surface, droplets are formed. Otherwise, a film is formed. Although dropwise condensation can yield much higher heat transfer coefficient than film condensation (in order of magnitude), it is extremely difficult to permanently sustain it. Therefore, filmwise condensation is the most common process in current industries and it dominates in plate condenser applications.

The evaporation process is sometimes also called boiling. Boiling is the change of phase from liquid to vapour, when heat is transferred to the liquid at or above the boiling temperature at the actual pressure. Evaporation is a more general term and can include phase changes at temperatures below the actual boiling point, e.g. evaporation of water from a water surface at ambient temperature and atmospheric pressure. Boiling can be further divided into two basic types: pool boiling and convective boiling. Pooling boiling occurs when a heat transfer surface is submerged in a relatively large body of quiescent liquid. The agitation is due to natural circulation and bubble motion. Convective boiling happens in a channel where a mixture of vapour and liquid flows. The heat transfer surface is generally the channel wall. A plate evaporator is obviously in the scope of convective boiling, and this chapter will concentrate on this type.

8.1 Flow patterns

During condensation and evaporation processes in channels, the liquid and the vapour take up a variety of configurations known as flow patterns. The flow pattern depicts the spatial and temporal distributions of vapour and liquid phases. The particular flow pattern depends on the conditions of pressure, channel geometry, relative flow rates of two phases, boundary conditions, etc. An important feature of a particular flow pattern is the direct relationship of the heat transfer and pressure drop characteristics to the pattern type, leading to a simple identification of important macroscopic heat transfer and pressure drop models [1]. This means that an approach to the selection of appropriate heat transfer and/or pressure drop correlations is preferred with an identification of the involved flow patterns. However, such a relationship is often difficult to establish due to the high complexity of flow phenomena. Experience has shown that reasonably accurate correlations for the pressure drop and the heat transfer coefficient can be obtained without consideration of the flow pattern [2]. Nevertheless, flow pattern is helpful in understanding basic flow mechanisms, and deserves particular attention.

Flow patterns in horizontal and vertical tubular channels have been studied for many years. Comprehensive information has already been established, although improvement is continuously undertaken through more advanced visualization technique. However, literature survey shows that the study of flow patterns in cross-corrugated channels of PHEs is very rare. Most visualizations for such complicated channels are for single-phase applications, reviewed in Chapter 7. The only study for the two-phase flow visualization in PHEs was conducted by Tribbe and Müller-Steinhagen [3]. Their results will be quoted later in this section.

In almost all practical applications, plate condensers and evaporators are placed in a vertical manner. Therefore, it is common to adopt the flow pattern results from vertical tubular pipes for the sake of rough estimation. The flow patterns encountered in vertical upwards co-current flow are shown schematically in Fig. 8.1 [4]. Five flow patterns have been identified:

1 Bubbly flow: the gas or vapour phase is distributed as discrete bubbles in a continuous liquid phase.

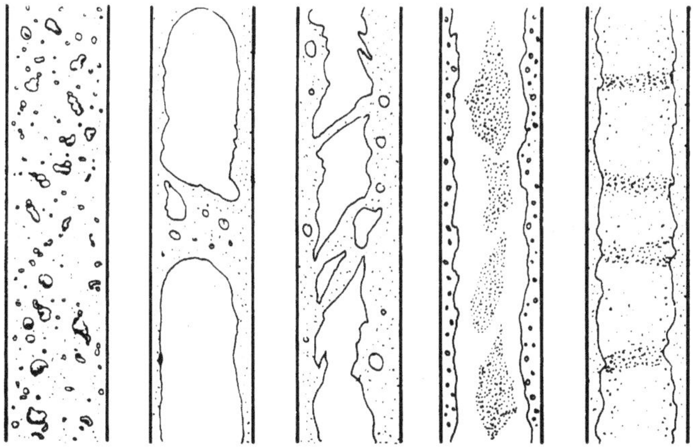

Figure 8.1: Flow patterns in vertical co-current flow [4]: (a) bubbly, (b) slug, (c) churn, (d) wispy-annular, and (e) annular.

2 Slug flow: the gas or vapour bubbles are approximately the diameter of the pipe.
3 Churn flow: churn flow is formed by the breakdown of the large vapour bubbles in the slug flow.
4 Wispy annular flow: the flow takes the form of a relatively thick liquid film on the walls of the pipe together with a considerable amount of liquid entrained in a central gas or vapour core.
5 Annular flow: a liquid film forms at the pipe wall with a continuous central gas or vapour core.

In general, at low vapour qualities and low mass flow rates, the flow usually obeys the bubbly flow pattern. At higher vapour qualities and mass flow rates, slug flow replaces the bubbly flow pattern. Further increase in vapour quality and mass flow rates leads to the appearance of the churn, wispy annular, and annular flow patterns. When condensation or evaporation occurs, the flow will experience the above flow patterns with the changes of vapour quality due to the heat flux on the wall. These changes occur due to two main reasons [4]. First, the departure from thermodynamic equilibrium coupled with the presence of temperature profiles in the channel. Second, the departure from local hydrodynamic equilibrium throughout the channel.

As Collier and Thome [4] pointed out, there is a widely felt need for simple methods to give some idea of the particular pattern likely to occur for a given set of local flow parameters. This is because a particular flow pattern usually has unique features of thermal and hydraulic performance. One method of representing the various transitions is the form of a flow pattern map. Flow pattern maps are entirely empirical and delineate the boundaries of transitions from one flow regime to another. Figure 8.2 is a flow pattern map constructed for a vertical upward flow by Hewitt and Roberts [5]. In this map, the respective patterns are represented as

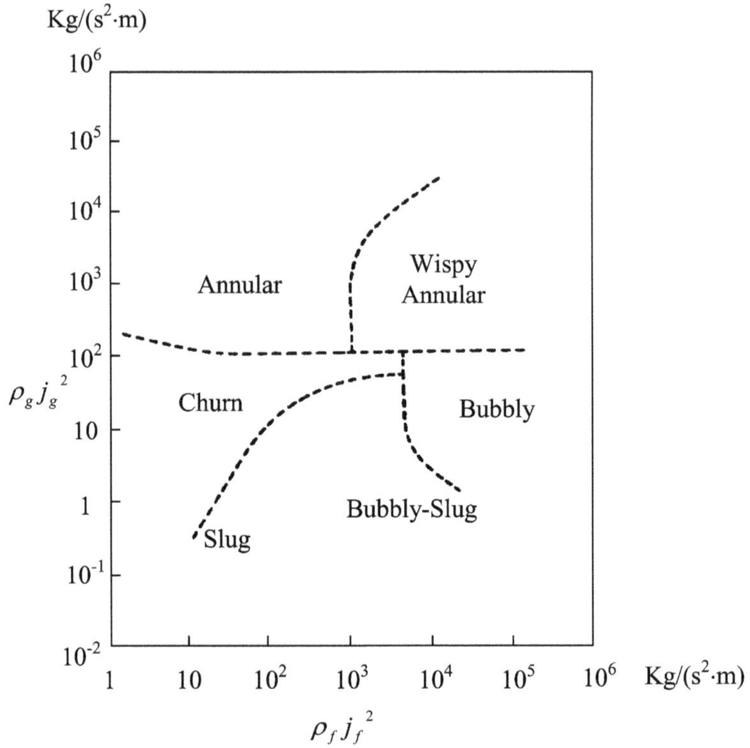

Figure 8.2: Flow pattern map for vertical flow [4].

areas on a graph, and the axes represent the superficial momentum fluxes of the liquid ($\rho_f j_f^2$) and vapour ($\rho_g j_g^2$) phases, respectively. Note that j is the volumetric flux (or sometimes 'superficial velocity'), which is defined as the rate of volumetric flow divided by the flow area (see eqn (8.1)).

$$j = Q/A \tag{8.1}$$

It should be pointed out that the momentum fluxes alone are not adequate to represent the influence of fluid physical properties or channel diameter, and therefore, such a map can only be regarded as a rough guide.

Recently, a two-phase flow visualization study was carried out by Tribbe and Müller-Steinhagen [3] for real PHEs. The visualization was for air–water downflow in replica channels with a channel gap of 6 mm and chevron angles of 30°, 45°, and 60°. Five flow patterns have been identified with the increase of gas flow rate: regular bubble flow, irregular bubble flow, churn flow, film flow, and partial film flow. The following description is based on the results from Tribbe and Müller-Steinhagen [3].

1 **Regular bubble flow:** Individual bubbles of approximately 3–5 mm in diameter flow along the furrows of both plates (see Fig. 8.3). Due to the shear stress, small bubbles are forced towards the surface contact point and large bubbles are split when they approach the contact point. One part continues along the same furrow while the other transfers to an opposite furrow and, therefore, changes direction. Similar to the single-phase flow, the tendency towards crossing flow diminishes as the chevron angle increases and is overtaken by longitudinal wavy flow character.

2 **Irregular bubble flow:** As gas flow rates increase, large irregularly shaped regions of gas appear among the regular bubble flow (Fig. 8.4). The bubbles spread over a number of furrows and exist over the entire depth of the channel.

3 **Churn flow:** The appearance of high-velocity liquid slugs indicates the transition to the churn flow region (Fig. 8.5). Initially, they are sporadic and often short-lived. As the gas flow increases, the stability of the intermittencies increases. They maintain their character along the length of the channel and occur with a regular frequency.

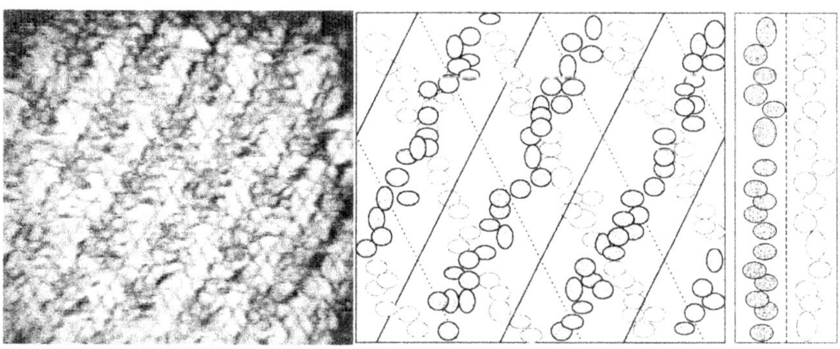

Figure 8.3: Regular bubbly flow [3]

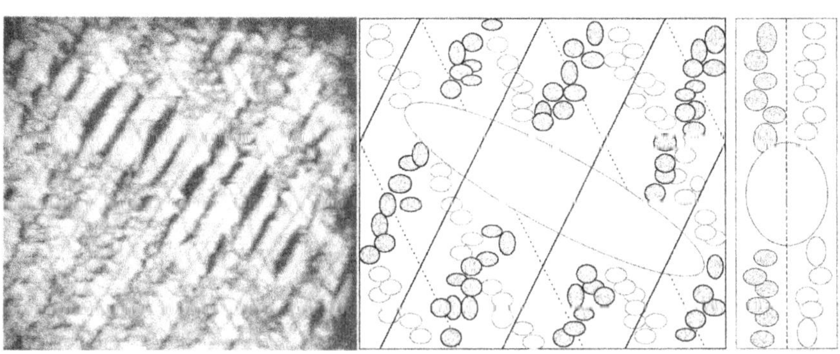

Figure 8.4: Irregular bubbly flow [3].

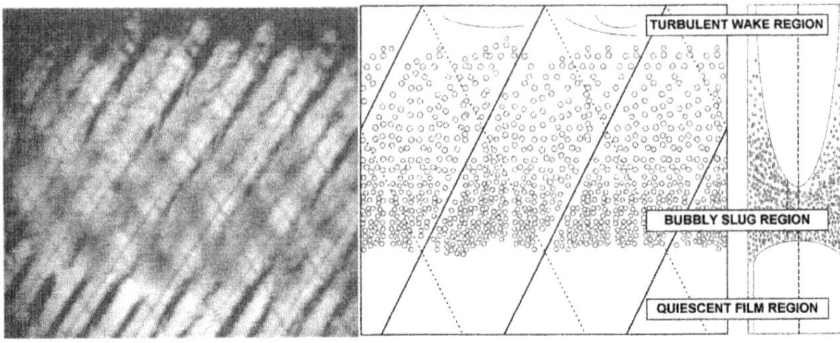

Figure 8.5: Churn flow [3].

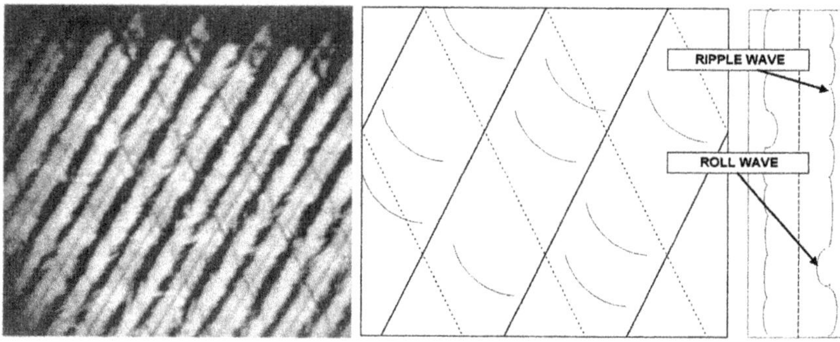

Figure 8.6: Film flow [3].

4 **Film flow:** The film flow pattern consists of a thin liquid film flowing along a furrow over which flows a fast-moving gas stream (Fig. 8.6). Although similar in nature to annular flow, the geometry of the channel along with the midplane shear means that the film does not form an annulus.

5 **Partial film flow:** At high vapour flow rates, the film no longer wets the entire surface and a region of dry surface appears (Fig. 8.7). Deposition of entrained liquid is observed at the gas/liquid interface, while liquid droplets are noted to appear on the dry edge.

Simple flow pattern maps for PHEs were constructed by Tribbe and Müller-Steinhagen [3], based on the superficial velocities. This is due to the fact that no physical property effects were investigated. In addition, the flow pattern maps derived by the authors are highly dependent on their individual test plates and have no universal validity. Therefore, those flow pattern maps are not adopted here.

In addition, it must be stressed that the above flow patterns identified are based on the experimental results of adiabatic gas and liquid mixture. Certainly, the fluid may experience significantly different flow patterns if either condensation or evaporation is involved. Therefore, the flow visualization for condensation and evaporation in PHEs is still not resolved and efforts have to be made in the future.

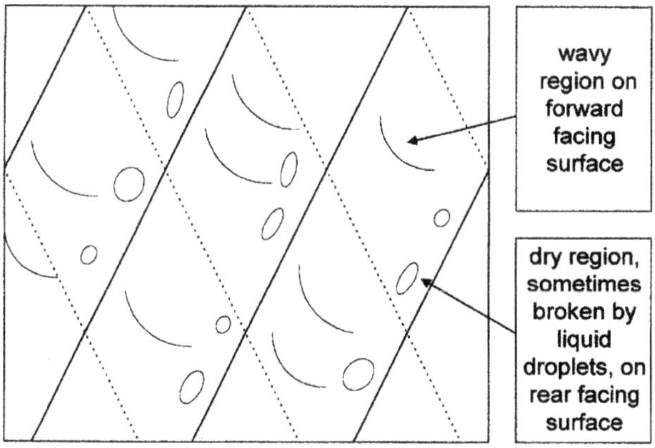

Figure 8.7: Partial film flow [3].

8.2 Performance of plate condensers

Because the application of condensation in PHEs has a relatively short history, the research on this specific subject is very limited so far. It is known that the condensation performance is dependent on many factors, such as fluid property, plate geometry, system pressure, mass flow rate, etc. This means that accurate prediction is somewhat difficult. Theoretical prediction appears very difficult at this stage, and hence most efforts have been placed on experimental investigation.

Common condensation applications are involved with steam, refrigerants, hydrocarbons, etc. In refrigerant applications, the working fluid can be either a pure substance (e.g. R22, R134a, and ammonia) or a mixture (e.g. R410a). In process industries, hydrocarbons and mixture of hydrocarbons (e.g. pentane, butane, and propane) are often encountered, e.g. in top condensers for distillations. The ideal correlation should be able to predict all situations for different fluids. However, a literature survey shows that this seems impossible, and specific correlations were developed for specific situations. The details will be explored in the following sections.

8.2.1 Fundamental mechanism of condensation

When vapour flows over a surface where the temperature is below the saturation temperature of the vapour, the vapour starts to condense, and consequently forms a thin film. The latent heat released during condensation must be transferred to the wall through the conduction of the condensate film. Therefore, a thin film will generally give a lower resistance and consequently a higher heat transfer coefficient than a thick one, all other things being equal. For films of the same thickness, a slow-moving laminar one will present a higher resistance than a fast-moving turbulent one. An important factor in producing fast-moving films is the vapour

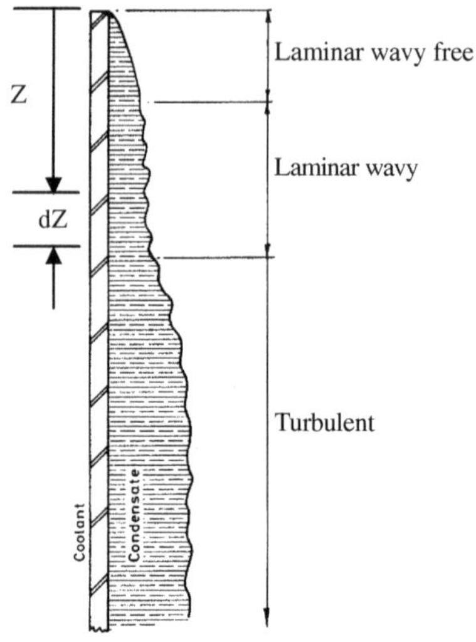

Figure 8.8: Condensation on a vertical surface.

shear stress. Some other factors that can affect the heat transfer coefficient are waves, splashing, and condensate sub-cooling.

Figure 8.8 illustrates the main features of condensation on such a surface when the vapour is stagnant, which means no shear stress at the interface. The condensate downward flow rate is zero at the top of the surface and increases with distance down the surface as the condensate accumulates. At the top of the surface, there is a region of very-low condensate Reynolds number, where the flow is both laminar and wave free. At some point down the surface, the Reynolds number grows to the point where instabilities form at the vapour–liquid interface, thus giving waves on the film. Still further down the surface, the Reynolds number grows to the point where turbulence occurs. In the laminar region, the heat transfer coefficient decreases with the thickening of the condensate, although the rate of decrease is less in the wavy region because of the mixing effect of the waves. In the turbulent region, the turbulence causes an extra thickening of the film in addition to that caused by the accumulated condensate. However, the thickening is usually more than compensated for by the better heat transfer in turbulent flow. The heat transfer coefficient therefore tends to increase in this region. In the presence of the vapour shear stress at the interface, the critical Reynolds number for the onset of turbulence is significantly reduced.

It is obvious that a downward vapour velocity will tend to increase the coefficient both by thinning the film and by increasing the likelihood of turbulence. This is the reason why most condensation applications occur in a downward style. However, an upward condensation does exist sometimes. An upward vapour velocity will tend to have the opposite effect, but this is not important because another phenomenon occurs at velocities well below those that would affect the heat transfer noticeably. This new phenomenon is flooding, which occurs when the condensate is prevented from damping from the bottom of the surface by the upward flow of vapour. Due to the limited applications for PHEs, the flooding phenomenon will not be discussed further in this book.

Condensation is controlled by gravity, vapour–liquid interfacial shear, inertia, surface tension force, etc. From this mechanism point of view, condensation can generally be categorized into two fundamental types: gravity controlled and shear controlled. Depending on the surface type, condensation can be divided into external condensation (plate or tube surface) or internal condensation (tube or other confined geometries). In addition, it should also be noticed whether the specific geometry is horizontal or vertical or inclined. Although previous correlations obtained from condensation in tubes may not be applicable to PHEs, they are often employed to make a rough estimation.

For gravity-controlled conditions, the pioneering work by Nusselt resulted in a simple equation for predicting the heat transfer coefficient on a vertical surface with laminar, wave free, no sub-cooling, linear temperature distribution conditions [6].

$$h = \left[\frac{\rho_l(\rho_l - \rho_g) g \gamma_{lg} k_l^3}{4\mu_l z (T_{gi} - T_w)} \right]^{1/4} \tag{8.2}$$

where z indicates the local position of the calculated place (see Fig. 8.8), and γ is the latent heat. The above Nusselt equation is the first theoretical analysis for condensation, and most other theoretical correlations developed later are improvements of this theory, such as Bromley [7] for sub-cooling consideration, Chen [8, 9] for wave consideration due to vapour drag force, Rohsenow [10] for the non-linear temperature distribution in condensate, etc. It should be pointed out that the Nusselt equation is valid for the local heat transfer coefficient prediction. The average heat transfer coefficient can be obtained by the integration along the condensation length, which is shown in the following equation:

$$\bar{h} = 0.925 \left[\frac{\rho_l(\rho_l - \rho_g) g k_l^3}{\mu_l \Gamma_z} \right]^{1/3} \tag{8.3}$$

where Γ_z is the condensate flow at a distance z from the top of the plane surface.

For shear-controlled conditions, there are many empirical correlations for various conditions, e.g. Carpenter and Colburn [11], Dukler [12], Akers et al. [13], Boyko and Kruzhilin [14], Shah [15]. All of these equations were obtained from experimental work either in vertical or horizontal tubes. Among them, the most common ones are the Boyko–Kruzhilin equation and the Shah equation.

The Boyko–Kruzhilin equation was proposed for steam condensation in a horizontal tube at elevated pressures ranging from 12.2 to 88 bar. The analogy between liquid film flow and single-phase flow in a pipe is used to correlate the experimental data, which resulted in the following equation:

$$h = h_l \left(\frac{\rho_l}{\rho_m} \right)^d \tag{8.4}$$

where the density correction exponent d is equal to 0.5; h_l is the heat transfer coefficient assuming all the mass is flowing as liquid; ρ_l is the density of the liquid phase, and ρ_m is the mean density of the liquid/vapour phase. This equation can be used to predict the local heat transfer coefficient as well as the average value.

The Shah equation was developed from a wide range of experimental data, which included water, refrigerants, and organics condensing in horizontal, vertical, and inclined pipes. Because many of the experimental data were obtained from literature, this equation has a wide acceptance in practical engineering calculations. It reads

$$h = h_l\left[(1-x)^{0.8} + \frac{3.8x^{0.76}(1-x)^{0.04}}{p_r^{0.38}}\right] \tag{8.5}$$

where h_l is the heat transfer coefficient assuming all the mass is flowing as liquid, and p_r is the local reduced pressure. This equation is applicable to the local heat transfer coefficient as well as the average value.

In PHEs, the condensing vapour usually has a very high speed, thus shear-controlled condensation usually prevails in most applications. Therefore, in the following sections, it always refers to the shear-controlled condensation, if not otherwise stated. The previous correlations developed for shear-controlled condensation in tubes could be employed as a rough estimation for PHEs. However, verification must be carried out and further development is certainly needed.

8.2.2 Condensation heat transfer

Condensation heat transfer in PHEs is influenced by flow rate, vapour quality, vapour pressure, medium properties as well as plate pattern. With the increase of flow rate, the vapour velocity is increased and turbulence is enhanced. Therefore, the condensation heat transfer is increased with the increase of flow rate. When vapour quality is high, vapour velocity as well as the interface shear stress is also high, which results in high condensation heat transfer. However, condensation heat transfer is decreased with regard to the increase of system pressure. The reason for this is that the volume flow rate is greater at lower pressure than at higher pressure under the same mass flow rate, so the vapour velocity is higher and consequently the condensation heat transfer is intensified. These influences have been verified by the experimental results and shown in Figs 8.9 and 8.10.

For the prediction of condensation heat transfer in PHEs, almost all efforts have been attempted through experimental tests, due to the complexity of cross-corrugated geometry in PHE channels. Tovazhnyanskiy and Kapustenko [18] conducted a steam condensation experiment in an extended-surface slot channel, and suggested that the Nusselt number was a function of Re_l, Pr_l, inlet and outlet steam qualities x_1, x_2, and density ratio ρ_l/ρ_g. Their correlation reads,

$$\mathrm{Nu} = 0.15\cdot \mathrm{Re}_l^{0.73}\cdot \mathrm{Pr}_l^{0.43}\cdot\left(\frac{1}{2}\right) \cdot\left(\sqrt{1+x_1\left(\frac{\rho_l}{\rho_g}-1\right)_1} + \sqrt{1+x_2\left(\frac{\rho_l}{\rho_g}-1\right)_2}\right) \tag{8.6}$$

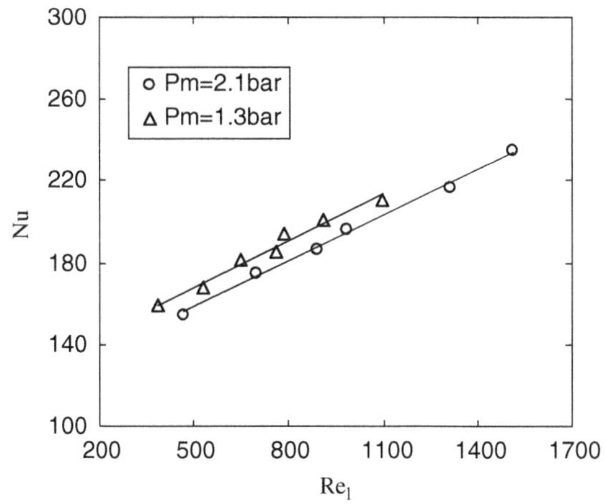

Figure 8.9: Complete steam condensation heat transfer – influence of mass flow rate and system pressure [16].

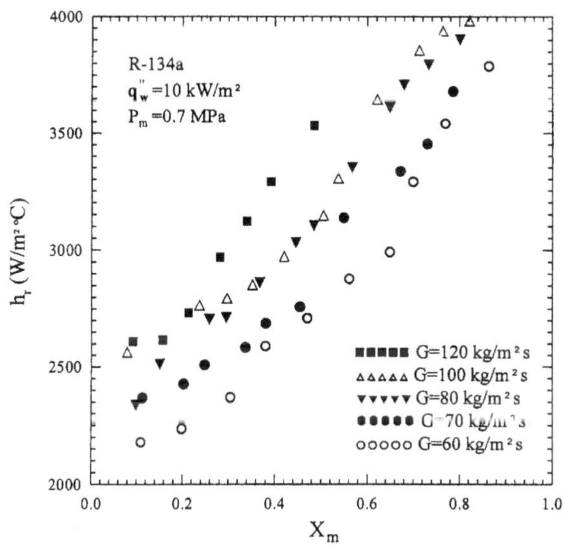

Figure 8.10: Condensation heat transfer coefficient of refrigerant R134a – influence of vapour quality and mass flow rate [17].

This correlation is an extension of the Boyko–Kruzhilin equation, which was applied to steam condensation in PHEs by Cooper [19]. However, as Wang et al. [16] pointed out, the Boyko–Kruzhilin equation overestimates the heat transfer coefficient of steam condensation in PHEs, as shown in Fig. 8.11. In the same figure, it is also shown that the Shah equation underestimates the real value in contrast. It is obvious that a large discrepancy between the predicted outlet steam quality and the test value exists.

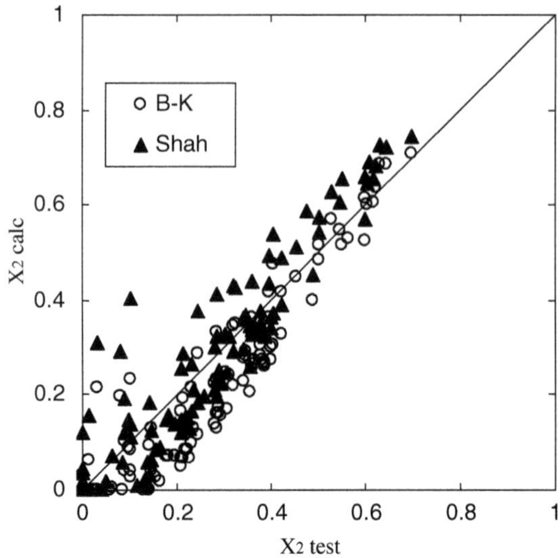

Figure 8.11: Shah and Boyko–Kruzhilin equations for steam condensation in PHEs [16].

It is easy to understand why the Boyko–Kruzhilin equation overestimates the heat transfer duty. This is due to the fact that the original Boyko–Kruzhilin equation was obtained through steam condensation at elevated pressures. If the density correction exponent is simply reduced to 0.45, the calculated condition will meet the experimental data relatively better, as shown in Fig. 8.12.

Wang and Zhao [20] performed steam condensation experiments in a PHE with a corrugation angle of 45°. They suggested the following correlation for predicting the average heat transfer coefficient:

$$\mathrm{Nu} = 0.00115 \left(\frac{\mathrm{Re}_{12}}{H}\right)^{0.983} \mathrm{Pr}_{\mathrm{l}}^{0.33} \left(\frac{\rho_{\mathrm{l}}}{\rho_{\mathrm{g}}}\right)^{0.248} \tag{8.7}$$

where Re_{12} is the Reynolds number of the condensate flow at the exchanger outlet, and the parameter $H = c_{\mathrm{p}} \cdot \Delta T/\gamma$ is claimed to account for the influence of the condensate film. Unfortunately, this equation suffers from a serious problem as explained below. It is easy to obtain the following relation:

$$\frac{\mathrm{Re}_{\mathrm{l}}}{H} = \frac{G(1-x_2)\cdot D_{\mathrm{e}}}{\mu_{\mathrm{l}} \cdot \boldsymbol{Cp}_{\mathrm{l}} \cdot \Delta T} \cdot \gamma = \frac{\dot{q}_w \cdot D_{\mathrm{e}}}{\mu_{\mathrm{l}} \cdot \boldsymbol{Cp}_{\mathrm{l}} \cdot \Delta T} \cdot \frac{A}{A_{\mathrm{c}}} = \frac{h \cdot D_{\mathrm{e}}}{\mu_{\mathrm{l}} \cdot \boldsymbol{Cp}_{\mathrm{l}}} \cdot \frac{A}{A_{\mathrm{c}}} = \frac{\mathrm{Nu}}{\mathrm{Pr}_{\mathrm{l}}} \cdot \frac{A}{A_{\mathrm{c}}} \tag{8.8}$$

where A_{c} and A are the channel-cross section area and the heat transfer area, respectively. The above equation shows the relationship between Nu, Re_{l}, and parameter H. This means that the correlation obtained by Wang and Zhao becomes meaningless.

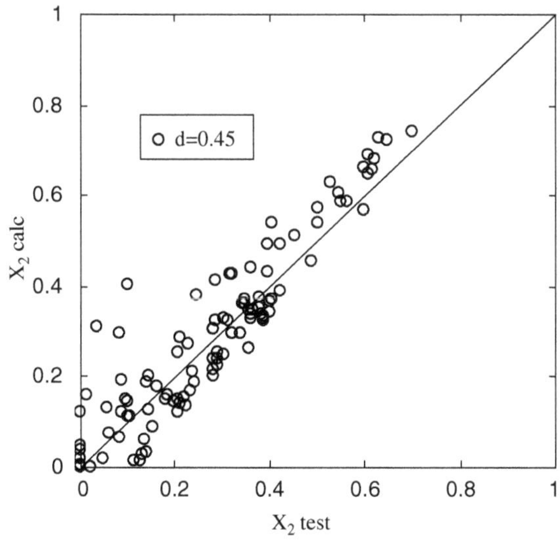

Figure 8.12: Modified Boyko–Kruzhilin equation for steam condensation in PHEs [16].

The latest experimental work on steam condensation in PHEs was carried out by Wang et al. [16]. In their experimental work, the steam inlet pressure ranged from 1.2 to 4.5 bar; the outlet steam quality ranged from 0 to 0.7; the steam mass velocity ranged from 20 to 180 kg/(m^2 · s). In total, seven different PHEs, including both gasketed heat exchangers and BHEs, have been tested. The detailed information about these exchangers is listed in Tables 8.1 and 8.2.

Based on the experimental values, a modified correlation of the Boyko–Kruzhilin equation is proposed to predict the local heat transfer coefficient. The structure of such an equation is

$$h = h_l \left(\frac{\rho_l}{\rho_m} \right)^{(a+b \cdot Re_l^c)} \tag{8.9}$$

where the constants a, b, and c are in ranges 0.3–0.37, 5.0–6.0, and −0.6 to −0.64, respectively. The comparison between the calculated and experimental values using the above equation is shown in Fig. 8.13, for tested PHEs. This figure shows that eqn (8.8) works very well for tested PHEs except for exchanger 7, which was blamed for maldistribution. Hence this equation is recommended for the calculation of steam condensation in PHEs.

For PHEs used in refrigeration applications, most research has concentrated on qualitative performance instead of searching for a widely accepted correlation. Kumar [21] reported some results for condensation of R22 and ammonia in several types of PHEs. Their results clearly indicated that the condensation was gravity controlled for low Reynolds numbers and shear controlled for higher Reynolds numbers. Typically, the condensation heat transfer coefficient of refrigerants in PHEs

Table 8.1: Tested PHEs by Wang et al. [16].

	Unit	PHE1	PHE2	PHE3	PHE4
Width	m	0.210	0.178	0.178	0.615
Height	m	0.484	0.448	0.448	0.678
Port diameter	m	0.06	0.054	0.054	0.21
Hydraulic diameter	m	0.006	0.005	0.005	0.0074
Corrugation angle	degrees	30	30	60	30
Surface area	m^2	0.122	0.095	0.095	0.5
Number of channels		21/7	19	19	9

Table 8.2: Tested PHEs by Wang et al. [16].

	Unit	PHE5	PHE6	PHE7
Width	m	0.615	0.615	0.1
Height	m	0.678	0.678	0.267
Port diameter	m	0.21	0.21	0.032
Hydraulic diameter	m	0.0074	0.0074	0.0048
Corrugation angle	degrees	45	60	30
Surface area	m^2	0.5	0.5	0.032
Number of channels		9	9	79

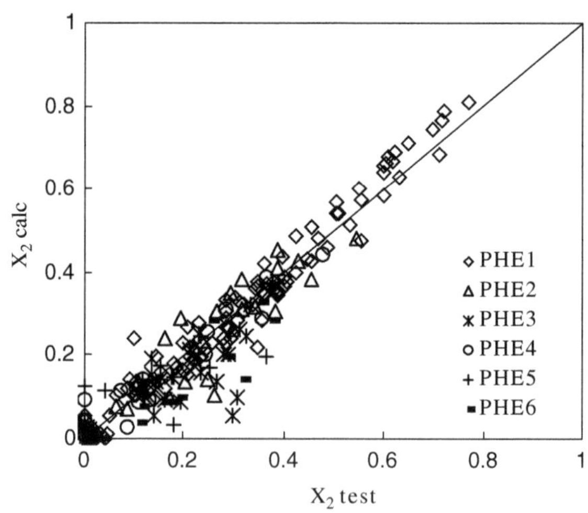

Figure 8.13: Comparison between the calculated and experimental values using the modified Boyko–Kruzhilin equation by Wang et al. for steam condensation in PHEs [16].

was 1.5–4 times that expected for vertical tubular condensation, and its value was a function of plate geometry, fluid properties, and flow condition. He also recommended the use of Nusselt equation [6] and the Carpenter and Colburn equation [11] for gravity- and shear-controlled regions, respectively.

Chopard [22] performed some tests with R22 at high Reynolds numbers in welded PHEs with three different geometries (plain rectangular, inline, or staggered studs). It was found that condensation heat transfer in the exchangers with inline or staggered studs was much higher than that in the plain rectangular channel. It was also found that the Shah equation underestimates the condensation heat transfer coefficient, which is a similar conclusion obtained from the steam condensation experiment by Wang et al. [16].

For ammonia applications, Panchal and Rabas [23] tested two plate-and-frame heat exchangers with corrugation angles of 30° and 60° as condensers. The results indicated that the plates with a corrugation angle of 60° gave higher heat transfer coefficients than the plates with a corrugation angle of 30°, as expected. No attempt was made to develop a correlation for predicting condensation heat transfer coefficient. In addition, Arman and Rabas [24] made a serious effort to develop a general computer program for heat transfer and pressure drop of a single component and binary mixtures during condensation in PHEs. A stepwise calculation method was used and a number of constants in the correlations were determined by comparison with experimental ammonia data. During the calculation, the Tovazhnyanskiy–Kapustenko [18] correlation and the Lockhart–Martinelli [25] model were used for heat transfer coefficient and pressure drop, respectively.

Yan et al. [17] conducted an experimental test for refrigerant R134a in a BHE with a corrugation angle of 60°. The influence effects of the mass flow rate, heat flux, system pressure, and vapour quality were investigated. The results indicated that at higher vapour quality the condensation heat transfer coefficient and pressure drop were significantly higher. A rise in the refrigerant mass flux only caused a mild increase in the heat transfer coefficient in most cases. It was also noted that the condensation heat transfer was only slightly better for a higher average imposed heat flux. In addition, at a higher system pressure the condensation heat transfer coefficient was found to be slightly lower. An attempt was made to summarize a correlation similar with Akers correlation for predicting heat transfer coefficient [26],

$$\mathrm{Nu} = 4.118\mathrm{Re}_{\mathrm{eq}}^{0.4}\mathrm{Pr}_{\mathrm{l}}^{1/3} \tag{8.10}$$

However, this correlation is only valid for their specific experimental condition, as the authors mentioned.

Recently, Gitteau et al. [27] studied the condensation heat transfer for pure and mixtures of hydrocarbons in a compact welded PHE. Three pure fluids (pentane, butane, and propane) and two mixtures (butane + propane) were tested. The operating pressure ranged from 1.5 to 18 bar. For pure fluids, two heat transfer mechanisms were identified. For low Reynolds numbers, condensation occurred almost filmwise and the heat transfer coefficient decreased with increasing Reynolds number. For higher values of the Reynolds number, the heat transfer increased gently.

For mixtures, the behaviour was different. For low Reynolds numbers, mass transfer affected heat transfer and reduced the heat transfer coefficient by a factor of 4. At higher Reynolds numbers, the heat transfer coefficient remained almost constant or increased slightly and the values were close to those of pure fluids. However, no correlation was proposed for the prediction of the condensation heat transfer coefficient.

The above reviews show that there has been quite a lot of research conducted for condensation heat transfer in PHEs. To summarize these results, the following conclusions are reached.

1 There are two fundamental mechanisms in this process: gravity controlled and shear controlled. Due to the high vapour velocity and corrugated geometry involved in PHEs, shear-controlled condensation is often the case.
2 In the shear-controlled region, the condensation heat transfer coefficient increases with the increase of mass flow rate and vapour quality, but decreases with the increase of system pressure. This relationship is valid for both steam and refrigerant.
3 For the gravity-controlled region, the condensation heat transfer coefficient can be predicted by the Nusselt equation. In the shear-controlled region, the shear-controlled equations should be used. However, the classical Nusselt theory gives a bottom limit for condensation in PHEs, although it is a very conservative estimation.
4 In the shear-controlled region, there is no single correlation, which can predict condensation heat transfer coefficient for different fluids in different plate channels. Equation (8.6) was developed from steam condensation in different PHEs, and probably can be recommended for general application. However, its use in refrigerant and hydrocarbon applications requires further verification.

8.2.3 Condensation pressure drop

As stated in Chapter 5, the total condensation pressure drop consists of the frictional pressure drop, the gravity pressure drop, the acceleration pressure drop, and the sum of additional pressure drops. Methods for calculating these different parts have been summarized there. Therefore, the frictional pressure drop can be obtained after the total pressure drop is measured during the experiments.

Frictional pressure drop of condensation is a function of flow rate, vapour quality, system pressure, etc. Figure 8.14 ($\mathrm{Re_{gin}}$ in the figure is the steam inlet Reynolds number, $\mathrm{Re_{gin}} = G_\mathrm{s} \cdot D_\mathrm{e}/\mu_\mathrm{s}$) was obtained from the experiment of steam condensation in PHEs by Wang et al. [16], and it shows that the condensation pressure drop decreases with the increase of saturation pressure, but increases with the increase of total mass flow rate (i.e. $\mathrm{Re_{gin}}$).

The pressure drop is increased with the increase of vapour quality. This has been demonstrated by Yan et al. [17] in an experimental investigation of R134a in a PHE, shown in Fig. 8.15.

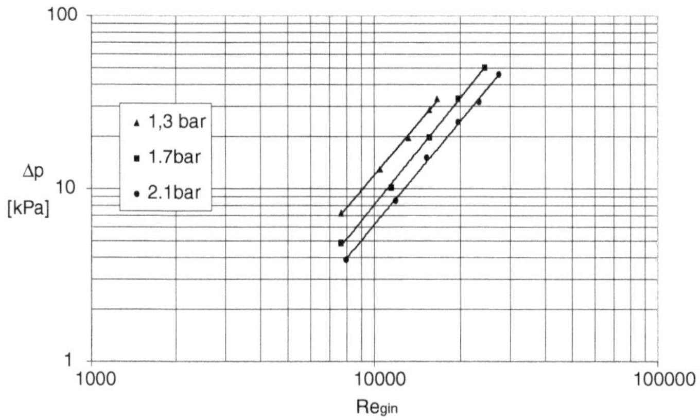

Figure 8.14: Frictional pressure drop as a function of mass flow rate (Re_{gin}) and system pressure of complete steam condensation in a PHE [16].

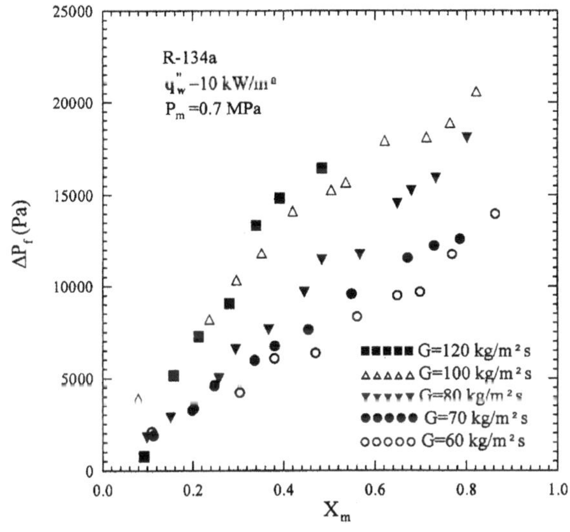

Figure 8.15: Variation of frictional pressure drop with mean vapour quality for various mass flow rates at constant system pressure ($Pm = 0.7$ MPa) for refrigerant R134a condensing in a PHE [17].

For the prediction of frictional pressure drop for condensation, a literature survey shows that a widely accepted method is to apply the Lockhart–Martinelli model [25]. The Lockhart–Martinelli parameter X and the two-phase friction multiplier ϕ_l for liquid are respectively defined as

$$X^2 = \frac{\Delta P_l}{\Delta P_g} \tag{8.11}$$

$$\phi_l^2 = \frac{\Delta P_{tp}}{\Delta P_l} \tag{8.12}$$

where ΔP_l is the liquid phase frictional pressure drop calculated as if the liquid phase were to be flowing alone, ΔP_g is the gas phase frictional pressure drop as if the gas phase were to be flowing alone, and ΔP_{tp} is the two-phase frictional pressure drop. The relationship between Lockhart–Martinelli parameter X and the two-phase friction multiplier ϕ_l for liquid can be illustrated graphically [25]. In practice, this model is often represented by the correlation suggested by Chisholm [28],

$$\phi_l^2 = 1 + \frac{C}{X} + \frac{1}{X^2} \tag{8.13}$$

The value of C is usually determined from experimental data for different situations. For plain tubes, it is usually selected according to the flow conditions as follows:

Liquid–vapour flow condition		C
Turbulent–turbulent	T–T	20
Viscous–turbulent	V–T	12
Turbulent–viscous	T–V	10
Viscous–viscous	V–V	5

For steam condensation in PHEs, a value of 16 was recommended for the constant C by Wang et al. [16] for predicting the average frictional pressure drop. However, a relatively big deviation between the predicted and experimental values existed. Later, they also pointed out that this single value could not perform well in the stepwise calculation procedure. In addition, different values were found for calculating frictional pressure drop for plate evaporators, which will be presented in later sections. These facts may suggest that the value of C is a function of geometry, fluid properties, and flow condition. Further work must be carried out for clarifying this issue.

8.3 Performance of plate evaporators

When a PHE is used for evaporation applications, it is called a plate evaporator. It can serve as a steam generator, which is used to produce steam. It can also serve as reboiler, which is used to supply heat and vapour to a distillation column. Not to mention, it is very often used in refrigeration systems as a refrigerant evaporator.

A literature survey shows there has been much experimental work on plate evaporators, but most of the work concentrated on refrigerant applications. Common working fluids include both pure fluids (e.g. R22, R134a, R142b, ammonia, propane, propene) and mixtures (e.g. R410a, R407c, mixture of propane and ethane). The large number of fluid types is attributed to the environment requirement by the international community. During a certain period, the synthetic chlorofluorocarbons (CFC) and hydrochlorofluorocarbons (HCFC) were widely used in refrigeration systems. However, it has been found that the use of such substances affects the ozone layer in the atmosphere and contributes to the global greenhouse effect. This has led to an international agreement to phase out these substances (the Montreal Protocol). This agreement has promoted a search for new and environmentally friendly refrigerants, which has resulted in a large number of working fluids for refrigeration systems. Because PHEs are increasingly used as evaporators in refrigeration systems, the working fluids of plate evaporators in most research work are refrigerant fluids. However, it must be realized that working fluids other than refrigerants are also found in plate evaporators. For example, plate evaporators are widely used as reboilers in process industries, where various working fluids (e.g. pure and mixtures of hydrocarbons) are present.

Literature survey also shows that most previous research work only resulted in qualitative conclusions, instead of refined general correlations for predicting the heat transfer coefficient and pressure drop. The reason for this is probably due to the fact that the evaporation process is very complicated and geometry dependent. To summarize a single correlation valid for different geometries and working fluids is a very difficult task. Therefore, many correlations are only valid within their respective experimental ranges.

8.3.1 Fundamental mechanism of evaporation

In the confined cross-corrugated channels in plate evaporators, the evaporation belongs to the forced convective evaporation category. The flow is a two-phase mixture of the liquid and its vapour, and it gains enthalpy and quality when it proceeds downstream. Heat transfer mechanisms are mostly complex functions of the flow regime, the geometry, the properties of the fluids and surfaces, and many other parameters. Previous studies of evaporation in smooth tubes can give relevant fundamental knowledge, although details can be different in plate channels. Therefore, general flow and heat transfer phenomena of evaporation in smooth tubes are described first in the following sections.

8.3.1.1 Evaporation regimes in smooth tubes

For a complete evaporation of upward flow in a vertical tube, the flow patterns experienced are depicted in Fig. 8.16.

During the first part, the liquid is being heated up to the saturation temperature and the wall temperature remains below that necessary for nucleation. The heat is transferred to a single phase liquid and the normal correlations for convection apply here. In the sub-cooled nucleate evaporation, vapour bubbles begin to form on the

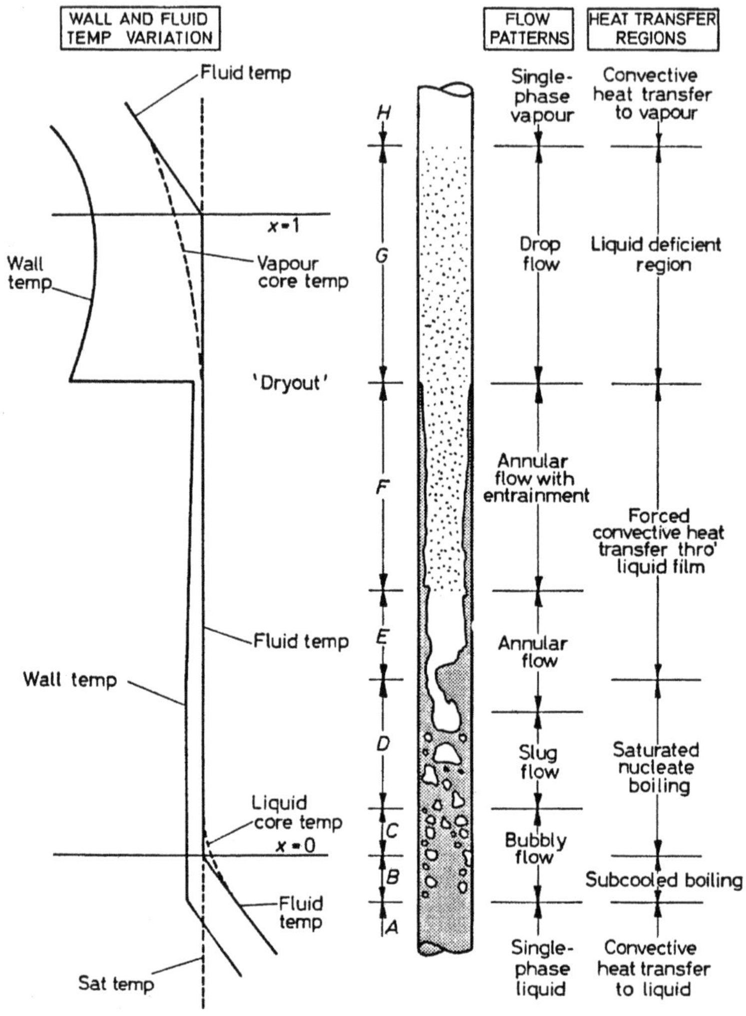

Figure 8.16: Flow regimes in a vertical evaporator tube [4].

wall from some nucleation sites, but the bulk of the fluid is still sub-cooled. This region is usually short, and the wall temperature remains essentially constant–a few degrees above the saturation temperature. After a certain distance of this region, the heat transfer regime becomes the saturated nucleate boiling, and the fluid core temperature equals saturation temperature. At low quality, the flow is bubbly flow, and then it becomes slug flow and annular flow with increase in vapour quality. In the annular flow region, there is a liquid annulus on the wall and a vapour core, which usually contains a significant amount of droplets. Note that nucleate evaporation may or may not exist in the annular flow region. If the liquid film on the wall is thin enough, heat is transferred by conduction through the liquid and vaporization occurs at the liquid–vapour interface. This region is referred to as the two-phase convective boiling region. At some critical point, the liquid film completely evaporates–a point called dryout. Although there are still liquid droplets in the bulk of the liquid, the heat transfer regime is termed liquid deficient region. Eventually, all the liquid evaporates and heat is transferred to a single-phase gas flow. The dryout point is very important in many practical designs. At this point, the surface temperature will arise abruptly in a heat flux controlled situation, due to the precipitous heat transfer coefficient drop. Attention must be given to this phenomenon.

Qualitatively, the progressive variation of the local heat transfer coefficient along the length of the tube as evaporation proceeds can be depicted by Fig. 8.17. As can be seen, the heat transfer coefficient is a function of heat flux and vapour quality, when the evaporation experiences different flow regimes. In this figure, heat flux is increased from curve i to vii. In the single-phase convective heat transfer region, the heat transfer coefficient is approximately constant and is influenced only by temperature on the liquid physical properties. In the sub-cooled nucleate boiling region, the heat transfer coefficient increases linearly with the length. But in saturated nucleate boiling region, the heat transfer coefficient remains constant with the length or vapour quality. In both regions, the heat transfer coefficient increases with regard to heat flux. In the two phase convective boiling region, the heat transfer increases with length or vapour quality, but the heat flux has a very small influence on the heat transfer coefficient in this region. At the dryout point, the

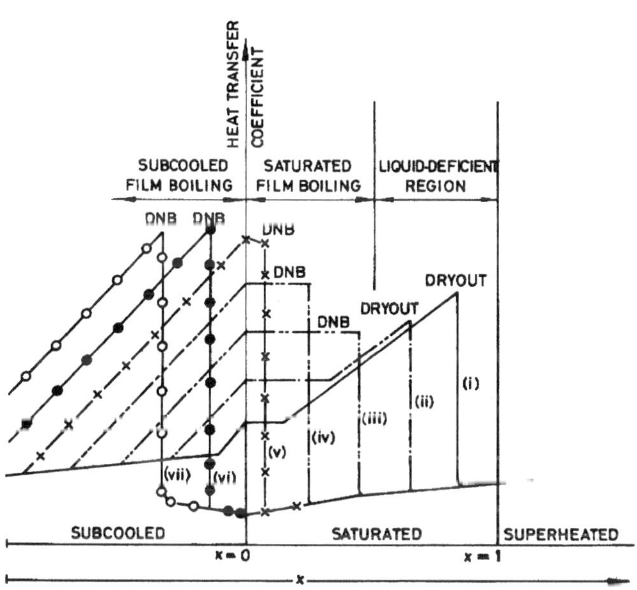

Figure 8.17: Variation of heat transfer coefficient with vapour quality with increasing heat flux as parameter [4].

heat transfer coefficient drops suddenly to that expected for heat transfer by forced convection to saturated steam. Finally, the heat transfer coefficient follows the trend of single-phase vapour convective flow. It should be pointed out that the above descriptions are restricted to the case where a relatively low heat flux is supplied to the walls of the tube. With higher heat flux, the flow regimes can be quite different and consequently the heat transfer coefficient. For example, when the heat flux is increased to iii in Fig. 8.17, the two-phase convective boiling region does not occur, and the flow goes to the liquid-deficient region directly, which is called departure from nucleate boiling (DNB). With even higher heat flux, DNB can occur even from a sub-cooled nucleate boiling region. The value of heat flux, which can initiate DNB, is often called the critical heat flux (CHF). At this point, the heat transfer coefficient precipitously drops to a much lower level. Under constant heat flux, such as in a nuclear reactor or a fossil fuel boiler, the wall surface temperature will increase sharply to compensate for the low heat transfer coefficient, and accidents may occur. However, the result of this phenomenon depends on the thermal boundary condition. In a temperature boundary condition, the local heat flux will decrease in response to the reduced overall heat transfer coefficient. In a two-fluid heat exchanger, the second fluid controls the wall temperature, thus the CHF condition does not exist.

8.3.1.2 General correlation for evaporation in tubes

The above description shows that three fundamental heat transfer mechanisms exist in the evaporation process, depending on the flow pattern. These are nucleate boiling (bubbly and slug flow), two-phase convective boiling (annular flow), and single-phase heat transfer to gas in the drywall region, downstream from the annular flow region. The third drywall region may or may not exist, depending upon the magnitude of the heat flux in the annular flow region. Therefore, in most cases, only the first two mechanisms are considered. In the annular flow region, the liquid velocity tends to suppress nucleate boiling. Therefore, the evaporation heat transfer can be divided into nucleate boiling dominated regime, two-phase convective boiling dominated regime, and a regime where both nucleate and two-phase convective boiling contribute.

Based on the fact that two mechanisms exist over the entire range of the evaporation and the contributions made by the two mechanisms are assumed additive, several correlations have been proposed, and the most commonly applied correlations are by Chen [29], Shah [30] and Steiner and Taborek [31], etc. Chen [29] carried out a comparison of several previous correlations and summarized the following correlation based on 594 data points of different fluids and upward tube flows with one exception (downflow). It reads,

$$h = h'_{\mathrm{nb}} + h'_{\mathrm{cb}} \tag{8.14}$$

where h'_{nb} is the contribution from nucleate boiling and h'_{cb} is the contribution from two-phase convective boiling. The equation of Forster and Zuber [32] was taken as the basis for the evaluation of the nucleate boiling component. Their pool boiling analysis was modified to account for the thinner boundary layer in forced convective boiling and the lower effective superheat that the growing vapour bubble experiences. The modified Forster–Zuber equation becomes

$$h'_{\text{nb}} = 0.00122\left(\frac{k_{\text{l}}^{0.79}c_{\text{pl}}^{0.45}\rho_{\text{l}}^{0.49}}{\sigma^{0.5}\mu_{\text{l}}^{0.29}i_{\text{lg}}^{0.24}\rho_{\text{g}}^{0.24}}\right)\Delta T_{\text{sat}}^{0.24}\Delta p_{\text{sat}}^{0.75}S_{\text{nb}} \tag{8.15}$$

where S_{nb} is a suppression factor defined as the ratio of the mean superheat experienced by the growing bubble to the wall superheat ΔT_{sat}, and Δp_{sat} is the change in vapour pressure corresponding to the wall superheat ΔT_{sat}. S_{nb} is a function of the local two-phase Reynolds number Re_{tp} and can be represented graphically or by the following curve-fitted correlation.

$$S_{\text{nb}} = \frac{1}{1 + 2.53 \times 10^{-6}\text{Re}_{\text{tp}}^{1.17}} \tag{8.16}$$

In the Chen correlation, the convective contribution h'_{cb} is calculated as the product of the single-phase convective heat transfer predicted by a Dittus–Boelter type correlation and the correction factor F_{cb}.

$$h'_{\text{cb}} = 0.023\left[\frac{\dot{G}(1-x)D_{\text{e}}}{\mu_{\text{l}}}\right]^{0.8}\left(\frac{\mu c_{\text{p}}}{k}\right)_{\text{l}}^{0.4}\left(\frac{k_{\text{l}}}{D_{\text{e}}}\right)F_{\text{cb}} \tag{8.17}$$

where the correction factor F_{cb} is a function of the Lockhart–Martinelli parameter X_{tt}, where the subscript '*tt*' means that both the liquid and gas phases are in turbulent condition. This relationship can be represented graphically or by the following curve-fitted correlation.

$$F_{\text{cb}} = 1 \quad \text{for } \frac{1}{X_{\text{tt}}} \leq 0.1 \tag{8.18}$$

$$F_{\text{cb}} = 2.35\left(\frac{1}{X_{\text{tt}}} + 0.213\right)^{0.736} \quad \text{for } \frac{1}{X_{\text{tt}}} > 0.1 \tag{8.19}$$

The Chen correlation fits the experimental data remarkably well, with a standard deviation of 11% on heat transfer coefficient. This correlation is regarded as one of the best available for the saturated forced convective boiling region in vertical ducts and is recommended for use with all single component non-metallic fluids.

Shah [30] developed a graphical chart which can be used for boiling in both vertical and horizontal channels. He also developed equations that represent his chart method [33]. In contrast to the Chen correlation, i.e. to add the two respective contributions, the Shah correlation takes the larger of his nucleate coefficient and his convective coefficient as the evaporation heat transfer coefficient. Kandlikar [34] has also proposed a Shah-like correlation that utilizes fluid-dependent constants.

Steiner and Taborek [31] proposed a general correlation for boiling in vertical tubes with upward (and downward) flow. They used a total number of data points of 12,067, which cover different fluids including water, refrigerants, hydrocarbons, cryogens, and ammonia. The advantage of this model is that it allows for continued improvement of its various parts without necessarily affecting other parts of the model. It reads,

$$h = [(h'_{\text{nb}})^n + (h'_{\text{cb}})^n]^{1/n} = [(h_{\text{nb,o}}F_{\text{nb}})^n + (h_{\text{l,o}}F_{\text{tp}})^n]^{1/n} \tag{8.20}$$

where $h_{\text{nb,o}}$ is the local 'normalized' nucleate pool boiling coefficient at a standard condition of heat flux and reduced pressure; F_{nb} is the nucleate boiling correction factor that accounts for the differences between pool and flow boiling; $h_{\text{l,o}}$ is the local liquid-phase forced convection coefficient based on the total flow as liquid; F_{tp} is the two-phase multiplier that accounts for the enhancement of convection in a two-phase flow. The details of how to calculate these individual variables can be found in the original publication. It should be pointed out that $n = 1$ means a simple addition (which is actually the form of the Chen correlation), $n = \infty$ means to choose the larger contribution (which is actually the form of the Shah correlation), and the other value means a power-law determination.

There are many correlations available, but the above-mentioned ones are the most often used correlations. Although all of them were developed from evaporation processes in tubes, they are often employed in PHEs. In such a case, it is logical to use single-phase correlations of PHEs instead of tubes, when the forced convective heat transfer is calculated. However, the correction factors may have significantly different values for PHEs and tubes, which will certainly affect the accuracy. Moreover, due to the different flow geometries, the onset of different flow regimes may occur quite differently. Therefore, verification and improvement must be made in order to accurately predict the evaporation process in PHEs.

8.3.2 Evaporation heat transfer

As indicated previously, research in the evaporation process in PHEs has been done mostly for refrigerant fluids. The working fluids cover both pure fluids (R22, R134a, R142b, ammonia, propane, propene) and mixtures (R410a, R407c, mixture of propane and ethane). Although some studies summarized some correlations for predicting evaporation heat transfer coefficient, many of them only came out with qualitative conclusions.

Engelhorn and Reinhart [35] tested a plate evaporator with R22 as working fluid. The PHE is 0.522 m high, 0.115 m wide, and the total heat transfer area is 3.02 m^2. The evaporation temperature ranged from −4.2°C to +9°C, and the heat flux ranged from 9 to 15 kW/m^2. The tests were conducted with and without distributor at the inlet. It was found that in both cases, the evaporation heat transfer coefficient increases with increase of the heat flux, which indicates that the nucleate boiling process is dominant. For the case without distributor, the film heat transfer coefficients were much higher (approximately 50%) as the refrigerant was saturated than when it was superheated at the outlet. Under the same heat flux, the evaporation heat transfer coefficient decreases with increasing evaporation temperature (i.e. increasing evaporation pressure). However, after the distributor was installed, the performance of the evaporator was changed considerably. The evaporation heat transfer coefficient increased significantly and no influence of evaporation temperature could be observed. This indicates that the maldistribution has considerable effect on the evaporation performance. The results for the case with distributor installed and saturated refrigerant at the outlet were compared with some correlations from the literature. The first comparison was made with a correlation of Gorenflo [36] for pool boiling on tube bundles. It was found that this correlation overpredicted the experimental results for the lower heat fluxes but underpredicted them for the higher heat fluxes; however, all results were within ±25% of the correlation. The second comparison was made with Slipcevic [37] also for pool boiling on tube bundles. The average heat transfer coefficient is the sum of the heat transfer coefficient for nucleate boiling h_{nb} and the contribution from the free convective heat transfer coefficient h'_{cb} according to

$$h = h_{nb} + h'_{cb} = h_{nb} + F_{cb}h_{cb} \tag{8.21}$$

where h_{nb} is calculated by the Stephan and Abdelsalam [38] equation, and h_{cb} is determined for the case of heat transfer with a turbulent boundary layer at the heated surface, i.e.

$$\mathrm{Nu} = 0.15 \cdot (\boldsymbol{Gr} \cdot \mathrm{Pr})^{1/3} \tag{8.22}$$

F_{cb} is the enhancement factor for considering the influence of free convection and can vary between 0.5 and 1. It was found that the deviation between experimental and predicted values was within ±10% when the enhancement factor F_{cb} was set equal to 0.5. The third comparison was made with Azarskov et al. [39], but the deviation between experimental and predicted values was very large.

The influence of flow distribution on the evaporation heat transfer of PHEs was also found by Dutto et al. [40]. The working fluid used was R142b. Two identical heat exchangers were used in the experiments, one with a distributor and one without a distributor. It was found that the boiling heat transfer coefficient was about 30% higher for the exchanger with distributor than for the one without. The values increased with increasing heat flux indicating that nucleate boiling was dominant. No correlations were suggested for the prediction of evaporation heat transfer coefficient.

Kumar [21] suggested the Chen-type correlation for predicting evaporation heat transfer coefficient for refrigerants. Because Kumar is an employee of APV (a manufacturer of PHEs), it is reasonable to believe that his correlation is used in practical design. The equation reads,

$$h = h_{micro+} + h_{macro} \tag{8.23}$$

where h_{micro} is the product of the nucleate boiling coefficient and a suppression factor arising from departure from pool boiling conditions. The suppression factor itself is a function of convective currents and decreases with increasing flow. The term h_{macro} is the forced convective single-phase heat transfer in a PHE corrected by a factor which is always greater than unity. It has been found that this correction factor varies greatly with corrugation inclination to the flow direction, and is a function of the Lockhart–Martinelli parameter. However, no details were given an how to calculate these terms. For any given channel configuration, the heat transfer coefficient was found to vary with the outlet quality, i.e. it increases with increasing outlet quality. However, it was observed that significant fall off in heat transfer coefficient could occur when the exit quality exceeded 0.7. In addition, Kumar suggested that the local heat transfer may differ considerably, and these changes can be computed in a stepwise fashion. Note that the mentioned working fluids in the related experiments were R22 and ammonia.

Syed [41], an employee of Alfa Laval, stated that correlations similar to those found in the open literature for circular tubes may be used for estimating the evaporation heat transfer coefficient in PHEs. The following Chen-type model was suggested:

$$h = h_{\text{nb}} S_{\text{nb}} + h_{\text{cb}} F_{\text{cb}} \tag{8.24}$$

where h_{nb} is the heat transfer coefficient for pool boiling; h_{cb} is the heat transfer coefficient for single-phase liquid flow; S_{nb}, the nucleate boiling suppression factor, is a function of heat flux and is always less than 1; F_{cb}, a two-phase multiplier, is a function of mass flow rate and is always larger than 1. The author indicated that the original correlation must be modified for use with PHEs, but no information was given to calculate these parameters.

Thonon et al. [42] gave a review of the research development of single- and two-phase flows in PHEs by that time. For evaporation, they suggested that annular flow pattern dominates in a large range of applications in such corrugated channels, unless the vapour quality is low. This flow pattern, in evaporation, allows a high heat transfer coefficient. As in tubes, the total heat flux in evaporation is assumed to have two components: nucleate boiling and convective boiling. But referring to Ohara et al. [43] and Marvillet [44], they concluded that the nucleate boiling has influence only at high heat flux or low vapour quality. This seems contrary to other investigations, e.g. Engelhorn and Reinhart [35] and Dutto et al. [40], who found a strong influence of the heat flux, indicating a dominating influence of nucleate boiling. The following criteria were suggested to distinguish whether the flow was nucleate boiling dominated or two-phase convective boiling dominated.

$$\text{Bo} \cdot X_{\text{tt}} > 0.00015 \quad \text{nucleate boiling}$$
$$\text{Bo} \cdot X_{\text{tt}} < 0.00015 \quad \text{convective boiling}$$

where Bo is the boiling number defined in eqn (8.25) and X_{tt} is the Lockhart–Martinelli parameter already defined.

$$\text{Bo} = \frac{\dot{q}}{G\gamma} \tag{8.25}$$

In the nucleate boiling dominated region, the Gorenflo correlation [35] is suggested (surface roughness is assumed as 0.4 μm),

$$h = h_0 \left(1.2P_{\text{r}}^{0.27} + \left(2.5 + \frac{1}{1 - P_{\text{r}}} \right) P_{\text{r}} \right) \left(\frac{\phi}{20{,}000} \right)^{0.9 - 0.3P_{\text{r}}^{0.15}} \tag{8.26}$$

In the above equation, h_0 is the heat transfer coefficient at standard condition and is provided in a table for specific condition (fluid and pressure); P_r is the reduced pressure; ϕ is the heat flux, W/m^2. In the two-phase convective boiling dominated region, the following enhancement factor method is used.

$$h = h_{lo} F_{tp} \tag{8.27}$$

where h_{lo} is the liquid heat transfer coefficient assuming all the fluid flows as liquid, and is calculated through the specific correlation for corrugated channels. The enhancement factor F_{tp} is calculated by

$$F_{tp} = 1 + 1.8 X_{tt}^{-0.79} \tag{8.28}$$

The evaporation heat transfer coefficient should then be taken as the greater value of the nucleate boiling term and the two-phase convective boiling term. In addition, they indicated that the critical vapour quality, at which the heat transfer coefficient drops, is lower than for a smooth geometry. A typical value of the critical quality is 0.7. This lower value may result from the fact that the corrugation tends to wash out the liquid film from the wall, but no experiments have clearly shown this. It is interesting to point out that this critical value is very similar to the finding from Kumar [21].

The evaporation heat transfer of refrigerant R134a was investigated in a single channel of a compact BHE by Margat et al. [45]. The tested heat exchanger had a 60° corrugation angle and 2 mm corrugation height. Electrical heating was used to vary the heat flux from 1 to 12 kW/m^2. The saturation pressure was kept at 6 bar, and the mass velocity ranged from 30 to 125 kg/(m$^2 \cdot$ s). The evaporation heat transfer coefficient was found to be independent of the heat flux but strongly dependent on the vapour quality. This indicates that nucleate boiling does not occur in such a corrugated channel, and the two-phase convective boiling is the sole mechanism. Based on this result, they proposed eqn (8.27) for predicting evaporation heat transfer. The enhancement factor F_{tp} in that equation was calculated by the following equation:

$$F_{tp} = \left(1 + \frac{C}{X_{tt}} + \frac{1}{X_{tt}^2}\right)^{0.5} \tag{8.29}$$

where $C = 3$. Comparison between eqns (8.28) and (8.29) shows that these equations give very similar values for the enhancement factor F.

Pelletier [46] tested three BHEs as evaporators with R22, propane, propene, and a commercial mixture of propane and ethane as working fluids. No exact geometry parameters were given for the tested heat exchangers, but the pictures show that they are very thin and long with high corrugation angle. All evaporators have the same heat transfer area by using different number of plates, and were tested at heat fluxes between 6 and 10 kW/m^2. Of the three tested heat exchangers, models A and B have increased heat transfer coefficients with increased heat flux, but model C has no such influence. This indicates that nucleate boiling is important for models A and B, but not important for model C. In addition, the heat transfer coefficients of R22 and propane were approximately the same for model C while for models A and B the heat transfer coefficient of propane was considerably higher. These findings suggest that different heat transfer mechanisms exist in the respective experiments, and they are probably the function of geometry, fluid property, and flow condition. Attempts were also made to compare the experimental values with the values predicted by different correlations, but only for model A. It was found that the correlations of Stephan and Abdelsalam [38], Cooper [47], and Gorenflo [36] could predict the experimental data of model A with R22, propane, and propene within $\pm 25\%$. It should be pointed out that all three correlations were developed for pool boiling, and the influence from the heat flux is well reflected. This means that these correlations are not applicable to model C because no heat flux effect on the heat transfer coefficient was found. Therefore, these correlations can not be regarded as general solutions for predicting evaporation heat transfer coefficients in PHEs. The author also compared the results within Slipcevic [37] and Gorenflo [36] for the situation of flooding tube bundles. It was found that the Slipcevic correlation predicted the heat transfer coefficient with $\pm 25\%$, although mostly this correlation overpredicted the experimental values. The Gorenflo correlation was found to overpredict the experimental values by up to 60%. In addition, two correlations for horizontal tubes, Pierre [48] and Steiner [49], were also compared. These could only predict one working fluid and fail for the other, or fail for all fluids. In addition, it is interesting to point out that the Slipcevic correlation was also found to be applicable in the experimental work by Engelhorn and Reinhart [35].

Sterner [50] performed measurements on a semi-welded PHE and various versions of the nickel BHE as evaporators, using ammonia as the working fluid. The corrugation angle was kept constant (59°), but the plate size and hydraulic diameter were different. All the tested heat exchangers had more than 60 channels, but distributors were used for some heat exchangers to minimize the maldistribution. No influence factors were identified, so that it is unknown whether nucleate boiling or convective boiling dominated. The following equation was suggested to predict the evaporation heat transfer coefficient:

$$\mathrm{Nu} = 436.592\,\mathrm{Re}_{\mathrm{lo}}^{1.076}\,\mathrm{Ja}^{-0.976}\,\mathrm{Bo}^{1.154}\,\mathrm{Pr}_{\mathrm{l}}^{0.333} \tag{8.30}$$

where Ja is the Jakob number. However, this equation is meaningless due to the same reason as eqn (8.8). In addition, performance comparison was carried out for the tested PHEs, with the heat transfer coefficient versus the pumping power. It was found that the thermal performances were affected by flow distribution and that adopting special distribution devices in the inlet port resulted in better heat transfer performance. This indicates that maldistribution is an influencing factor for the thermal performance of plate evaporators.

The group of Lin [51, 52] carried out evaporation tests for R134a and R410a in a vertical PHE, respectively. The maldistribution was eliminated due to the single channel arrangement for the refrigerant. The same plate with a corrugation angle of 60° was used for all the tests. The effects of vapour quality, mass flow rate, heat flux, and pressure on the evaporation heat transfer and pressure drop were investigated in detail. It was found that the evaporation heat transfer coefficient in PHEs was much higher than that in circular tubes, particularly in the two-phase convective boiling dominated regime of high vapour quality. In their experimental range, the following influences were found:

1 For R134a, the evaporation heat transfer coefficient increases significantly with the vapour quality. No critical vapour quality, after which the heat transfer coefficient drops significantly, was identified during the experiments, at least when the vapour quality was lower than 0.9. No such information was given to R410a.
2 For R134a, the mass flow rate exhibits significant effects on the heat transfer coefficient only at high vapour quality. For R410a, the effect of the refrigerant mass flux on the saturated flow boiling heat transfer coefficient is significant only at higher imposed heat flux.
3 For R134a, the heat flux does not have significant effects on the heat transfer at high qualities, but it shows some influences at low quality. In comparison, the boiling heat transfer coefficient increases almost linearly with the heat flux for R410a.
4 For R134a, the increase in the system pressure results in a low heat transfer coefficient in the high vapour quality regime. In the low vapour quality regime, this influence becomes very small. For R410a, the system pressure has very slight influences on the evaporation heat transfer coefficient.

This investigation indicates that the influences of different parameters on the evaporation heat transfer are very complex. These parameters are interrelated and are difficult to separate during the experiments. Based on their experimental data, the following equation was suggested for predicting the evaporation heat transfer coefficient of R134a.

$$\mathrm{Nu} = 1.926\mathrm{Pr}_1^{1/3}\mathrm{Bo}_{\mathrm{eq}}^{0.3}\mathrm{Re}_{\mathrm{eq}}^{0.5}[(1-x)+x(\rho_1/\rho_\mathrm{g})^{0.5}] \tag{8.31}$$

where $\mathrm{Re}_{\mathrm{eq}}$ and $\boldsymbol{Bo}_{\mathrm{eq}}$ are, respectively, the equivalent Reynolds number and boiling number in which an equivalent mass flux is used in the definitions first proposed by Akers [13]

$$\mathrm{Re}_{\mathrm{eq}} = \frac{G_{\mathrm{eq}}D_\mathrm{e}}{\mu_1} \tag{8.32}$$

$$\mathrm{Bo}_{\mathrm{eq}} = \frac{\dot{q}}{G_{\mathrm{eq}}\cdot\gamma} \tag{8.33}$$

where

$$G_{\mathrm{eq}} = G[(1-x)+x(\rho_1/\rho_\mathrm{g})^{0.5}] \tag{8.34}$$

For R410a, the following equation was suggested:

$$h = h_{\mathrm{lo}}(88\mathrm{Bo}^{0.5}) \tag{8.35}$$

where h_{lo} is the heat transfer coefficient assuming all the fluid flows as liquid. These equations were claimed only valid for their specific situations. Obviously, no uniform equation was developed, even for two fluids flowing in the same channel. This indicates the difficulty in reaching a general correlation for predicting the evaporation heat transfer coefficient for PHEs.

Boccardi et al. [53] have studied the thermal performance of two compact BHEs as both evaporator and condenser using two pure refrigerants (R22 and R134a) and two mixtures (R407c and R410a). The tested PHEs had 40 channels and distributors were used to minimize the maldistribution during experiments. It was found that the thermal performance depended on the refrigerants, the thermal load, and the heat transfer process (evaporation or condensation). The heat exchanger comparison is based on the overall heat transfer coefficient under similar operating conditions. For evaporation, comparison shows that R410a has the highest heat transfer coefficient, and R407a the lowest. Based on their results, the R22 replacement aspects in PHEs for air conditioning were discussed. They also pointed out that the performance rank for different refrigerants could be reversed on the chosen PHEs. It should be pointed out that comparison of their results with others was not possible because no information about plate size and corrugation pattern was presented because parts of the investigation had to be classified as confidential. In addition, no attempts were given to develop correlations for predicting evaporation heat transfer coefficients.

Additional works in the literature about plate evaporators are exemplified by Panchal et al. [54], Kedzierski [55], etc. However, their conclusions are more or less similar to the above-cited references. Based on the above information, the following conclusions can be summarized:

1 Although there has been much work on evaporation in PHEs, most research works only published qualitative results on overall performance of individual heat exchangers for specific fluids. Attempts to summarize a universal correlation predicting the evaporation heat transfer coefficient are very rare, even if some authors tested several PHEs on their own.
2 There are contradictory conclusions about whether the evaporation process is nucleate boiling dominated or two-phase convective boiling dominated. The influences on the evaporation heat transfer coefficient from heat flux, mass flow rate, vapour quality, and pressure are sometimes contradictory for different experiments. This may be due to the maldistribution or the accuracy of individual experiments. On the other hand, this may also indicate that the influence is a function of geometry and fluid type.
3 It appears that flow distribution in the inlet manifold has an important influence on the overall performance. In almost all related experiments, the difference in overall performance with and without distributors was observed. Therefore, the influence of maldistribution was not eliminated for those experiments with a large number of flow channels but without distributor. This absolutely increases the difficulty in finding a more general correlation. Moreover, this may explain the contradictory conclusions between different researchers.
4 Although there is no universal correlation available in the literature for predicting evaporation heat transfer coefficient, it seems that the Chen-type correlation

is the most frequently used format for practical calculation. However, the accurate prediction of correction factor and suppression factor must be developed. The discovery of nucleate boiling dominated or convective boiling dominated will certainly help this development. At this stage, probably the correlations suggested from Syed or Thonon et al. may be recommended for calculating evaporation heat transfer coefficient for PHEs.

Therefore, in order to fully understand the evaporation process in PHEs, further work must be carried out. The above literature survey shows that it is particularly important to conduct systematic investigations with wide variations of plate geometry, working fluid, mass flow rate, vapour quality, heat flux, and pressure. In addition, the influence of maldistribution must be investigated.

8.3.3 Evaporation pressure drop

In the above-cited literature, investigating the evaporation heat transfer in PHEs, most of them did not cover the aspects of pressure drop. The main work on this subject was conducted by Margat et al. [45], Sterner [50], and Lin [51, 52]. Their studies demonstrated that the frictional pressure drop of evaporation increased by increasing the vapour quality and the mass velocity, but it decreased with increasing system pressure. This conclusion is similar to that of condensation, and the explanation is also similar. Figures 8.18 and 8.19 from these studies clearly show this effect.

For the prediction of evaporation friction pressure drop, the most widely accepted correlation is probably the Lockhart–Martinelli model, similar to the condensation case. This means that eqn. (8.13) is still valid. However, the selection of the

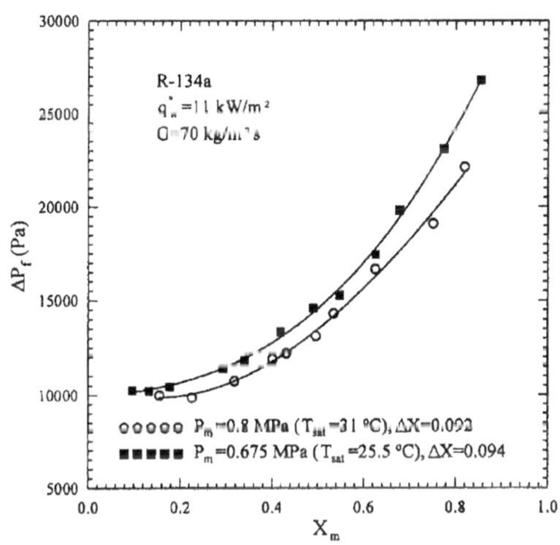

Figure 8.18: Variation of frictional pressure drop with mean vapour quality for various system pressures for R134a in a PHE [51].

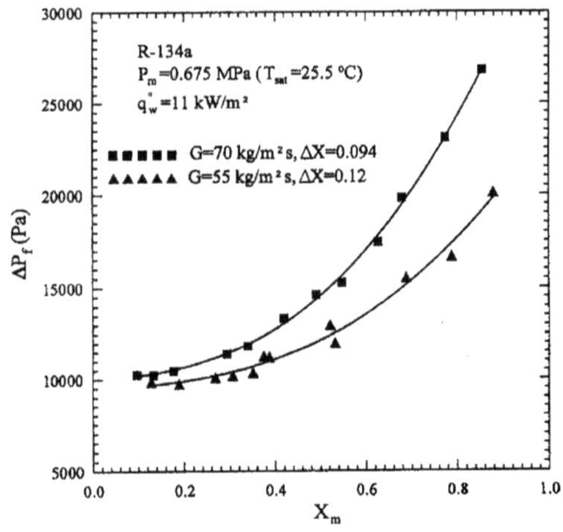

Figure 8.19: Variation of frictional pressure drop with mean vapour quality for various system pressures for R134a in a PHE [51].

C value is quite contradictory in the literature. A value of eight was suggested by Thonon et al. [42], but a value of three was suggested by the same group, however, based on their own experimental work [45]. In the experimental work conducted by Sterner [50], the value of *C* is function of the Reynolds number, and may vary from 1 to 200, well beyond the value recommended for plain tubes (between 5 to 20). One interesting finding by Holt et al. [56] is that the value of *C* is a function of the hydraulic diameter, from their experimental results in the ducts with different cross-sections (circular, trapezoidal, and rectangular) for different mixtures (air/water, helium/water, and air/60% aqueous glycerol). However, putting the findings from other researchers in the same figure shows that there is still a large deviation between different results (see Fig. 8.20). These contradictory findings may suggest that the value of *C* is a function of fluid properties and flow condition in addition to geometry. Further work is thus obviously necessary to clarify this issue.

Nomenclature

A	heat transfer area, m^2
A	flow area, m^2
A_c	channel cross area, m^2
Bo	boiling number
C	constant in Chisholm correlation
c_p	specific heat, J/(kg · K)
d	density correction exponent
D_e	equivalent diameter, m
D_h	hydraulic diameter, m
F	correction factor
g	gravity acceleration, m^2/s

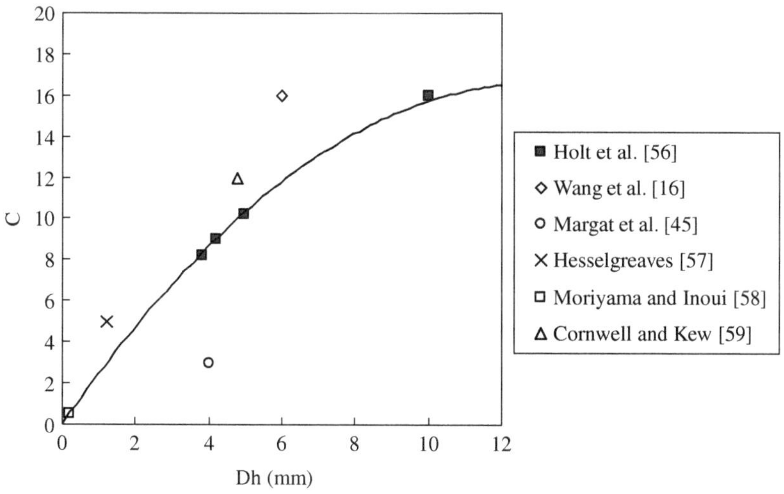

Figure 8.20: Correlation of Holt et al. [56] of Chisholm's C parameter with hydraulic diameter, with additional data (From Hesselgreaves [60]).

G	mass velocity, kg/(m^2 · s)
Gr	Grashof number
h	heat transfer coefficient, W/(m^2 · K)
H	dimensionless parameter, $H = c_p \cdot \Delta T/\gamma$
j	volumetric flux (or superficial velocity), m/s
Ja	Jakob number
k	thermal conductivity, W/(m · K)
Nu	Nusselt number
p	pressure, Pa
p_c	critical pressure, Pa
p_r	reduced pressure $= p/p_c$
Pr	Prandtl number
Q	volumetric flow rate, m^3/s
$\dot{q}$	heat flux, W/m^2
Re	Reynolds number
S	suppression factor
T	temperature, K
x	vapour quality
X	Lockhart Martinelli parameter
z	length, m
ΔT	temperature difference, K

Greek symbols

ϕ	heat flux, W/m^2
ϕ_l	two-phase friction multiplier for liquid

γ	latent heat, J/kg
μ	dynamic fluid viscosity, kg/(m · s)
ρ	fluid density, kg/m^3
σ	surface tension, N/m
Γ	condensate flow rate per unit width, kg/(m · s)

Subscripts

1	inlet
2	outlet
cb	two-phase convective boiling
eq	equivalent
f	frictional
g	gas or vapour
i	interface
l	liquid
lo	liquid only
m	mean value
nb	nucleate boiling
o	normalized or standard condition
tp	two phase
tt	turbulent–turbulent
w	wall
z	distance from the top, m

References

[1] Shah, R.K. and Sekulic, D.P., Heat Exchangers, in *Handbook of Heat Transfer*, 3rd edn, ed. W. Rohsenow, J. Hartnett, and Y. Cho, McGraw-Hill, New York, chapter 17, 1998.

[2] Kakac, S. and Liu, H., *Heat Exchangers: Selection, Rating and Thermal Design*, CRC Press, Boca Raton, FL, 1998.

[3] Tribbe, C. and Müller-Steinhagen, H.M., Gas/Liquid Flow in Plate-and-Frame Heat Exchangers, *Heat Transfer Engineering*, vol. 22, no. 1, pp. 5–21, 2001.

[4] Collier, J.G. and Thome, J.R., *Convective Boiling and Condensation*, 3rd edn, Oxford University press, Oxford, 1994.

[5] Hewitt, G.F. and Roberts, D.N., Studies of Two-phase Flow Patterns by Simultaneous X-ray and Flash Photography, AERE-M 2159, HMSO, 1969.

[6] Nusselt, W., The Condensation of Steam on Cooled Surfaces, *Z. Ver. Deutsch. Ing.*, vol., 60, pp. 541–546, 569– 575, 1916. (Translated into English by D. Fullarton, *Chemical Engineering Fundamentals.*, vol. 1, no. 2, pp. 6–19, 1982.)

[7] Bromley, L.A., Effect of Heat Capacity of Condensate, *Industrial Engineering Chemistry*, vol. 44, 2966, 1952.

[8] Chen, M.M., An Analysis Study of Laminar Film Condensation, Part I – Flat Plates, *ASME Journal of Heat Transfer*, Series C, vol. 83, pp. 48–55, 1961.

[9] Chen, M.M., An Analysis Study of Laminar Film Condensation, Part II – Single and Multiple Horizontal Tubes, *ASME Journal of Heat Transfer*, Series C, vol. 83, pp. 55–60, 1961.

[10] Rohsenow, W.M., Heat transfer and Temperature Distribution in Laminar Film Condensation, *Transactors of ASME*, vol. 79, pp. 1645–1648, 1956.

[11] Carpenter, E.F. and Colburn, A.P., The Effect of Vapour Velocity on Condensation Inside Tubes, *Proceedings of the General Discussion on Heat Transfer*, pp. 20–26, IMechE/ASME, 1951.

[12] Dukler, A.E., Fluid Mechanics and Heat Transfer in Vertical Falling Film Systems, *Chemical Engineering Progress Symposium Series No. 30*, 1960.

[13] Akers, W.W., Deans, H.A., and Crosser, O.K., Condensation Heat Transfer within Horizontal Tubes, *Chemical Enginearing Progress Symposium Series*, vol. 55, pp. 171–176, 1959.

[14] Boyko, L.D. and Kruzhilin, G.N., Heat Transfer and Hydraulic Resistance during Condensation of Steam in a Horizontal Tube and in a Bundle of Tubes, *International Journal of Heat Mass Transfer*, vol. 10, no. 2, pp. 361–373, 1967.

[15] Shah, M.M., A General Correlation for Heat Transfer during Film Condensation inside Pipes, *International Journal of Heat Mass Transfer*, vol. 22, no. 3, pp. 547–556, 1979.

[16] Wang, L., Christensen, R., and Sundén, B., An Experimental Investigation of Steam Condensation in Plate Heat Exchangers, *International Journal of Heat Exchangers*, vol. 1, no. 2, pp. 125–150, 2000.

[17] Yan, Y.Y, Lio, H.C. and Lin, T.F., Condensation Heat Transfer and Pressure Drop of Refrigerant R134a in a Plate Heat Exchanger, *International Journal of Heat Mass Transfer*, vol. 42, no.6, pp. 993–1006, 1999.

[18] Tovazhnyanskiy, L.L. and Kapustenko, P.A., Heat Transfer from Steam Condensing in an Extended-surface Slot Channel, *Heat Transfer Soviet Research*, vol. 12, no. 4, pp. 34–36, 1980.

[19] Cooper, A., Condensation of Steam in Plate Heat Exchangers, *AIChE Symposium*, vol. 70, no. 138, pp. 172–177, 1987.

[20] Wang, Z. and Zhao, Z., 1993, Analysis of Performance of Steam Condensation Heat Transfer and Pressure Drop in Plate Condensers, *Heat Transfer Engineering*, vol. 14, no. 4, pp. 32–41.

[21] Kumar, H., Design of Plate Heat Exchangers for Refrigerants, *Proceedings of Conference of the Insitiue Refrigeration*, 1991–1992, 5.1–5.2, 1992.

[22] Chopard, F., Marvillet, C. and Pantaloni, J., Assessment of Heat Transfer Performance of Rectangular Channel Geometries, *Proceedings of the First European Conference on Thermal Science*, IChemE, vol. 1, no. 129, pp. 725–733, 1992.

[23] Panchal, C.B. and Rabas, T.J., Thermal Performance of Advanced Heat Exchangers for Ammonia Refrigeration Systems, *Heat Transfer Engineering*, vol. 14, no. 4, pp. 42–57, 1993.

[24] Arman, B. and Rabas, T.J., Condensation Analysis for Plate-frame Heat Exchangers, ASME HTD-vol. 134, *Proceedings of the National Heat Transfer Conference, vol. 12*, pp. 97–104, 1995.

[25] Lockhart, R.W. and Martinelli, R.C., Proposed Correlation of Data for Isothermal Two-phase, Two-component Flow in Pipes, *Chemical Engineering Progress*, vol. 45, no. 1, pp. 39–48, 1949.

[26] Akers, W.W., Deans, H.A., and Crosser, O.K., Condensation Heat Transfer within Horizontal Tubes, *Chemical Engineering Progress Symposium Series*, vol. 55, pp. 171–176, 1959.

[27] Gitteau, J., Thonon, B., and Bontempts, A., Condensation of Pure Hydrocarbons and Their Mixture in a Compact Welded Heat Exchanger, in *Compact Heat Exchangers and Enhanced Technologies for the Process Industries*, ed. R.K. Shah, A.W. Deakin, H. Honda, and T.M. Rudy, Begell House, New York, pp. 487–494, 2001.

[28] Chisholm, D., A Theoretical Basis for the Lockhart–Martinelli Correlation for Two-phase Flow, *International Journal of Heat Mass Transfer*, vol. 10, pp. 1767–1778, 1967.

[29] Chen, J.C., A Correlation for Boiling Heat Transfer to Saturated Fluids in Convective Flow, *6th National Heat Transfer Conference*, Boston, August 11–14, 1963.

[30] Shah, M.M., A New Correlation for Heat Transfer during Boiling Flow through Pipes, *ASHRAE Transactions*, vol. 82, no. 2, pp. 66–96, 1976.

[31] Steiner, D. and Taborek, J., Flow Boiling Heat Transfer in Vertical Tubes Correlated by an Asymptotic Model, *Heat Transfer Engineering*, vol. 13, no. 2, pp. 43–69, 1992.

[32] Forster, H.K. and Zuber, N., Dynamics of Vapour Bubbles and Boiling Heat Transfer, *AIChE Journal*, vol. 1, no. 4, pp. 531–535, 1955.

[33] Shah, M.M., Chart Correlation for Saturated Boiling Heat Transfer: Equations and Further Study, *ASHARE Transactions*, vol. 88, no. 1, pp. 185–196, 1982.

[34] Kandlikar, S.G., A General Correlation for Saturated Two-phase Flow Boiling Heat Transfer Inside Horizontal and Vertical Tubes, *ASME Journal Heat Transfer*, vol. 112, pp. 219–228, 1990.

[35] Engelhorn, H.R. and Reinhart, A.M., Investigations on Heat Transfer in Plate Evaporator, *Chemical Engineering Process*, vol. 28, pp. 143–146, 1990.

[36] Gorenflo, D., *Pool Boiling*, VDI-Verlag, Düsseldorf, Germany, 1993.

[37] Slipcevic, B., Wärmeübergang bei Verdampfern in Näturlicher Strömung and Bemessung von überfluteten Verdampfern, *Handbuch der Kältetechnik*, Vok. VI/B, Springer, Berlin, 1988.

[38] Stephan, K. and Abdelsalam, M., Heat Transfer Correlations for Natural Convection Boiling, *International Journal of Heat Mass Transfer*, vol. 23, pp. 73–87, 1980.

[39] Azarskov, V.M., Danilowa, G.N., and Zemskov, B.B., Heat Transfer in Plate Evaporators of Different Geometries, *Kholodilnaja Technika* (in Russian), vol. 58, no. 4, pp. 25–31, 1981.

[40] Dutto, T., Blaise, J.C., and Benedic, T., Performance of Brazed Plate Heat Exchanger Set in Heat Pump (in French), *Proceedings of the 18th International Congrers Refrigeration*, vol. III, pp. 1284–1288, Montreal, Quebec, Canada, August 10–17, 1991.

[41] Syed, A., Use of Plate Heat Exchangers as Evaporators and Condensers in Process Refrigeration, in *Heat Exchanger Engineering*, ed. E.A. Foumeny and P.J. Heggs, Ellis Horwood Series in Chemical Engineering, vol. 1, Design of Heat Exchangers, pp. 139–157, 1992.

[42] Thonon, B., Vidil, R., and Marvillet, C., Recent Research and Development in Plate Heat Exchangers, *Journal of Enhanced Heat Transfer*, vol. 2, nos. 1–2, pp. 149–155, 1995.

[43] Ohara, T., Yamamoto, T., and Fujita, H., Heat Transfer and Pressure Drop of Boiling Flow in Cross-ribbed Flat Channels, *International Communication of Heat Mass Transfer*, vol. 15, no. 5, pp. 556–557, 1990.

[44] Marvillet, C., Welded Plate Heat Exchangers as Refrigerants Dey-Ex Evaporator, in *Design and Operation of Heat Exchangers*, ed. W. Roetzel, P.J. Heggs, and D. Butterworth, Springer Verlag, Germany, pp. 255–268, 1992.

[45] Margat, L., Thonon, B., and Tadrist, L., Heat Transfer and Two-phase Flow Characteristics during Convective Boiling in a Corrugated Channel, in *Compact Heat Exchangers for the Process Industries*, ed. R.K. Shah, K.J. Bell, S. Mochizuki, and V.V. Wadekar, Begell House, New York, pp. 323–330, 1997.

[46] Pelletier, O., Propane as Refrigerant in Residential Heat Pumps, *Engineering Licentiate Thesis*, Royal Institute of Technology, Department of Energy Technology, Stockholm, Sweden, 1998.

[47] Cooper, M.G., Heat Flow Rates in Saturated Nucleate Pool Boiling – A Wide Ranging Examination Using Reduced Properties, *Advances in Heat Transfer*, vol. 16, ed. J.P. Harnett and T.F. Jr. Irvine, Academic Press, Orlando, Florida, 1984.

[48] Pierre, B., Värmeövergång vid Kokande Köldmedier I Horisontella Rör, *Kylteknisk Tidskrift* (in Swedish), vol. 28, no. 5, pp. 3–12, 1969.

[49] Steiner, D., Nucleate Boiling in Vertical Tubes, *VDI-Heat Atlas*, VDI-Verlag GmbH, Düsseldorf, Germany, 1993.

[50] Sterner D., Experiments and Analysis of Performance of Plate Heat Exchangers as Evaporators in Refrigerator Systems, *Engineering Licentiate Thesis*, Publ. 99/2007, Division of Heat Transfer, Department of Heat and Power Engineering, Lund Institute of Technology, Lund, Sweden, 1999.

[51] Yan, Y.Y. and Lin, T.F., Evaporation Heat Transfer and Pressure Drop of Refrigerant R-134a in a Plate Heat Exchanger, *ASME Journal of Heat Transfer*, vol. 121, no. 1, pp. 118–127, 1999.

[52] Hsieh, Y.Y and Lin, T.F., Saturated Flow Boiling Heat Transfer and Pressure Drop of Refrigerant R-410A in a Vertical Plate Heat Exchanger, *International Journal of Heat Mass Transfer*, vol. 45, no. 5, pp. 1033–1044, 2002.

[53] Boccardi, G., Celata, G.C., Cumo, M., Gerosa, A., Donati, F.M., and Zorzin, A., R22 Replacement Aspects in Compact Heat Exchangers for Air

Conditioning, *International Journal of Heat Exchangers*, vol. 1, no. 1, pp. 77–91, 2000.

[54] Panchal, B.B., Hillis, D.L., and Thomas, A. Convective Boiling of Ammonia and Freon 22 in Plate Heat Exchangers, *ASME/JSME Thermal Engineering Conference*, Honolulu, pp. 261–268, 1983.

[55] Kedzierski, M.A., Effect of Inclination on the Performance of a Compact Brazed Plate Condenser and Evaporator, *Heat Transfer Engineering*, vol. 18, no. 3, pp. 25–38, 1997.

[56] Holt, A.J., Azzopardi, B.J., and Biddulph, M.W., Two-phase Pressure Drop and Void Fraction in Channels, *5th UK National Heat Transfer Conference on Heat Transfer*, Session D, IChemE, London, 1997.

[57] Hesselgreaves, J.E., Single Phase and Boiling Performance of a Novel Highly Compact Heat Exchanger Surfaces, *5th UK National Heat Transfer Conference*, Session D, IChemE, London, 1997.

[58] Moriyama, K. and Inoue, A., The Thermohydraulic Characteristic of Two-phase Flow in Extremely Narrow Channels, Heat Transfer – Japanese Research, vol. 21, no. 8, 1992.

[59] Cornwell, K. and Kew, P.A., Compact Evaporators, in Convective Flow and Pool Boiling, *Proceeding of the International Engineering Foundation 3rd Conference*, Taylor & Francis, Irsee, Germany, 1999.

[60] Hesselgreaves, J.E., *Compact Heat Exchangers: Selection, Design and Operation*, Elsevier, Amsterdam, 2001.

CHAPTER 9

Fouling, corrosion, and erosion

The most frequent causes of failure of a well-designed and fabricated PHE at the designed conditions are the fouling deposits over the plates and the corrosion and erosion damages to the plates. The fouling of heat exchangers may be defined as the accumulation of undesired deposits on the heat transfer surface. It is the consequence of various heat, mass, and energy transfer phenomena involved with heat exchanger operation. The foulant layer results in narrowing of the flow area, which increases the flow velocity for a given volumetric flow rate. This in general leads to increased pressure drop, and consequently the operating pumping power. Although the convective heat transfer coefficient may be slightly increased due to increased velocity, the foulant deposit imposes an additional resistance to the overall heat transfer, which results in the decrease of heat transfer rate. Corrosion is defined as the degradation of a material because of reaction with the environment. It is an electrochemical process by which a metal returns to its nature state, i.e. iron oxide or rust. This ultimately results in the failure of heat exchangers, which consequently may cause the failure of the entire engineering system. Erosion is defined as the loss of material due to high velocity impact of liquid streams, suspended drops in gaseous flow, suspended vapour bubbles in liquid flow, or suspended solid particles in gaseous or liquid streams. For PHEs, both corrosion and erosion must be kept at a minimum in order to protect very thin plates. Failure to do so will result in short operating lifetime. It should also be pointed out that fouling, corrosion, and erosion do not act independently, and instead, they can influence reciprocally. An example is the corrosion fouling, which is caused by the corrosion products sticking to the surface reducing the heat transfer due to their low thermal conductivity and also providing nucleation sites for the fouling to deposit by other mechanisms. Another example is the erosion corrosion, which is the corrosion accelerated as a result of an increase in the relative motion between the corrosive fluid and a metal wall. Therefore, fouling, corrosion, and erosion must be considered together when a new heat exchanger is designed. Failure to do so will result in performance deterioration or even a total fail of the exchanger after a certain period of operation. Therefore, research on fouling, corrosion, and erosion has been extensive during the past years for different heat exchangers. For PHEs, considerable research efforts have been made on fouling formation characteristics and the methods for diminishing fouling. Studies on corrosion in PHEs are also extensive, and various guidelines on selecting plate materials in different corrosive environments have been established. For the erosion in PHEs, the common way is to avoid critical velocities for erosion formation at some particular places, and the fundamental studies are limited. Therefore, the main part of this chapter will be devoted to fouling and corrosion phenomena, and only a brief description of erosion will be provided at the end.

9.1 Fouling

Fouling is the collection and growth of unwanted material on heat transfer surface. It has great influence on heat exchanger performance. Therefore, fouling must be considered in heat exchanger design. Improper consideration of fouling will result in the failure of heat exchanger performance after a certain period, and consequently the entire system.

9.1.1 Basic consideration

In a heat exchanger, the process of heat transfer from a hot fluid stream and a cold fluid stream involves various conductive and convective processes. These can be individually represented in terms of a thermal resistance. The summation of these individual resistances is the total thermal resistance, and its inverse is the overall heat transfer coefficient U. For PHEs where the heat transfer surface is flat, the overall heat transfer coefficient U is calculated from

$$\frac{1}{U} = \frac{1}{h_1} + \frac{1}{h_2} + \frac{\delta_p}{k_p} + R_{f,1} + R_{f,2} \tag{9.1}$$

where h_1 and h_2 are the heat transfer coefficients of the two fluid streams, respectively, $W/(m^2 \cdot K)$; δ_p is the plate thickness, m; k_p is the thermal conductivity of the plate, $W/(m \cdot K)$; $R_{f,1}$ and $R_{f,2}$ are the fouling resistances on the plate surfaces of the two streams, respectively, $m^2 \cdot K/W$. When the heat exchanger is new, the fouling resistance is zero, and the overall heat transfer coefficient in this case can be designated as U_c. For two cases with and without fouling under a constant heat duty, the heat transfer rate is the same as well as the mean temperature difference. Therefore, the following equation can be derived to estimate the heat transfer surface ratio of the clean service and the fouled service [1].

$$\frac{A_f}{A_c} = 1 + U_c R_f \tag{9.2}$$

where A_f is the heat transfer surface in the fouled case, m^2; A_c is the heat transfer surface in the clean case, m^2; R_f is the total fouling resistance ($R_f = R_{f,1} + R_{f,2}$). In eqn (9.2), $U_c R_f$ represents the additional heat transfer area due to the heat exchanger fouling. For a range of fouling resistances, Fig. 9.1 shows the percentage of increase

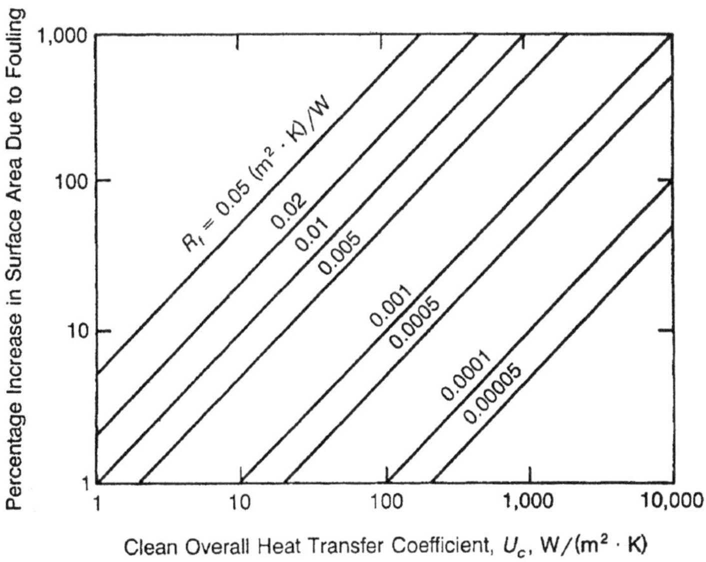

Figure 9.1: Effects of fouling on surface area [2].

in the heat transfer surface area due to fouling. It is obvious that with the increase of U_c, the fouling influence becomes more significant. For example, the total fouling resistance R_f of 0.0001 ($m^2 \cdot K/W$) gives the added surface 1% of the original surface if the clean overall heat transfer coefficient U_c is 100 ($W/(m^2 \cdot K)$). However, the same value of R_f gives the added surface 100% of the original surface if U_c is 10,000 ($W/(m^2 \cdot K)$). From this perspective, the fouling influence is more severe for PHEs than shell-and-tube heat exchangers because of the higher heat transfer coefficients in PHEs. The common range of overall heat transfer coefficients is between 1000 ($W/(m^2 \cdot K)$) and 10,000 ($W/(m^2 \cdot K)$), depending on specific process conditions.

Typically, the main effects of fouling resistance are the reduced heat transfer and increased pressure drop. Equation (9.1) shows that fouling resistances result in reduced overall heat transfer coefficients, consequently increased heat transfer surface from the basic equation $A = Q/(U \cdot \Delta T_m)$. Equation (9.2) further demonstrates that even a small value of the fouling resistance can result in significant heat transfer surface increase if the clean overall heat transfer coefficient is high. On the other hand, when fouling is created on the heat transfer surface, the cross-flow area is reduced. Hence the mean fluid velocity is increased for a constant volumetric flow rate. If constant friction factor is assumed, the total pressure drop is certainly increased. It must be recalled that the pressure drop is a function of the square of the fluid velocity. Therefore, fouling can result in marked pressure drop increase, and consequently, increased pumping power, i.e. increased operating cost. In addition, fouling may create nucleate sites, which can cause different forms of corrosion. This further deteriorates the performance of heat exchangers.

Because of the effects of fouling on the thermal and hydraulic performance of heat exchangers, a big additional cost is cast to related industrial sectors. These costs may include the increased capital expenditure (due to oversized heat transfer surface), increased operating cost (due to increased pressure drop), loss of production (due to heat exchanger failure), increased maintenance cost (due to cleaning requirement), etc. Therefore, the cost of fouling is considerable, and it was estimated that the fouling of all heat exchangers represented an annular expense in the United States of somewhere between \$4.2 and \$10 billion [3]. This has stimulated the research on the fouling of heat exchangers, but until now it still remains an unresolved issue due to the complexity of its mechanism [4].

9.1.2 Fundamental mechanism

Fouling is a time-dependent process, and the newly installed heat exchanger only encounters the designed constant fouling value after a certain operating period. The fouling buildup usually experiences five successive stages, namely initiation, mass transport, deposition, removal and auto-retardation, and ageing. Short descriptions are provided below, and detailed information can be found in Epstein [5] and Kho [6].

Initiation Following the start-up of a clean heat exchanger, a certain period of time is required to pass before any appreciable fouling is recorded. This period is known as the initiation or incubation period. During the initiation stage, the surface is conditioned for the fouling that will take place later. Surface temperature, material, roughness, and others strongly influence the initial delay.

Transport During the mass transport stage, the conditions required for deposition are established, and the fouling substances from bulk fluid are transported to the heat transfer surface. This transport is accomplished by a number of phenomena including diffusion, sedimentation, and thermophoresis.

Deposition During this period, the species responsible for fouling (which have already been transported to the heat transfer surface) attach to the heat transfer surface. The rate of deposition is dependent on the rate of the diffusion-driven mass transfer to the surface, and the chemical reaction rate at the surface. Properties of the material such as density, size, and surface conditions are important during this stage.

Removal Some material is removed from the surface immediately after deposition and some is removed later. The removal rate is dependent on the deposit strength and the fluid shear forces. Dissolution, erosion, and spalling have been proposed as plausible mechanisms for removal.

Ageing Once the foulant is deposited, it undergoes an ageing process via chemical or physical changes, which may either strengthen or weaken the deposit.

The overall result of the above stages is the net deposition of material on the heat transfer surface. Obviously, the amount of fouling material is time dependent. Accurate prediction of this behaviour is important to determine oversizing percentage as well as the period of the cleaning cycle. The following model, originally proposed by Kern and Seaton [7], regards the net fouling deposition rate $\mathrm{d}m_\mathrm{f}/\mathrm{d}t$ as a combination of the deposition rate $\dot{m}_\mathrm{d}$ and the removal rate $\dot{m}_\mathrm{r}$.

$$\frac{\mathrm{d}m_\mathrm{f}}{\mathrm{d}t} = \dot{m}_\mathrm{d} - \dot{m}_\mathrm{r} \tag{9.3}$$

Over the years, this model has been widely accepted, with the recognition that the corresponding equations for the two mass fluxes, $\dot{m}_\mathrm{d}$ and $\dot{m}_\mathrm{r}$, must be sophisticatedly modelled. As a result, various models have been developed in the literature to describe these two fluxes specifically for each fouling mechanism [8, 9], but they are not further explored here.

Depending on the operating conditions, different types of deposition and removal processes occur simultaneously, resulting in various forms of fouling curves. In general, there are four types of fouling curves and they are illustrated in Fig. 9.2. The respective type of fouling curve develops depending on the differential rates of the deposition and removal. Curve A is a linear fouling curve, where the fouling resistance develops progressively at a constant rate. In this situation, the removal rate of deposits is either negligible or constant but slower than its corresponding deposition rate. Curve C is an asymptotic fouling curve, which is the most commonly reported type. In this case, fouling develops with progressively diminishing rate asymptotically to an ultimate constant value. This curve is obtained when the deposition rate is constant and the removal rate is proportional to the fouling layer thickness. Curve B lies between the linear and the asymptotic fouling curves. Most likely, this behaviour results from the deposition rate inversely proportional to the fouling thickness. Curve D is a sawtooth fouling curve, which has a mean trend

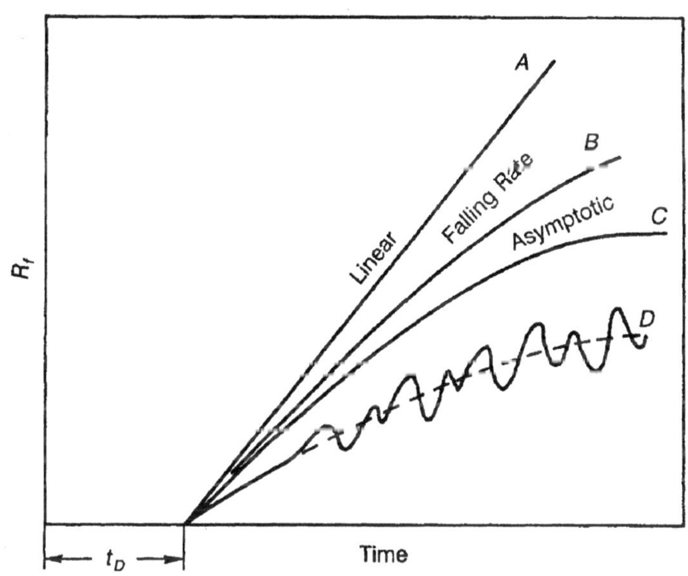

Figure 9.2: Development of fouling resistance as a function of time [2].

similar to curve C, but fluctuation of the fouling resistance is observed with time. This curve is obtained when relatively large chunks of the fouling deposit are periodically removed from the heat transfer surface due to increased shear stress and other reasons.

For PHEs, the influence factors of fouling formation include operating conditions, fluid properties, plate designs, etc. Usually, six different types of fouling are identified in practice. These will be described in the next section. The effects of the influence factors are certainly different for the specific type of fouling. However, different fouling types often occur simultaneously, and sometimes it is difficult to separate them. Therefore, some studies have been carried out without specific targeting of which fouling type in present. Cooper et al. [10] investigated cooling water fouling using an APV model R405 PHE. The water was chemically treated before entering the tested exchanger. Some of the important results from that investigation are shown in Fig. 9.3.

It is obvious that the fouling resistance in PHEs is significantly lower than that in shell-and-tube heat exchangers, despite the typical lower flow velocities. If the flow velocity is increased, the fouling resistance decreases similarly as it is found in shell-and-tube heat exchangers. This influence of velocity can also be seen in Fig. 9.4, where the influence of surface temperature is illustrated. With the increase of plate surface temperature, the fouling resistance is increased.

In addition to the overall fouling resistance, the local distribution of fouling is also of interest. This is usually dependent on the local velocity and temperature distribution, which results from flow arrangement, plate pattern, etc. Figure 9.5 is from the experimental work of Cooper et al. [10], and it clearly indicates that fouling occurred only in the upper third of the plates. This is the region near the hot

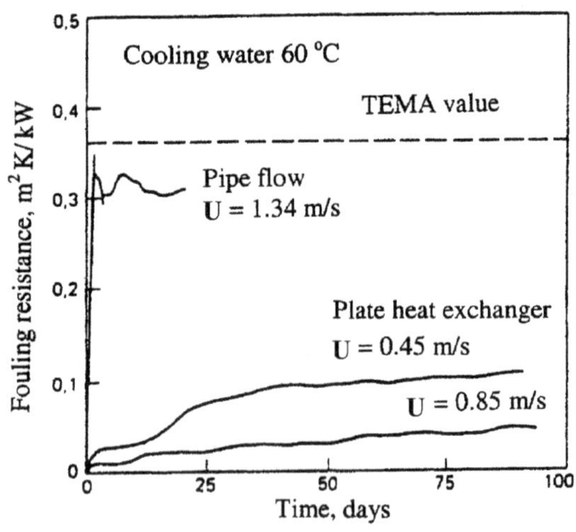

Figure 9.3: Comparison of fouling in a PHE and a shell-and-tube heat exchanger [10].

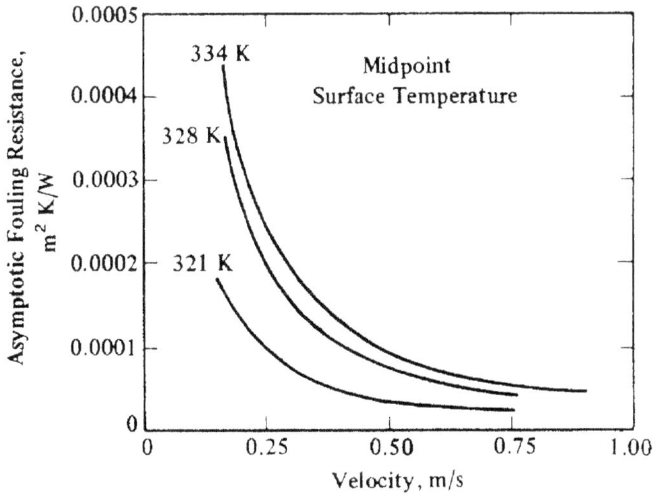

Figure 9.4: Fouling resistance in a PHE as a function of flow velocity and temperature [10].

Figure 9.5: Fouling distribution on plates [10].

water inlet and the cold water outlet, a region of high surface temperature profile for their experimental condition.

9.1.3 Forms of fouling

Based on various mechanisms, fouling is usually categorized into six main branches: particulate, crystallization, biological, chemical reaction, corrosion, and solidification. Different names may appear in the literature. To name a few, particulate fouling may also be called sedimentation fouling; crystallization fouling may also be called precipitation fouling; polymerization fouling and coking fouling belong to chemical reaction fouling. In addition, scale is an often-used word in practice, which sometimes refers to the deposits from water systems caused by the super-saturation of inorganic salts. But as pointed by Gupta [11], generally no distinction is made between fouling and scale in the literature, and they can be used interchangeably.

In this part, these six fouling categories will be explored. The respective mechanism and the influence factors will be described. Because of the scope of this book, concentration will be on fouling in PHEs. The influence factors (e.g. fluid velocity, surface and bulk temperatures, surface material, plate pattern, etc.) will be discussed where applicable. A literature survey shows that most research on the fouling of PHEs was conducted for particulate, crystallization, and biological fouling, which are in fact the most frequently encountered fouling types in practice.

9.1.3.1 Particulate fouling

Particulate fouling is the accumulation of particles suspended in the process stream onto the heat transfer surface. Water streams often contain mud, sand, dust, or rust fragments that aggregate in regions of low velocity. These particles sometimes act as catalysts precipitating chemical reactions and combine with crystallization processes occurring simultaneously. For PHEs, the turbulence introduced by the plate corrugations is prone to keep the suspension of the particles. Therefore, particulate fouling is less severe in PHEs than in shell-and-tube heat exchangers. In addition, specially designed PHEs with wide gaps can certainly further reduce the particulate fouling.

The particulate fouling in PHEs was investigated by Müller-Steinhagen and Middis [12] and Thonon et al. [13]. The following conclusions were reached from their work, and detailed information can be found in their original publications.

1 The fouling resistance was found to increase rapidly without delay at the beginning of the experiment and approached an asymptotic value after some time. With increasing concentration, the rate of deposition and the value of the asymptotic fouling resistance are increased.
2 Similar with other fouling types, the asymptotic fouling resistance decreased with increasing fluid velocity due to the increase in wall shear stress.
3 The particle type and flow conditions have great influences on the particulate fouling rate. This is particularly true for particles Al_2O_3 or TiO_2, where the solution pH has a strong influence on the particle size distribution.
4 The effect of plate geometry is important for the particulate fouling development. It has been demonstrated that for a given velocity, high corrugation angle leads to low fouling resistance.
5 No influence of temperature is identified for particulate fouling, but its influence on other fouling mechanisms may be substantial.

Figure 9.6 shows the particulate fouling distributions on plates with corrugation angles of 30° and 60°, from the experimental work of Thonon et al. [13]. They

(a) (b)

Figure 9.6: Particulate fouling view of plates [13]: (a) 30° and (b) 60° corrugation angle.

concluded that at low and intermediate velocities (less than 0.75 m/s), for a plate with a corrugation angle of 30°, the deposit layer is quite uniform and seems to be homogeneous. For a plate with a corrugation angle of 60°, the deposit is scarce and it is located mainly at the downstream of the contact points. If the velocity is low (less than 0.1 m/s), which corresponds to a Reynolds number of 800, the deposit layer becomes homogeneous. They further linked this phenomenon to the flow distribution, which depends on the corrugation angle. Interested readers should consider their original publication for further details.

9.1.3.2 Crystallization fouling

Crystallization fouling mostly takes place in cooling water systems, when water soluble salts, predominantly calcium carbonates, become supersaturated and crystallize on the heat transfer surface to form scaling. Normally the concentration increases with temperature, but some salts typically found in brackish water or seawater exhibit the inverse solubility characteristic (sometimes above a certain temperature). These salts with inverse solubility will crystallize out when the liquid is heated. Common examples of this category include $CaSO_3$, $CaSO_4$, etc. Obviously, temperature has the largest influence on crystallization fouling.

The crystallization fouling from $CaSO_4$ solutions in PHEs has been investigated by Bansal and Müller-Steinhagen [14]. It was found that the rate of deposition increased with increasing wall temperature, increasing bulk concentration, and decreasing flow velocity. With increasing flow velocity, both the initial fouling rate and the fouling resistance decreased. The initial fouling rate decreased in case of lower wall temperature at high flow velocities whereas the fouling resistance decreased due to a higher removal rate in conjunction with a lower interface temperature. It was also found that the $CaSO_4$ concentration was only important in the initial stages of the fouling process. Later, the channel gap was reduced considerably by the deposit, thus causing other factors to dominate such as higher shear stresses and low interface temperatures. Due to blockage of the outlet flow distribution area the increase in pressure drop may be significantly higher than the increase in thermal fouling resistance. In addition, the local distribution of crystallization fouling was dependent on the local velocity and temperature distribution. In general, low velocity and high temperature resulted in more crystallization fouling.

Figure 9.7 provides a close look at the crystal formation in a certain area of the heat exchanger plate [14]. It reveals that the crystals are varying in size and amount. Around the contact points between plates, the flow velocity is low, resulting in higher wall temperature and lower removal rates. Due to the particular geometry at these locations, the temperature profile in the vicinity of the contact points is different from that in the flow channel because a volume of liquid is heated from several directions. As a result of these effects, most crystal formation is initiated near the contact points. Obviously, this does not apply for the stagnant flow zones in the heat exchanger. Further examination of the crystal formation near a contact point (shown in Fig. 9.8) shows that the crystals can be divided into two different size categories. An almost homogeneous deposit of small crystals is observed directly on the plate surface whereas long single crystals are formed on top of this initial layer.

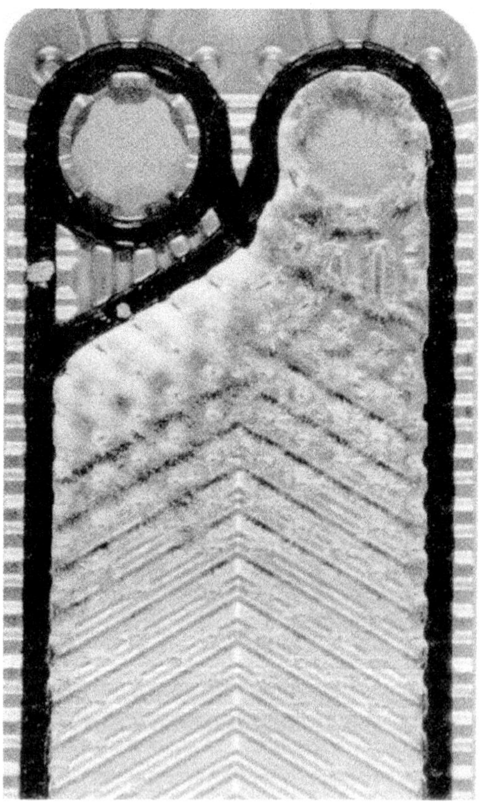

Figure 9.7: Crystal deposition pattern on plate [14].

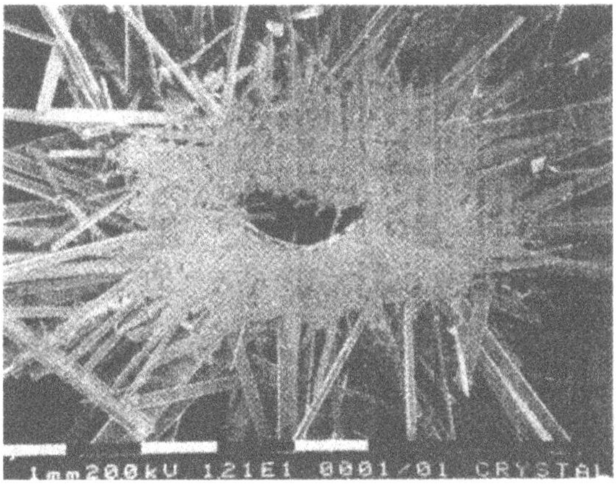

Figure 9.8: Crystal growth near contact point on plates [14].

In addition to pure crystallization fouling, the effect of suspended particles on crystallization fouling in PHEs was investigated by Bansal et al. [15]. It was found that the suspended particles settle on the heat transfer surface and act as nuclei. The availability of extra nucleation sites increases the crystallization rate significantly.

9.1.3.3 Biological fouling

Deposition or growth of material of a biological origin on a heat transfer surface results in biological fouling. Such material may include micro-organisms (e.g. bacteria, algae, and fungi) and macro-organisms (e.g. seaweed, water weed, and barnacle). Biological fouling is frequently encountered where the cooling water is used in asdrawn condition from river, lake, sea, coastal water, etc. On contact with heat transfer surfaces, these organisms can attach and breed, sometimes completely clogging the fluid passages, and sometimes entrapping silt or other suspended solids and giving rise to deposit corrosion. Because PHEs are generally designed with cooling mediums in all kinds of natural water sources, biological fouling obviously becomes an important issue.

Novak [16] studied the fouling behaviour of the Rhine River water near Mannheim in Germany and of the Öresund seawater in Sweden. For both waters, mainly biological fouling was observed. In most cases, it was found that the fouling resistance increased almost linearly over the observed period. The influence factors include fluid velocity, surface temperature, and wall shear stress. With the increase of fluid velocity, the fouling resistance is decreased, which is very similar to other fouling mechanisms. For constant flow velocity, it was found that maximum fouling occurred at a surface temperature of about 35°C, due to the preferred living conditions of biological matter. The influence of wall shear stress is more or less related to fluid velocity. Higher fluid velocity usually gives higher wall shear stress, leading to lower fouling resistance.

9.1.3.4 Chemical fouling

Deposits formed by chemical reactions at the heat transfer surface in which the surface material itself is not a reactant are known as chemical reaction fouling. Polymerization, cracking, and coking of hydrocarbons are prime examples of chemical reaction fouling. Chemical fouling can be found in food industry, chemical industry, and petroleum refineries, where PHEs are frequently used. A critical variable is the surface temperature that determines the reaction rate. A study of gas oil application by Shibuya et al. [17] demonstrated that fouling resistance of PHEs is much lower than that of shell-and-tube heat exchangers after a long period of operation. There have been some efforts to develop theoretical models to predict chemical fouling in food industries, e.g. Paterson and Fryer [18], De Jong [19], but it is very difficult to apply them to real industries due to the complexity of food process plants. Therefore, the current solution to chemical fouling in PHEs is to monitor the fouling, and consequently determine the appropriate cleaning period to perform the cleaning task [20].

9.1.3.5 Corrosion fouling

A heat transfer surface may react with a corrosive fluid through an electrochemical process, producing corrosion products. Part of these corrosion products are sheared off and become suspended in the flowing stream only to sediment out downstream. The other part may stick to the surface, reducing the heat transfer due to their low thermal conductivity and providing nucleation sites for fouling to deposit by other mechanisms, e.g. crystallization and particulate fouling.

It should be noted that there are no systematic studies on corrosion fouling in PHEs. This is due to the fact that the effect of corrosion fouling is very hard to separate from other fouling mechanisms. However, it is widely believed that the methods of decreasing corrosion rates should be applicable for reducing corrosion fouling consequently. The possible methods for reducing corrosion in PHEs will be described later in this chapter.

9.1.3.6 Solidification fouling

When the surface temperature is lower than the solidification or freezing temperature of one or more components present in the fluid, such components solidify on the surface leading to solidification fouling. An example is the deposition of wax from crude oil in pipelines and heat exchangers. Because solidification fouling is encountered only in specific applications for PHEs, the research on this fouling type is really limited. Obviously, the surface temperature is very critical in this situation, and the control of solidification fouling is thus dependent on the control of plate surface temperature.

It must be recognized that most fouling situations involve a number of different fouling mechanisms. Moreover, some of the fouling processes may complement each other. For instance, corrosion fouling provides nucleation sites for crystallization fouling and particulate fouling. On the other hand, corrosion is often accelerated by the presence of other deposits such as crystallization deposits or bio-films, and consequently the corrosion fouling is promoted. Therefore, different fouling mechanisms should be considered simultaneously where applicable in practice.

9.1.4 Mitigation of fouling

In order to maintain or restore the heat transfer efficiency, different techniques for mitigating fouling have been developed. Some of the methods are to prevent or slow down the fouling formation. Others are to clean the fouled PHEs. These mitigation methods can be classified into two groups: off-line and on-line cleaning methods. Off-line cleaning requires a shutdown of process operations, while on-line cleaning is performed during operation; thus, continuous process operation can be maintained. The comprehensive description of the various cleaning methods for general heat exchangers can be found in the literature, e.g. Bott [4] and Kuppan [21], and the current section concentrates on the cleaning methods for PHEs only.

9.1.4.1 Off-line cleaning

For off-line cleaning methods, there are two-types available: cleaning-off-place and cleaning-in-place.

Cleaning-off-place The cleaning-off-place method is to remove the plates from the PHE, and the plates are cleaned outside the exchanger by either mechanical or chemical cleaning. This method requires the heat exchanger to be dismantled, and thus is only applicable to plate-and-frame heat exchangers. For mechanical cleaning, non-metallic brushes or high-pressure washing can be used. For chemical cleaning, various agents can be used, depending on the nature and degree of fouling. Common cleaning agents for encrusted scales and sedimentation include hot water, nitric-, sulfuric-, sulfamic-, citric- or phosphoric-acid, sodium polyphosphates, etc. For biological growth and slime, alkaline cleaning agents such as sodium hydroxide and sodium carbonate are usually effective. If more extensive service is needed, the plates can be sent to the manufacturer for total reconditioning. In this case, the plates can meet 'as good-as-new performance standards'.

Cleaning-in-place Cleaning-in-place does not require the dismantlement of the plates, and thus it is applicable to all types of PHEs. The principle of this method is to circulate external chemical solution inside the exchanger, to dissolve and take away the fouling. This is an economical method of maintaining the exchanger at peak performance and extending its operating life. The selection of chemical solution depends on the nature of fouling as well as plate materials. Under no circumstance should hydrochloric acid be used to clean stainless steel plates, nor should hydrofluoric acid be used to clean titanium plates. If these acids are used for these types of plates, the plates will corrode and need replacement. Cleaning agents containing ammonia and oxidizing acids such as nitric acid must not be used to clean plates in nickel, monel, Hastelloy B, and copper-brazed heat exchangers. These agents will result in corrosion of the plates and brazed joints and ultimate failure of the exchanger.

9.1.4.2 On-line cleaning

For on-line cleaning methods, there are several types available, including automatic strainer, backflushing, chemical injection, etc. The principles of these methods are described as follows.

Automatic strainer Strainers are used in open systems (such as cooling tower systems or central cooling systems), where a significant amount of debris in suspension is contained in the cooling water. They can protect heat exchangers and other equipment from clogging when using low-quality water. Automatic strainers are available with pneumatic or electrical control actuators, which automatically backflush and clean the filter basket at pre-set intervals. It should be pointed out that the installation of an automatic strainer is only valid for particles with relatively large diameters.

Backflushing Backflushing means reversal of the flow in the heat exchanger for a short period of time in order to flush out accumulated debris from the heat exchanger inlet channel. This is done by installing a nozzle with a back flush valve in the pipes and reversing the flow, which carries the dirt out of the unit. Regular backflushing of a heat exchanger may be effective in removing particles or debris from the inlet port area. However, it works only if the particles have not yet reached the PHEs' heat transfer surface. Usually, this system is less costly than the automatic strainer.

Chemical injection The chemical injection method is to use chemical additives to minimize fouling. The selection of additives depends on the nature of fouling. For instance, minerals from water are removed by softening, which is beneficial to crystallization fouling reduction. Another example is to use continuous injection of chlorine or other types of biocides to suppress biological fouling. It should be pointed out that this method usually does not depend on the type of heat exchangers, and it is more up to the selection of appropriate chemical additives. Therefore, relevant knowledge of shell-and-tube heat exchangers can be adopted, with the consideration of plate materials.

There are also some new developments in on-line cleaning methods, e.g. electrical anti-fouling technology and surface modification. In electrical anti-fouling technology, an oscillating electric field using time-varying magnetic field generated in a solenoid is wrapped around a feed pipe. Cho et al. [22] have demonstrated in their experimental work that it is effective in reducing fouling in PHEs. However, the effect of this method needs further verification, and many practical issues have to be solved before general application in industries. For the surface modification method, it has been found that plate surface treated by ion beam implantation, magnetron, sputtering, etc., can mitigate fouling significantly [23]. These effects have already been demonstrated at the laboratory level, and their applications in practice can be anticipated, although significant further work has to be done.

Overall, both on-line and off-line mitigation methods are used for PHEs in practice. The selection of the specific method depends on the degree of fouling, nature of foulant, material of plates, regulations of environment, cost factors, exchanger type, etc. For a specific application in practice, it is advisable to consult the PHE manufacturer for appropriate fouling mitigation methods.

9.1.5 Design of PHEs subject to fouling

Because fouling is an inevitable process for PHEs, it must be taken into account during the design. The common solution is to oversize the exchanger by setting fouling margins. However, one must be careful in selecting appropriate values of the fouling resistance. The introduction of fouling margins usually, although not always, leads to the increase of the number of plates and therefore, also to an increase of the cross-sectional area for fluids. For the same mass flow rate, the fluid velocity as well as the shear stress in the exchanger channels is certainly reduced. This will increase the fouling rate, as indicated by the fundamental mechanism of fouling formation. Consequently, heat exchangers designed with a high fouling resistance can foul more rapidly than those designed with a low fouling resistance. This increase in fouling rate may be so extensive that exchangers designed with high fouling margins must be cleaned more frequently than those designed with low fouling margins [24]. Therefore, the selection of fouling resistance for PHE applications is very important. As indicated previously, PHEs are usually less prone to fouling than shell-and-tube heat exchanger and their fouling factors are normally 20–25% of those used in shell-and-tube heat exchangers. For design purposes, fouling values not greater than one-fifth of the published tubular figures are recommended by Cooper et al. [10] Table 9.1 was suggested by Marriot [25] for selecting the fouling factors for plate-and-frame heat exchangers.

Table 9.1: Fouling resistances for plate-and-frame heat exchangers [24].

Fluid	Fouling resistance (m^2K/W)
Water	
Demineralized or distilled	0.000009
Towns (soft)	0.000017
Towns (hard) heating	0.000043
Cooling tower (treated)	0.000034
Sea (coastal) or estuary	0.000043
Sea (ocean)	0.000026
River, canal, borehole, etc.	0.000043
Engine jacket	0.000052
Oils, lubricating	0.000017–0.000043
Oils, vegetable	0.000017–0.000052
Solvents, organic	0.000009–0.000026
Steam	0.000009
Process fluids, general	0.000009–0.000052

Table 9.2: Effect of plate arrangement on fouling [26].

Over-design	Expected effect on		
	Flow velocity	Wall temperature	Fouling[a]
Longer plates	↑	↓	↑
Wider plates	↑	↑	↑
Same aspect ratio	←	←	←
Additional plates	↓	↑	↑
Additional plates in second pass	↑	↓	↓

Note: [a]Also depends on type of fouling.

In addition to selecting appropriate values of fouling resistance, the arrangement of oversized heat transfer area is also important. This arrangement will influence flow velocity and wall temperature, and consequently the fouling rate. A sensitivity study on the various possibilities of providing excess heat transfer surface area for fouling was carried out by Zettler et al. [26], and the approximate effects are displayed in Table. 9.2. They concluded that minimum fouling occurs if the excess surface area is provided by a two-pass arrangement of the original plates, followed by the use of larger plates with the same width, followed by larger plates with standard width/height ratio. The poorest performance is obtained when the excess surface is simply added as parallel plates.

9.2 Corrosion

Corrosion is a part of the cycle of growth and decay that is the natural order to things. Today's tanks, battleships, and automobiles are the iron deposits of tomorrow. Although corrosion is generally undesirable, one must remember that it is nature's way to disposing of garbage. Without corrosion, our homes would be old decrepit places filled with ancient, broken-down items that are impossible to get rid of. From this perspective, corrosion is a good thing – in fact essential [27]. However, corrosion ultimately results in the failure of the equipment, and consequently the failure of the engineering systems. It was estimated by Fontana and Greene [28] that the annular cost of corrosion in the United States was up to $30 billion in 1977. It should be pointed out this figure was for the general corrosion cost (not limited to heat exchangers only), but it did not include the loss of production, plant downtime, etc. Therefore, an understanding of corrosion principles and corrosion control is of great interest to industry as well as the general public.

For heat exchangers, corrosion is an important issue, which must be taken into account in the design. For shell-and-tube heat exchangers, corrosion is allowed to some extent because of the large thickness of tube wall. But for PHEs, the corrosion margin is very small due to the small thickness of plates, and hence corrosion is not acceptable. However, there are many reasons why PHEs are subject to some particular corrosion possibilities [29, 30]. Plates must be sealed around their edges to contain process streams, by either gaskets or welding or brazing. Consequently, crevice corrosion may occur under gaskets or adjacent to seal welds. Localized corrosion may be either initiated or aggravated by the leaching of harmful ionic species from polymer gasket materials. For instance, chloride ions can cause stainless steels and nickel alloys to corrode, and fluoride ions can cause titanium to corrode. Stressed corrosion failures are also encountered in PHEs, particularly at cold-formed corrugations. Therefore, corrosion in PHE design is an important subject and deserves special attention.

9.2.1 Fundamental mechanism

Corrosion is an electrochemical process by which a metal returns to its natural state, i.e. iron oxide or rust. It is based on the universal laws of nature and is just a question of how quickly. For PHEs, some examples are chloride attack on stainless steel, fluoride attack on titanium and ammonia attack on copper in a copper brazed heat exchanger. A corrosion cell consists of an anode, a cathode, and an electrolyte (see Fig. 9.9). Metal ions dissolve into the electrolyte (aqueous solution) at the anode. Electrically charged particles (electrons) are left behind. These electrons flow through the metal to other points (cathodes) where electron-consuming reactions occur. The result of this activity is the loss of metal at the anode and often the formation of a deposit. The conductivity of the electrolyte is the key to the speed of the corrosion process. A solution with low conductivity produces a slow corrosion reaction, whereas a solution with high conductivity produces rapid corrosion. In the absence of an electrolyte, little or no corrosion takes place. An

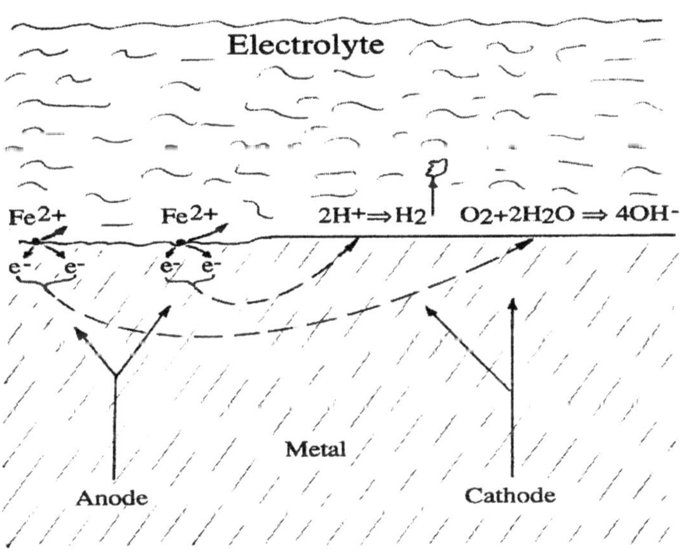

Figure 9.9: The electrochemical erosion cell.

example is that iron exposed to dry desert air remains bright and shiny because the water necessary to the rusting process is not available. PHEs are primarily served in liquid applications, and the basic circuit environment for corrosion always exists.

The corrosion mechanism can be illustrated using iron immersed in water. The chemical reactions involved are the following:

- At the anode, the neutral metal is dissolved and converted to electrically charged metal ions and electrons.

$$Fe \rightarrow Fe^{2+} + 2e^- \tag{9.4}$$

- At the cathode, the electrons are utilized through the reduction of dissolved oxygen to hydroxyl ions according to

$$0.5O_2 + H_2O + 2e^- \rightarrow 2OH^- \tag{9.5}$$

  or the reduction of hydrogen ions to hydrogen gas in the absence of oxygen

$$2H^+ + 2e^- \rightarrow H_2 \tag{9.6}$$

- The above reactions are followed by

$$Fe^{2+} + 2OH^- \rightarrow Fe(OH)_2 \tag{9.7}$$

In practice, however, the electrochemical reactions are much more complex than the above equations. There are various influence parameters on the corrosion process, and the important ones are

1 Environment factors such as concentration of the corroding species (H^+, O_2, Cl^-, SO_4^{2-}, HCO_3^-, Ca^{2+}), pH values, velocity, impurities, formation of a protective film, deposits, and temperature of the medium;
2 material factors such as composition of the alloy, structure of the alloy (single phase, multiple phase, presence of precipitates), mechanical state of the alloy (degree of cold work, presence of defects), and tendency of fouling, etc.;
3 operational factors such as presence of dissimilar metals, equipment design features (sharp bends, stagnant regions, etc.), stresses present (tensile or compressive, static or dynamic, applied or residual), thermal and mechanical history, etc.

These influence factors on corrosion formation are generally illustrated in Fig. 9.10. It is obvious that the corrosion is a result of a combined effect of environment and material. As will be described later, there are different forms of corrosion in terms of specific mechanisms. Therefore, the detailed influence factors are different for a specific corrosion type, and these will be described in the next section.

9.2.2 Forms of corrosion

It is widely accepted by international scientists and engineers that corrosion can be categorized in to eight forms, depending on the specific mechanism [28]. They are general corrosion (uniform corrosion), pitting corrosion, crevice corrosion, stress

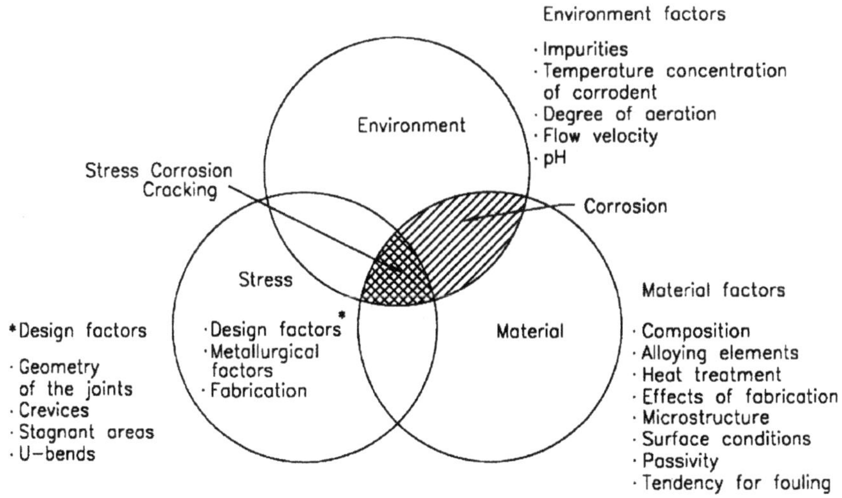

Figure 9.10: Factors influencing corrosion [21].

corrosion, erosion corrosion, intergranular corrosion, galvanic corrosion, and selective leaching. General corrosion takes place uniformly and has an even distribution over a broad area. The other corrosions are localized and have an uneven distribution. In the following section, detailed description of the respective corrosion type will be given, and concentration will be placed on PHEs instead of general heat exchangers. Parts of the following description are based on the materials from Kuppan [21] Walker [27], Fontana and Greene [28], Novak [31], and more detailed information can be found in these sources. The pictures of various corrosion forms are provided by Alfa Laval, which is greatly acknowledged.

9.2.2.1 General corrosion

General corrosion is the most common form of corrosive attack on metals. It is characterized by a chemical or electrochemical reaction that proceeds uniformly over the entire exposed surface or a substantial portion of that surface. General corrosion proceeds without appreciable localized attack. It results from prolonged contact with strong acidic environment when the passivity cannot be maintained. The metal becomes progressively thinner and eventually fails because of the stress loadings imposed on it. General corrosion in PHEs is shown in Fig. 9.11a (overall picture) and b (detailed structure).

In comparison with other forms of corrosion, general corrosion is the easiest to handle [27]. The rate of decomposition can be determined by comparatively simple immersion tests of a specimen in the fluid. The life of equipment can therefore be predicted and extended to the degree required by the addition of a corrosion allowance to the metal wall thickness as necessary to sustain the pressure or other stress loading applied. For PHEs, general corrosion has to be kept as low as possible, due to the very small corrosion allowance.

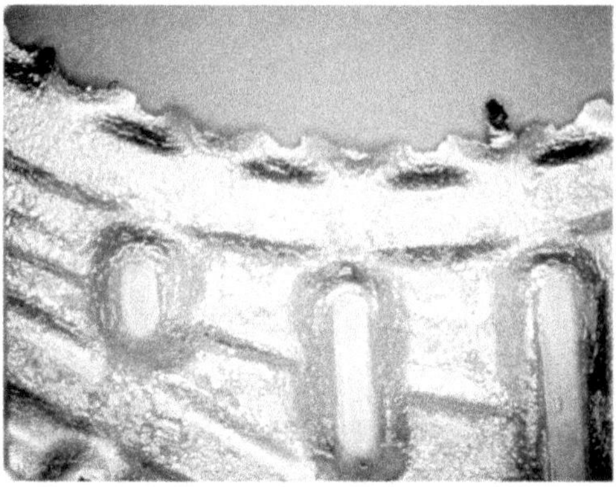

(a)

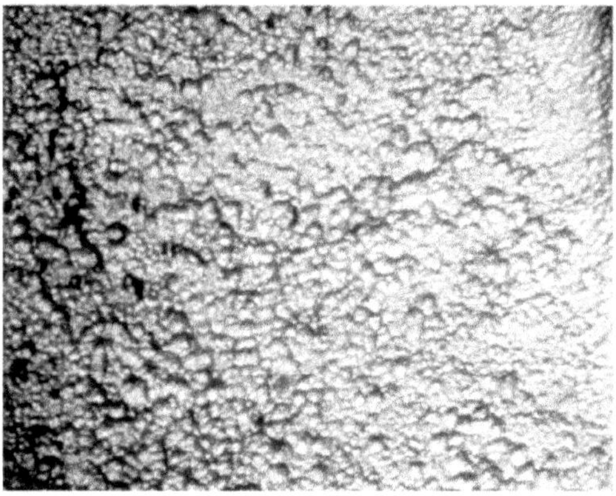

(b)

Figure 9.11: General corrosion of plates [31]: (a) overall and (b) detailed view.

9.2.2.2 Pitting corrosion

Pitting corrosion is the phenomenon whereby an extremely localized attack results in the formation of holes in the metal surface that eventually perforate the wall. Pitting usually occurs on metals that are covered with a very thin adherent protective surface film that forms on the metal surface during the surface treatment process or is produced by reaction with an environment [21]. The pits can have various sizes and they can be isolated or grouped together. Pitting corrosion usually occurs in slightly acidic to slightly alkaline environment, where the most common promoters are halides and oxidants. Mostly frequently, pitting corrosion is caused by high chloride content and is proportional to the chloride concentration, the fluid temperature, and inversely proportional to the fluid pH value. There is usually an incubation period required for pitting corrosion starts. Thereafter, the speed of pitting corrosion will increase, and the pitting hole will spread out. The pitting corrosion is illustrated in Fig. 9.12.

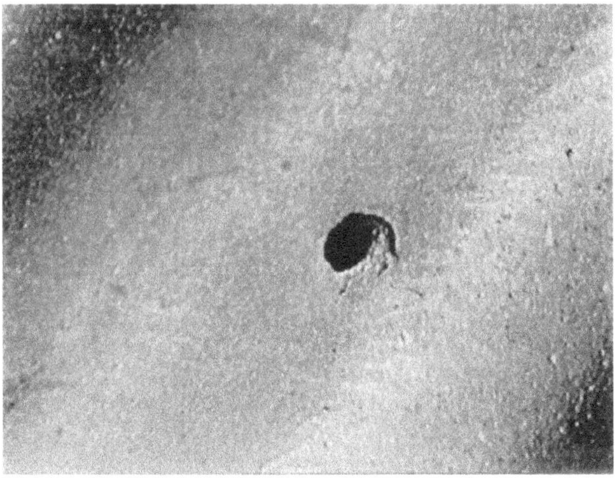

Figure 9.12: Pitting corrosion of plates [31].

The influence factors for pitting corrosion include metallurgical and structural factors, environmental factors, polarization phenomena, presence of corrosion products, etc. However, the mechanism of pitting corrosion, particularly the initiation of pitting at any particular point is not clear, although studies suggest that the sites for initiation of pits on passive metal surfaces may be generally related to defect structures of the underlying metal such as dislocations, grain boundaries, or non-metallic inclusions.

9.2.2.3 Crevice corrosion

Crevice corrosion, similar to pitting, is a localized form of corrosion that takes place at localized environments that are distinctly different from the bulk environments. It is characterized by intense local corrosion in crevices and other shielded areas on metal surfaces exposed to stagnant corrosive liquids. For PHEs, it can occur where any undisturbed liquid film exists, such as beneath gaskets, plate contact points, and beneath deposits. Such a corrosion is shown in Fig. 9.13.

The mechanism of crevice corrosion is a very complex electrochemical process. It is widely believed that crevice corrosion is associated with the depletion of oxygen in the stagnant liquid pool, which results in corrosion of the metal walls adjacent to the crevice. The corrosion rate increases as the crevice mouth narrows and as the external cathode area is increased. Metals and alloys that depend on oxide films or passive surface layers for corrosion resistance are most susceptible to crevice corrosion. The films are often destroyed by the chloride or hydrogen ions resulting from the electrochemical process in the stagnant film. Contact between a metallic and any non-metallic surface is particularly prone to the crevice corrosion of the metal part. This is exactly what usually happens at the joint place of plate and gasket for plate-and-frame heat exchangers.

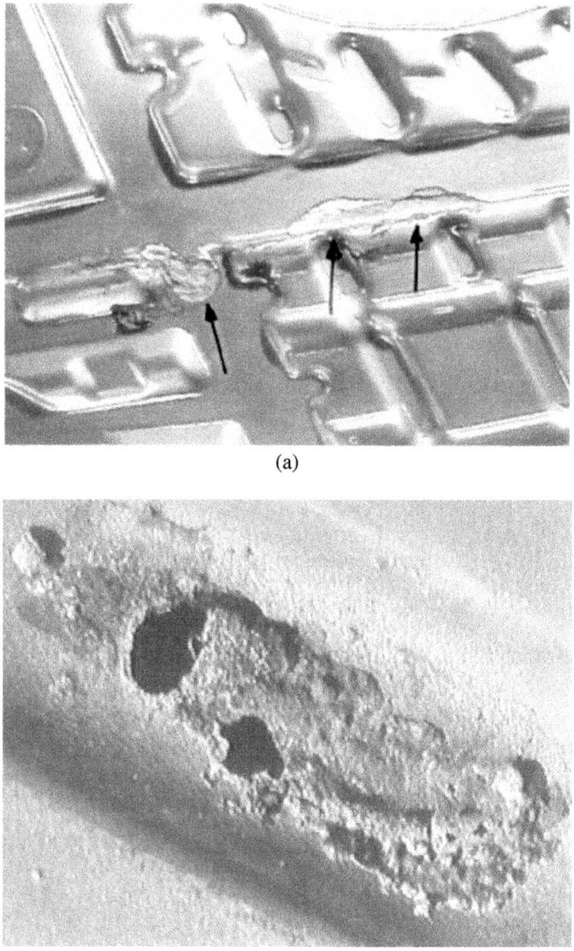

Figure 9.13: Crevice corrosion of plates [31]: (a) overall and (b) detailed view.

Compared to pitting corrosion, crevice corrosion has the same mechanism of propagation, but the mechanisms of initiation are different. Crevice corrosion is initiated by differential concentration of oxygen or ions in the electrolyte, where pitting is initiated by metallurgical factors and structural factors only. In general, the level of crevice corrosion occurring at crevices such as under deposits or gaskets is significantly greater than that of pitting on an open surface.

9.2.2.4 Stress corrosion

Stress corrosion is created by simultaneous action of corrosive environment and stress. It is often divided into two categories: stress corrosion cracking (SCC)

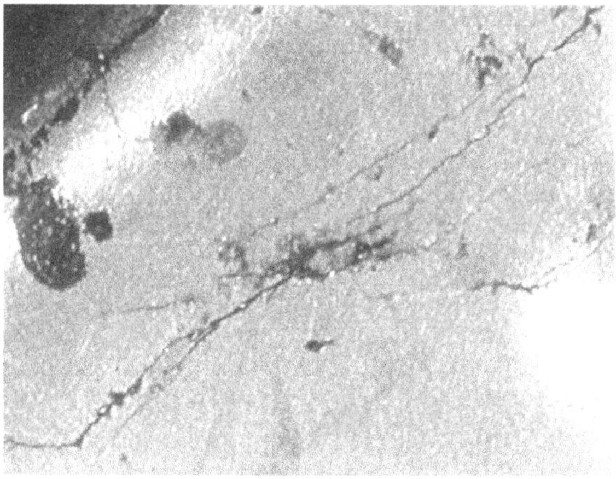

Figure 9.14: SCC of plates [31].

involving the effects of static stresses and corrosion, and corrosion fatigue involving variable stresses and corrosion. Cracks resulting from the first phenomenon are predominantly intergranular, and cracks resulting from the latter phenomenon are transgranular.

SCC develops in specific metal–fluid combinations when the stress level is above the minimum level that depends on the temperature, alloy structure, and the environment. In simple terms, SCC requires the simultaneous occurrence of a susceptible material, a corrosive environment, and tensile stress. Most frequently, SCC occurs in environment containing halides, strong alkali, hydrogen sulfide, and ammonia. It should be pointed out that welded joints are particularly prone to SCC because a residual tensile stress and stress concentration usually are present. This certainly applies to welded PHEs. For the initialization, SCC frequently starts from small corrosion pits. The base of the pit acts as a stress raiser so that the local stress concentration is very high and fosters continuing development of the crack. When the crack penetrates further into the metal, the remaining wall section has to take care of the load. As a result of continuous propagation, the metal may eventually fail suddenly when the stress in the remaining metal exceeds the ultimate strength. An illustrated picture is shown in Fig. 9.14, where the crack starts from pits.

Fatigue is the process whereby a metal subject to a fluctuating load fails at stress levels substantially below the yield point. It is well recognized that the fatigue characteristics of metals are drastically affected when the fluctuating load is imposed in the presence of a corrosive medium. When a metal is exposed to a corrosive environment with a fluctuating load, a corrosion fatigue crack likely starts at the base of a corrosion pit, and the crack propagates via simultaneous mechanical and electrochemical action. Usually, the fatigue crack is straight, not branched (see Fig. 9.15). It has been found that the frequency of stress fluctuation is important to corrosion fatigue, whereas it is not significant in conventional fatigue failures.

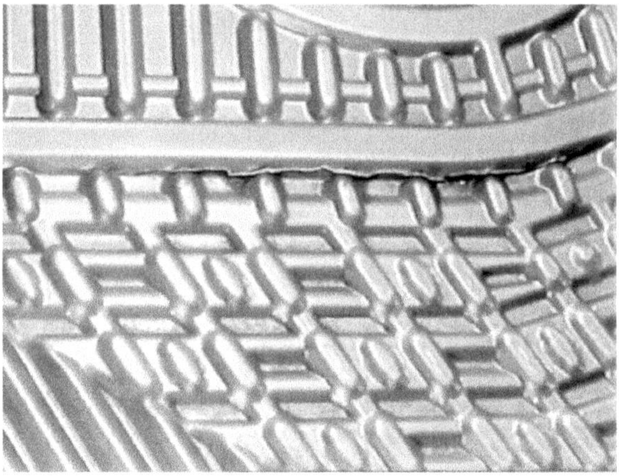

Figure 9.15: Corrosion fatigue of plates [31].

9.2.2.5 Erosion corrosion

Erosion corrosion is the term used to describe corrosion that is accelerated as a result of an increase in the relative motion between a corrosive fluid and a metal wall. Usually, when a metal corrodes, the corrosion film stays on the surface and reduces the rate of further corrosion. However, when that corrosion film is removed by erosion, fresh metal surface gets exposed and corroded. Thus, the cycle repeats continuously. The effect of the two is synergistic and far greater than if either erosion or corrosion were present alone. It is obvious that erosion corrosion is attributed by the removal of protective surface film or adherent corrosion products by the fluid shear stress under high turbulence condition. Therefore, erosion corrosion differs markedly from most other forms of corrosion, where the rate of attack is highest under stagnant or low velocity conditions. Examples include copper alloys in seawater, titanium in caustics, nickel alloys in sulfuric acid, etc. For PHEs, it is more severe at the inlet than at the outlet. Figure 9.16 is an illustration of erosion corrosion in PHEs.

The main governing factors for erosion corrosion are metal materials and flow conditions. For metal materials, metals whose corrosion resistance depends on the formation of a protective surface film are particularly susceptible to be attacked by erosion corrosion. Aluminium and stainless steels are in this category. The protective film is eroded by mechanical scrubbing, exposing the soft core to chemical or electrochemical attack in addition to the continued mechanical wear. For flow conditions, erosion corrosion is usually accelerated when the fluid is entrained with air or abrasive solid particles, such as sand. It is also generally agreed that the velocity increase can increase the rate of erosion corrosion. For many materials there appears to be a critical velocity value above which the rate of attack increases rapidly. These critical values can be found in many engineering books (e.g. Walker [27]), although they are relatively subjective. These critical values should not be exceeded during the course of heat exchanger designs.

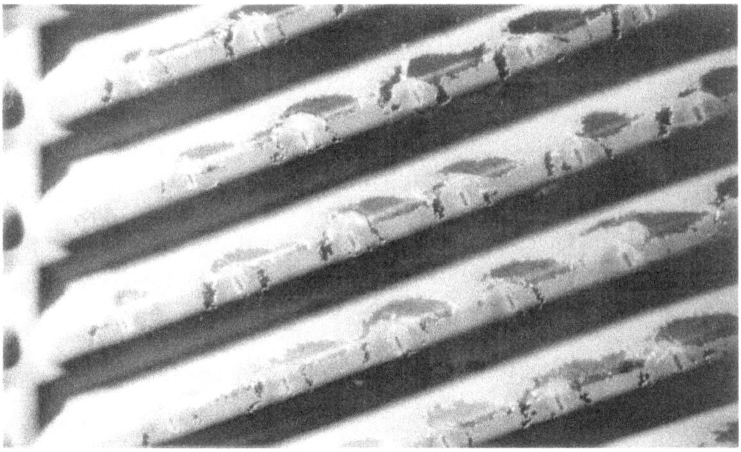

Figure 9.16: Erosion corrosion in plates [31].

9.2.2.6 Intergranular corrosion

A localized and preferential form of corrosion attack in a narrow region along the grain boundaries or closely adjacent regions without appreciable attack on the grains is called intergranular corrosion (see Fig. 9.17). Due to this form of corrosion, the metal loses its strength. Intergranular corrosion generally takes place when the corrodent preferentially attacks the grain boundary phase. The boundaries between individual grains tend to be more chemically active than the grains themselves and are attacked more rapidly when exposed to a corrosive. The depletion of a particular alloying element along the grain boundaries is usually caused by improper heat treatment or heat from welding or any other high-temperature operation that causes the precipitation of certain alloying element at the grain boundary. In austenitic

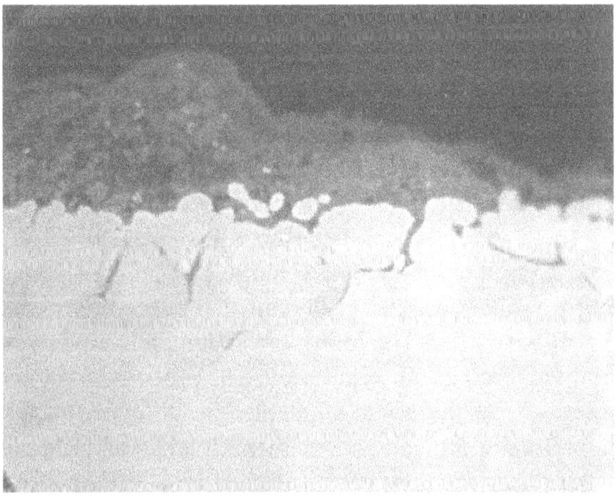

Figure 9.17: Intergranular corrosion [31].

stainless steel type-304 18-8 composition (18% chromium and 8% nickel), intergranular corrosion is most common and is of particular interest. Numerous failures of this type of stainless steel have occurred because of intergranular corrosion in environments where excellent corrosion resistance was anticipated.

The mechanism of intergranular corrosion is an electrochemical process as a result of local cell action in the grain boundaries. A galvanic cell is formed due to potential difference between second-phase micro-constituents and the depleted solid solution from which the constituents are formed. The carbide precipitate and the grain matrix are cathodic to the locally depleted grain boundary region. The high cathode to anode area ratio results in rapid corrosion of the grain boundary material and the metal disintegrates.

9.2.2.7 Galvanic corrosion

When dissimilar metals are immersed in a corrosive or electrically conductive solution, a galvanic cell is formed and a voltage (potential difference) is established between them. This results in corrosion of one of the metals, known as the anode of the couple. The cathode is usually not affected by galvanic corrosion, which is known as a noble metal. Galvanic corrosion can also take place even within the same group of metals due to local imperfections or heterogenities on the metal surfaces or due to variation in local solution chemistry. Most often, galvanic corrosion shows up as furrows or troughs on the corroded metal at its point of contact with the more noble metal. The initiation of galvanic corrosion requires four essential components: the anode, the cathode, the electrolyte, and a metallic path between anode and cathode, which completes the circuit. For PHEs, galvanic corrosion often takes place on welded joints, brazed joints, and soldered joints.

The parameter to determine the degree of galvanic corrosion is the so-called galvanic series. Commonly, various metals and alloys can be arranged in the order of decreasing (or increasing) corrosion resistance. Typical galvanic series for plate materials in seawater is shown in Table 9.3 and the more comprehensive list can be found in Walker [27].

9.2.2.8 Selective leaching

Selective leaching is the term used to describe a corrosion process wherein one element is removed preferentially corroded over others from the parent alloy, leaving behind a weak structure. This phenomenon occurs principally in brasses with a high zinc content (dezincification) and in other alloys from which aluminium, iron, cobalt, chromium, and other elements are removed. The mechanism for selective leaching is likely a multi-stage electrochemical process in which the metal first dissolves in the corrosive fluid. The more noble element is placed back on the surface, while the anodic ions remain in solution. The process is characterized by soft, porous surface deposits of the noble element. The influence factors for selective leaching are mainly metallurgical, environmental, and water chemistry. However, they are not further discussed here because selective leaching corrosion is usually ignored in PHEs because of the common plate materials.

In addition, it is worthwhile to mention hydrogen damage, which is also related to a corrosion process. Hydrogen damage is the generic term applied to the variety of unfortunate consequences following exposure of metal to hydrogen. Hydrogen embrittlement is one of the typical forms encountered by PHEs. Usually, it involves two issues: the crack has been formed by hydrogen, and the toughness changes because the material has been embrittled. This is due to the fact that atomic hydrogen dissolved in steel can interface with the normal process of plastic deformation. For titanium, the following conditions must exist simultaneously for hydrogen embrittlement to occur [31]: pH must be below 3 or above 12, or the passive film must be damaged by mechanical means; the temperature must be above 80°C and there must be some mechanism for generating nascent hydrogen.

9.2.3 Control of corrosion

Corrosion is not acceptable for PHEs due to the thin plates. This makes corrosion control very important to keep PHEs working properly. There are many different ways to control corrosion for general heat exchanger applications. The details of all those methods are beyond the scope of this book, and readers are referred to specific books on corrosion (e.g. Fontana and Greene [28]). Here, only some brief information will be given and focus is on PHEs.

Generally speaking, there are two fundamental methods for corrosion control. One category is to use corrosion-resistant metals and the other category is either by isolating the corroding metal from the environment or by modifying of the environment so that either the anodic or the cathodic reaction is brought under control. The typical corrosion control methods of the second category are:

1 Anodic reaction reduction If anodic reaction is reduced, corrosion is reduced or eliminated consequently. One way is to make a metal cathode by means of

Table 9.3: Galvanic series in seawater [31].

Metal or alloy	Galvanic series
Graphite	Most noble
Stainless steel (passive)	
Hastelloy C	
Titanium	
Copper–Nickel 90–10	
Copper	
Stainless steels (active)	
Cast iron	
Aluminium	
Zinc	
Magnesium	Least noble

an impressed current or attachment to a more anodic metal (sacrificial anode) than the metal in the galvanic couple. The other way is to form a passive film on the metal surface that blocks corrosive ions from reaching the metal.

2 Cathodic reaction reduction Prevention of cathodic reaction can be reached through the change of corrosive environment, e.g. reduction of oxygen concentration, reduction of H^+ (increase of pH value), formation of protective deposits, etc. These can be achieved by appropriate water treatment by use of certain inhibitors.

3 Reduction of both anodic and cathodic reactions Coatings are generally relatively thin films separating the two reactive materials or a metal from an environment. Applying a protective barrier between a corrosive environment and the material to be protected is a fundamental method of corrosion control. Coatings can be metallic, inorganic, or paints.

For the second category, selecting corrosion-resistant materials, it is important to know the types of corrosion that a certain material is subjected to, and under which conditions the corrosions can form. Table 9.4 from Novak [31] lists the common corrosion types and formation reasons for stainless steels, titanium, and nickel, which are usually used in PHEs.

In practice, to select appropriate materials to minimize corrosion for a specific condition is never an easy task. It can be accumulated by application experience and fundamental research, which means that material experts and PHE manufacturers know more details. Here, a partial list from literature gives some suggestions for some typical applications. In alloys selection, there are several natural metal and environment combinations. These combinations usually represent the maximum corrosion resistance at the lowest cost and should always be considered. A partial list includes: stainless steels/nitric acid; copper–nickels/sea and brackish water; nickel and nickel alloys/caustic soda; monel/hydrofluoric acid; hastelloys/hot hydrochloric acid; titanium/hot strong oxidizing solutions [32]. In addition, there are some metal and environment combinations that are readily attacked by corrosion and must be avoided. A partial list is ammonia and ammoniacal solutions/copper and copper alloys; halogens and halides/austenitic stainless steels; mercury/copper and copper alloys, monel; sulfur and sulfides (at high temperatures)/nickel and nickel alloys, etc. [32]. In summary, the best way in practice to prevent corrosion is to use a material that is not corroded by the heat transfer medium, as suggested by Cross [33].

9.3 Erosion

Erosion is the wearing away of mass from the surface of a body by a flow of a fluid containing particles. High velocity is usually the main reason for erosion, and obstruction of the flow creates high local velocities. For shell-and-tube heat exchangers, it is primarily encountered at the inlet of the tubes and on the outside surface of the tubes under the shell-side nozzles. For PHEs, it can be found typically in the inlet neck and distribution area (see Fig. 9.18). In severe cases, the inlet port of the plate can be eroded and this can lead to leakage. For two-phase flows, it is usually

Table 9.4: Corrosion [31].

Reason	Promoted by	Corrosion type
Corrosion of stainless steel		
Halogenides (chlorides, bromides)	Oxidants, low pH, temperature	C, P, SCC
Strong inorganic acids	Temperature, halogenides	G
Strong organic acids	Halogenides, oxidants, temperature	G, P, C
Strong alkalies (NaOH, KOH)	Temperature, stress	G, SCC
Sensitization in HAZ (chromium carbides)	Acids, oxidants, temperature	SCC, IG
Action of micro-organisms	No flow condition	C, P
Corrosion of titanium and titanium alloys		
Hydroflouric acid	Not suitable at all	G
Fluorides	Acidic environment, temperature	G, C, SCC
Strong alkalies (NaOH, KOH)	High concentration, high velocity, high temperature	G, EC
Non-oxidizing, inorganic acids	Concentration, velocity, temperature	G
Strong organic acids	Absence of oxidants	G
Hydrogen (in nascenti)	Other corrosion types	HE
Corrosion of nickel alloys		
Strong inorganic acids (sulfuric, chromic, phosphoric)	Halogenides, temperature	G, EC
Chlorides, bromides, fluorides (cooling water, brine)	Oxidants, low pH, temperature	C, P, G
Strong alkalies	Temperature, oxidants	G, S

Notes: C = crevice corrosion, G = general corrosion, IG = intergranular corrosion, S = selective corrosion, EC = erosion corrosion, HE = hydrogen embrittlement, P = pitting corrosion, SCC = stress corrosion cracking

more common because of the high speed due to low density. In general, erosion is dependent on the following factors: fluid velocity, whether single- to two-phase flow, fluid properties, material properties, and strength [11]. In addition, erosion often takes place together with corrosion, and they influence each other. This is called erosion corrosion, which is a combination of electrochemical decomposition and mechanical wear action. This has been already described in discussing various

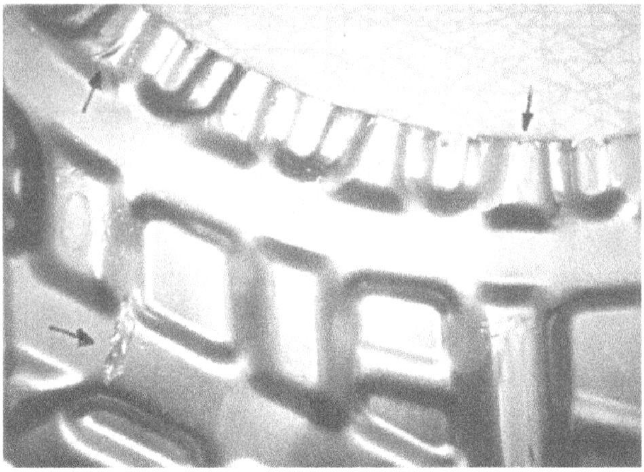

(a)

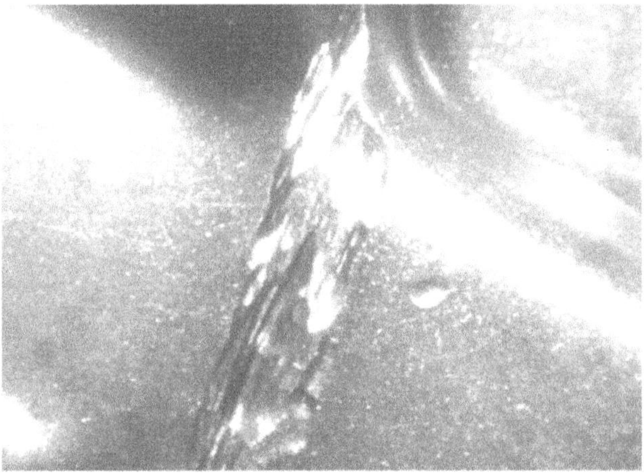

(b)

Figure 9.18: Typical erosion in PHEs [31]: (a) overall and (b) detailed view.

corrosion types. Unlike pure erosion, erosion corrosion is not confined to the high speed zone, and it can also be found in the relatively low speed zone where corrosion is present.

One common way to prevent erosion is to eliminate the abrasive solids from the heat transfer fluid by filtration or by straining using a strainer or similar device to remove any large particles causing maldistribution. During the design stage of PHEs, care must be taken to monitor the fluid velocity through the inlet port and neck area of the plates. Adjustments may be necessary to reduce the velocity in these high velocity areas by either reducing the fluid flow and/or by increasing the number of plates in the heat exchanger.

Nomenclature

h convective heat transfer coefficient, $W/(m^2 \cdot K)$
k thermal conductivity, $W/(m \cdot K)$
m mass, kg
Q heat transfer rate, W
R fouling resistance, $(m^2 \cdot K)/W$
t time, s
U overall heat transfer coefficient, $W/(m^2 \cdot K)$
ΔT temperature difference, K

Greek symbol

δ plate thickness, m

Subscripts

1 fluid stream 1
2 fluid stream 2
c clean
d deposition
f fouling
m average
p plate
r removal

References

[1] Kakac, S. and Liu, H., *Heat Exchangers: Selection, Rating and Thermal Design*, CRC Press, Boca Raton, FL, 1997.
[2] Chenoweth, J.W., Fouling Problems in Heat Exchangers, in *Heat Transfer in High Technology and Power Engineering*, ed. W.K. Yang and Y. Mori, Hemisphere, New York, pp. 406–419, 1987.
[3] Chenoweth, J.M., Final Report of the HTRI/TEMA Joint Committee to Review the Fouling Section of the TEMA Standards, *Heat Transfer Research, Inc.*, Alhambra, CA, 1988.
[4] Bott, T.R., *Fouling of Heat Exchangers*, Elsevier, Amsterdam, 1995.
[5] Epstein, N., Thinking about Heat Transfer Fouling: A 5 × 5 Matrix, *Heat Transfer Engineering*, vol. 4, no. 1, pp. 43–54, 1983.
[6] Kho, T., Effect of Flow Distribution on Scale Formation in Plate Heat Exchangers, *PhD Thesis*, University of Surrey, Surrey, UK, 1998.
[7] Kern, D.Q. and Seaton, R.E., A Theoretical Analysis of Thermal Surface Fouling, *British Chemical Engineering*, vol. 4, no. 5, pp. 258–262, 1959.

[8] Epstein, N., Fouling in Heat Exchangers, in *Low Reynolds Number Flow Heat Exchangers*, ed. S. Kakac, R.K. Shah, and A.E. Bergles, Hemisphere, New York, 1981.

[9] Watkinson, A.P. and Epstein, N., Particulate Fouling of Sensible Heat Exchangers, *Proceeding of the 4th International Heat Transfer Conference*, Paris, vol. 1, HE 1.6, 1970.

[10] Cooper, A., Suitor, J.W., and Usher, J.D., Cooling Water Fouling in Plate Heat Exchangers, *Heat Transfer Engineering*, vol. 1, no. 3, pp. 50–55, 1980.

[11] Gupta, J.P., *Working with Heat Exchangers*, Hemisphere, New York, 1990.

[12] Müller-Steinhagen, H. and Middis, J., Particulate Fouling in Plate Heat Exchangers, *Heat Transfer Engineering*, vol. 10, no. 4, pp. 584–591, 1989.

[13] Thonon, B., Grandgeorge, S., and Jallut, C., Effect of Geometry and Flow Conditions on Particulate Fouling in Plate Heat Exchangers, *Heat Transfer Engineering*, vol. 20, no. 3, pp. 12–20, 1999.

[14] Bansal, B. and Müller-Steinhagen, H., Crystallization Fouling in Plate Heat Exchangers, *ASME Journal of Heat Transfer*, vol. 115, pp. 584–591, 1993.

[15] Bansal, B., Müller-Steinhagen, H., and Chen, X.D., Effect of Suspended Particles on Crystallization Fouling in Plate Heat Exchangers, *ASME Journal of Heat Transfer*, vol. 117, pp. 568–574, 1997.

[16] Novak, L., Comparison of the Rhine River and the Öresund Sea Water Fouling and its Removal by Chlorination, *ASME Journal of Heat Transfer*, vol. 104, pp. 663–669, 1982.

[17] Shibuya, H., Morohashi, M., Levy, W., and Costa, C., Fouling Tests Using Pilot-Scale Packinox Heat Exchangers with Untreated Straight-rum Gas Oils, in *Fouling Mitigation of Industrial Heat Exchanger*, ed. C. Panchal, T. Bott, E. Somerscales, and S. Toyama, Begell House, New York, pp. 525–536, 1997.

[18] Paterson, W.R. and Fryer, P.J., A Reaction Engineering Approach to the Analysis of Fouling, *Chemical Engineering Science*, vol. 43, pp. 1714–1717, 1988.

[19] De Jong, P., Bouman, S., Van Der Linden, J.J.L.J., Fouling of Heat Treatment Equipment in Relation to the Denaturation of β–Lactoglobulin, *Journal of the Society of Dairy Technology*, vol. 45, pp. 3–8, 1992.

[20] Zubair, S.M. and Shah, R.K, Fouling in Plate-and-Frame Heat Exchangers and Cleaning Strategies, in *Compact Heat Exchangers and Enhanced Technologies for the Process Industries*, ed. R.K. Shah, A.W. Deakin, H. Honda, and T.M., Rudy, Begell House, New York, pp. 553–565, 2001.

[21] Kuppan, T., *Heat Exchanger Design Handbook*, Marcel Dekker, New York, 2000.

[22] Cho, Y.I., Choi, B.G., and Drazner, B.J., Electronic Anti-fouling Technology to Mitigate Precipitation Fouling in Plate and Heat Exchangers, *International Journal of Heat Mass Transfer*, vol. 41, no. 17, pp. 2565–2571, 1998.

[23] Bornhorst, A., Zhao, Q., and Müller-Steinhagen, H., Reduction of Scale Formation by Ion Implantation and Magnetron Sputtering on Heat Transfer Surfaces, *Heat Transfer Engineering*, vol. 20, no. 2, pp. 6–14, 1999.

[24] Novak, L., Fouling in Plate Heat Exchangers and Its Reduction by Proper Design, in Heat Exchangers: Theory and Practice, Hemisphere, ed. D. Yugosl, Washington, pp. 871–883, 1983.

[25] Marriott, J., Where and How to Use Plate Heat Exchangers, *Chemical Engineering*, vol. 78, no. 8, pp. 127–134, 1971.

[26] Zettler, H., Kho, T., and Müller-Steinhagen, H., Parametric Study of Fouling and Excess Heat Transfer Area in Plate Heat Exchangers, in *Proceedings of an International Conference on Mitigation of Heat Exchanger Fouling and Its Economic and Environmental Implications*, ed. T.R. Bott, A.P. Watkinson, and C.B. Panchal, Banff, Begell House, Redding, CT, pp. 50–58, July 1999.

[27] Walker, G., *Industrial Heat Exchangers: A Basic Guide*, Hemisphere, New York, 1982.

[28] Fontana, M.G. and Greene, N.D., *Corrosion Engineering* 2nd edn., McGraw-Hill, New York, 1978.

[29] Turissini, R.L., Bruno, T.V., Dahlberg, E.P., and Setterlund, R.B., Corrosion Failures in Plate Heat Exchangers, *Materials Performance*, vol. 37, no. 1, pp. 59–61, 1998.

[30] Turissini, R.L., Bruno, T.V., Dahlberg, E.P., and Setterlund, R.B., Prevent Corrosion Failures in Plate Heat Exchangers, *Chemical Engineering Progress*, vol. 93, no. 9, pp. 44–50, 1998.

[31] Novak, L., Corrosion in Compact Heat Exchangers, *Internal Presentation at Alfa Laval Lund AB*, Lund, Sweden, 2001.

[32] Moore, R.E., Selecting Materials to Meet Environmental Conditions, in *Materials Engineering: Selecting Materials for Process Equipment*, ed. K.J. McNaughton, McGraw-Hill, New York, pp. 19–22, 1980.

[33] Cross, P.H., Preventing Fouling in Plate Heat Exchangers, *Chemical Engineering*, vol. 86, no. 1, pp. 87–90, 1979.

CHAPTER 10

Extended design and operation issues

10.1 Flow distribution

There are many reasons why flow distributions in PHEs are important. Ideally, the flow is distributed evenly in every path channel of the same pass. Even flow distribution includes two parts: even mass flow rate for different channels and even flow velocity distribution over the plate. Failure of realizing even flow distribution is called maldistribution, which will result in less than optimal thermal operation and consequently lower product quality. Moreover, maldistribution will result in local low-velocity zones, where fouling and corrosion are prone to form. The formation of fouling and corrosion will certainly reduce the overall performance, and eventually lead to total failure of the exchanger.

There are mainly three types of maldistribution occurring in PHEs [1]:

1 Port flow maldistribution. The port pressure distribution in a given pass is the result of the fluid friction in the manifold and the momentum changes due to the velocity change. The variation of port pressure produces flow rate variations from channel to channel along the port.

2 Flow maldistribution resulting from non-identical channel design. As described earlier, non-identical channel design is to use parallel arrangement of two different channel types to match the thermal and hydraulic requirements simultaneously. However, the mass flow rates of different channels will be different, primarily caused by the difference of hydraulic flow resistance between different channel types. Obviously, fluid passes more easily through the channel with lower hydraulic flow resistance.

3 Channel flow maldistribution. Channel flow maldistribution results from vertical and diagonal flow arrangements. The study of Kho [2] shows that the channel flow maldistribution is more likely to occur in vertical flow arrangement than is diagonal flow arrangement. Proper design of the inlet and outlet flow distribution areas of plates can minimize this problem.

Although the channel flow maldistribution is an important issue for the operation of PHEs, it is difficult to estimate, as suggested by Kho [2]. In addition to experimental investigations, numerical simulation is probably an efficient alternative way of analysis. However, due to the current numerical simulation technology and available computer capacity, investigation of full plates by numerical simulation is still an unsolved problem [2, 3]. Hence this issue is not further explored here. The current section will focus on the port flow maldistribution and the flow maldistribution due to non-identical channel design.

10.1.1 Mathematical modelling

The distribution of flow rates in channels depends on the pressure variation in port manifolds and channels. These variations are a function of the total flow rate, friction property of the port, plate pattern, flow arrangement, etc. In order to predict the pressure variation in the manifolds and the consequent flow rate distribution throughout the plate pack, an analytical mathematical model must be established. There are generally two theories for this modelling, i.e. work-energy theorem and conservation of momentum theorem. In applying the work-energy equation to the branching process, it was argued that the mechanical energy before the branching should be equal to the mechanical energy after branching plus some losses due to friction. However, it is difficult to estimate the energy loss term due to friction, as suggested Bajura and Jones [4]. Therefore, many authors avoided this difficulty by applying the momentum theorem, e.g. Acrivos *et al.* [5], Wilkinson [6], Bassiouny and Martin [7, 8], Shen [9], Heggs and Scheidat [10]. In this section, the momentum conservation theorem will be adopted, and the following modelling equations are mostly based on the results from Wilkinson [6].

For flow arrangement, there are two basic types for PHEs, i.e. U arrangement and Z arrangement, which have already been illustrated in Fig. 5.3. Figure 10.1 illustrates the flow arrangement for one fluid stream in the U arrangement. The fluid flows out of the lower (inlet) manifold and into the channels, where it exchanges heat with the other stream (not shown in the figure). Afterwards, it flows out of the channels, and then combines in the upper (exit) manifold. In the inlet manifold, fluid friction gives rise to a pressure drop in the x direction, i.e. in the direction of flow, but the deceleration of the fluid, caused by the outflow, causes a pressure rise from the momentum consideration in the x direction. In this case, the effects of fluid friction and fluid momentum changes are counter-acting. The net result depends on many factors, in particular the plate hydraulic characteristic and the friction loss in the manifold. The plate hydraulic characteristic determines the rate at which the fluid leaves the manifold, and the friction loss in the manifold is affected by the port size and inlet fluid flow rate. In the exit manifold, fluid friction gives rise to an increase in pressure in the x direction, and the momentum contribution is in the same sense because the fluid is accelerating towards the open end as fluid flows into the manifold. The two effects therefore reinforce each other.

In the example shown in Fig. 10.1, the inlet and outlet velocities are u_0 at $x = 0$ and the corresponding pressures are $p_1(0)$ and $p_2(0)$. The difference between $p_1(0)$

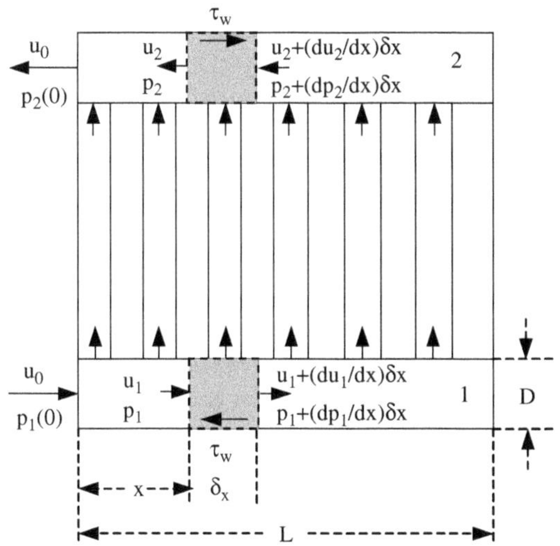

Figure 10.1: U arrangement of PHEs.

and $p_2(0)$ is the total pressure drop. In the following derivation, it is assumed that the inlet and exit port manifolds have the same dimensions and the fluid properties are constant. A control volume can be selected in the way that the width δx is equal to the sum of channel widths of the considered stream and the other stream, shown in shaded rectangle in Fig. 10.1. A momentum balance over the control volume between x and $x + \delta x$ of the inlet manifold gives

$$\frac{\pi D^2}{4}\frac{\mathrm{d}p_1}{\mathrm{d}x}\delta x + \pi D \tau_{\mathrm{w}} \delta x + k_1 \frac{\pi D^2}{4}\frac{\mathrm{d}}{\mathrm{d}x}(\rho u_1^2)\delta x = 0 \tag{10.1}$$

where k_1 is the momentum correction factor for considering that the fluid leaving the manifold may not emerge normal to the main direction of flow. This is similar to the procedure of Acrivos *et al.* [5] for manifold problems. The wall shear stress τ_{w} is calculated by

$$\tau_{\mathrm{w}} = \tfrac{1}{2}\rho u_1^2 f \tag{10.2}$$

where f is the Fanning friction factor. Incorporation of the above equation in eqn (10.1) results in

$$\frac{\mathrm{d}p_1}{\mathrm{d}x} + \frac{2f\rho u_1^2}{D} + 2k_1\rho u_1 \frac{\mathrm{d}u_1}{\mathrm{d}x} = 0 \tag{10.3}$$

A similar procedure for the exit manifold leads to the following equation:

$$\frac{\mathrm{d}p_2}{\mathrm{d}x} - \frac{2f\rho u_2^2}{D} + 2k_2\rho u_2 \frac{\mathrm{d}u_2}{\mathrm{d}x} = 0 \tag{10.4}$$

The profile of the pressure drop over the plate channels can be obtained by subtracting eqn (10.4) from eqn (10.3).

$$\frac{\mathrm{d}(p_1 - p_2)}{\mathrm{d}x} + \frac{2f\rho(u_1^2 + u_2^2)}{D} + 2k_1\rho u_1\frac{\mathrm{d}u_1}{\mathrm{d}x} - 2k_2\rho u_2\frac{\mathrm{d}u_2}{\mathrm{d}x} = 0 \tag{10.5}$$

For U arrangement, u_1 is equal to u_2 because of mass continuity. Therefore, eqn (10.3) becomes

$$\frac{\mathrm{d}(p_1 - p_2)}{\mathrm{d}x} + \frac{4f\rho u_1^2}{D} + 2(k_1 - k_2)\rho u_1\frac{\mathrm{d}u_1}{\mathrm{d}x} = 0 \tag{10.6}$$

The above equation is the governing equation for the velocity and pressure distribution in U arrangement in PHEs. In order to solve this equation, an additional equation must be established to link the pressure drop and the velocity. This can be done in the following way. The channel flow rate $\dot{V}$ at x can be calculated through the mass conservation of the control volume.

$$\dot{V} = -\frac{\pi D^2}{4}\frac{\mathrm{d}u_1}{\mathrm{d}x}\delta x \tag{10.7}$$

Therefore, the velocity in the plate channel is calculated by

$$\mathrm{v} = \frac{\dot{V}}{A_\mathrm{c}} = -\frac{\pi D^2}{4A_\mathrm{c}}\frac{\mathrm{d}u_1}{\mathrm{d}x}\delta x \tag{10.8}$$

where A_c is the cross-section area of the plate channel. For a specific plate channel, the pressure drop is a function of the channel velocity, and its relationship often follows the formula given below.

$$p_1 - p_2 = 2f_\mathrm{p}\frac{L_\mathrm{p}}{D_\mathrm{e}}\rho\mathrm{v}^2 \tag{10.9}$$

where f_p, L_p, and D_e are the Fanning friction factor of the plate channel, the plate length, and the equivalent diameter of the plate channel, respectively. For a specific plate, the friction factor f_p usually can be correlated by [11]

$$f_\mathrm{p} = a\mathrm{Re}^{-b} = a\left(\frac{\rho\mathrm{v}D_\mathrm{e}}{\mu}\right)^{-b} \tag{10.10}$$

Therefore, eqn (10.9) becomes

$$p_1 - p_2 = 2a\frac{\mu^b L_\mathrm{p}}{D_\mathrm{e}^{1+b}\rho^{b-1}}\left(\frac{\pi D^2}{4A_\mathrm{c}}\right)^{2-b}(\delta x)^{2-b}\left(-\frac{\mathrm{d}u_1}{\mathrm{d}x}\right)^{2-b} = a'\left(-\frac{\mathrm{d}u_1}{\mathrm{d}x}\right)^{b'} \tag{10.11}$$

Combination of eqns (10.6) and (10.11) results in the governing equation of the U arrangement as follows:

$$a'b'\left(-\frac{\mathrm{d}u_1}{\mathrm{d}x}\right)^{b'-1}\frac{\mathrm{d}^2u_1}{\mathrm{d}x^2} - 2(k_1 - k_2)\rho u_1\frac{\mathrm{d}u_1}{\mathrm{d}x} - \frac{4f\rho u_1^2}{D} = 0 \tag{10.12}$$

For Z arrangement, the illustration picture is shown in Fig. 10.2.

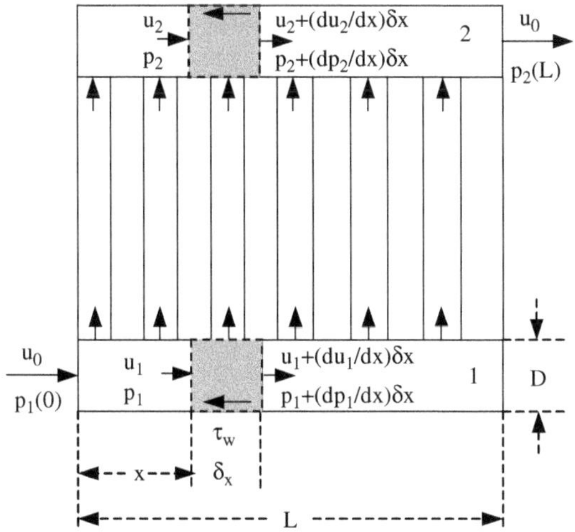

Figure 10.2: Z arrangement of PHEs.

The inlet velocity at $x = 0$ and the outlet velocity at $x =$ L are u_0, and the corresponding pressures are $p_1(0)$ and $p_2(L)$, respectively. The difference between $p_1(0)$ and $p_2(L)$ is the total pressure drop. Similar to the U arrangement, the inlet and exit manifolds are governed by the following equations of momentum conservation, respectively.

$$\frac{\mathrm{d}p_1}{\mathrm{d}x} + \frac{2f\rho u_1^2}{D} + 2k_1\rho u_1 \frac{\mathrm{d}u_1}{\mathrm{d}x} = 0 \tag{10.13}$$

$$\frac{\mathrm{d}p_2}{\mathrm{d}x} + \frac{2f\rho u_2^2}{D} + 2k_2\rho u_2 \frac{\mathrm{d}u_2}{\mathrm{d}x} = 0 \tag{10.14}$$

For Z arrangement, the following relationship between u_1 and u_2 always exists:

$$u_1 + u_2 = u_0 \tag{10.15}$$

After a derivation similar to that conducted for the U arrangement, the governing equation of the Z arrangement is obtained as follows:

$$a'b'\left(-\frac{\mathrm{d}u_1}{\mathrm{d}x}\right)^{b'-1}\frac{d^2u_1}{\mathrm{d}x^2} - 2\rho(k_1u_1 - k_2u_1 + k_2u_0)\frac{\mathrm{d}u_1}{\mathrm{d}x} - \frac{2f\rho u_0}{D}(2u_1 - u_0) = 0 \tag{10.16}$$

Equations (10.12) and (10.16) are governing equations for U and Z arrangements, respectively. These equations are subject to the appropriate boundary conditions, i.e. the velocity u_1 is equal to u_0 and 0 at $x = 0$ and at $x =$ L, respectively. Obviously, the governing equation is a non-linear second-order differential equation, which can be solved numerically. After the velocity profile is obtained, the pressure distribution can be calculated by eqn (10.11).

It must be noted that the above treatment has been simplified to illustrate the main feature of the problem. For instance, k_1 and k_2 are the momentum correction factors for considering that the fluid leaving the manifold may not emerge normal to the main direction of flow. Obviously, accurate prediction of the velocity distribution depends on the values of these two factors, which are a function of geometry, flow rate, etc. Unfortunately, there is still lack of experimental or theoretical work to determine such values. In addition, the practical situation may have to consider many additional factors, e.g. non-isothermal flow and multi-pass arrangement.

10.1.2 Typical results

Typical results of flow distribution are presented in this section. Most of these are adopted from literature, and detailed information will thus be found in the original source. Huang [1] used the computer program of Heat Transfer Research Incorporation (HTRI: http://www.htri.net) to investigate the port flow distribution for both Z and U arrangements. The results are very similar to those from Wilkinson [6], and the accuracy was found to be good by comparison with experimental data. However, no information about the model was provided, probably due to commercial confidentiality. The main results regarding the port flow distribution by Huang [1] are as follows.

The port pressure distribution is shown in Figs 10.3 and 10.4 for U and Z arrangements, respectively. As can be seen, the inlet port pressure increases along the port for both arrangements, although with different curve shapes. This increase suggests that momentum gain from the decrease in flow rate is higher than the sum of friction and turnaround losses. As a result of friction and momentum losses, pressure in the

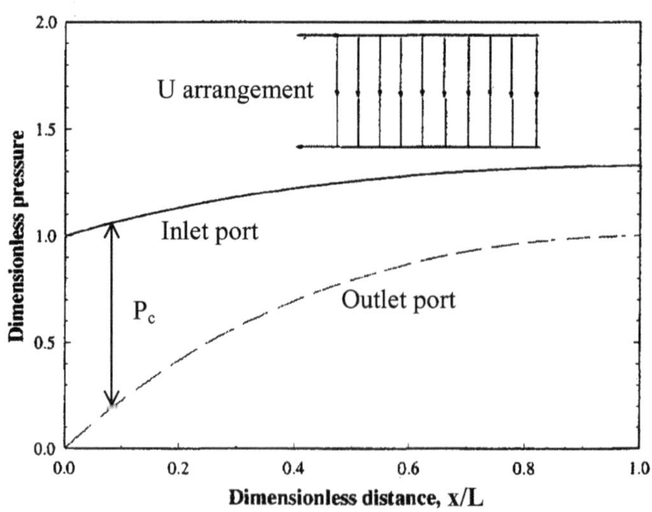

Figure 10.3: Port pressure and channel pressure drop distribution for U arrangement [1].

outlet port decreases in the direction of fluid flow for both U and Z arrangements. However, because the fluid flow direction in the outlet port differs for these two arrangements, the channel pressure drop distributions also differ. Channel pressure drop decreases in the direction of inlet port fluid flow for a U arrangement and increases for a Z arrangement, which can also be seen in Figs 10.3 and 10.4.

The channel velocity distribution for U and Z arrangements are shown in Figs 10.5 and 10.6, respectively. Note that the channel velocity distribution corresponds to

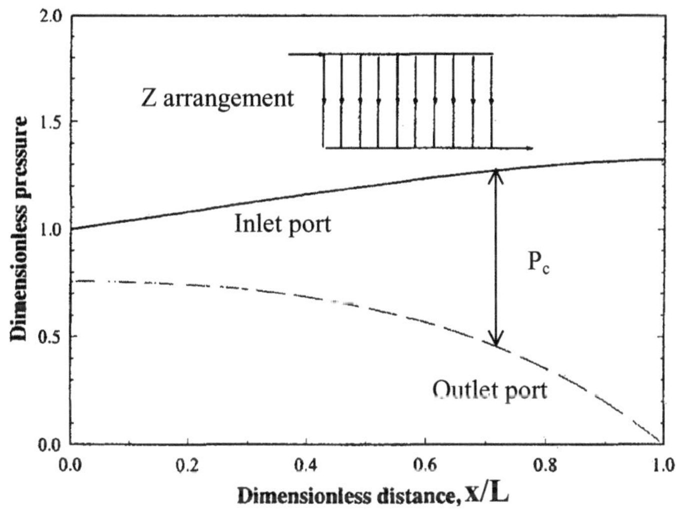

Figure 10.4: Port pressure and channel pressure drop distribution for Z arrangement [1].

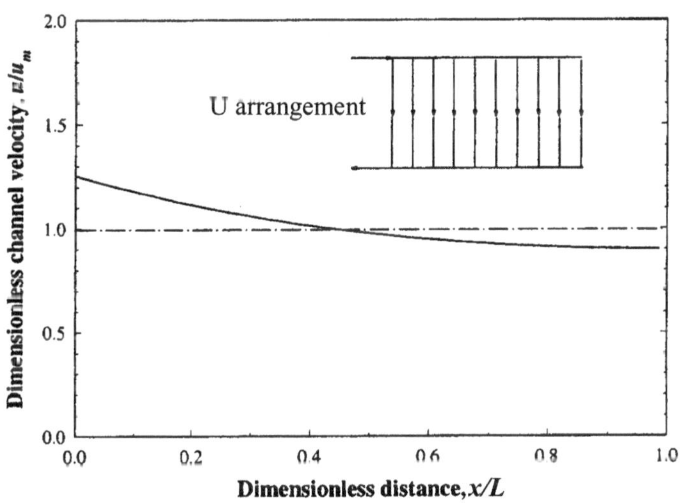

Figure 10.5: Channel velocity distribution for U arrangement [1].

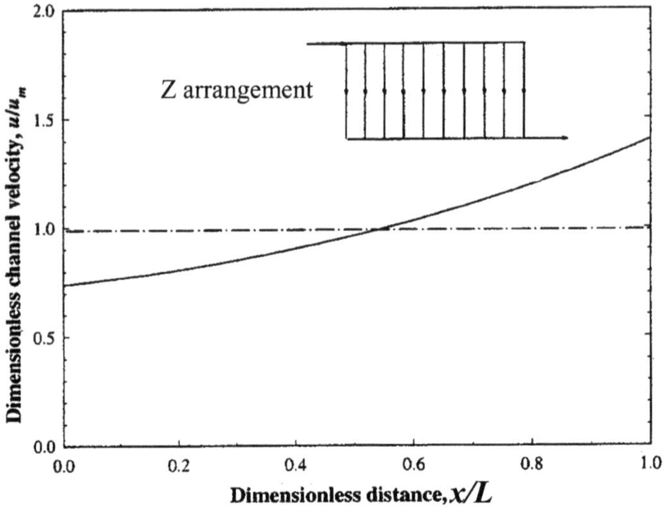

Figure 10.6: Channel velocity distribution for Z arrangement [1].

the channel flow rate distribution. It is interesting to point out that higher channel pressure drop means higher channel velocity, for both U and Z arrangements. Comparison also shows that Z arrangements have higher flow maldistribution than U arrangements, for the same geometry and inlet conditions. This is because the Z arrangement's inlet port momentum gain and exit port momentum loss are in the same direction causing channel pressure drop and channel velocity to vary more along the port.

For non-identical channel design, PHEs experience another type of maldistribution. In Figs 10.7 and 10.8 for the first arrangement type, the plates with higher chevron angle are placed in the first group (close to stream inlet). In Fig. 10.7, the pressure distribution of U arrangement is shown. Due to the port pressure distribution, the channel pressure drop decreases along the port of respective groups. This results in the decrease of channel velocity within these two groups. However, the channel velocity jumps at the connection point from group one to group two, as shown in Fig. 10.8. This is due to the lower hydraulic resistance of group two, which means that it is easier for the fluid to pass through.

In the second arrangement type, plates with lower chevron angle are placed in the first group, resulting in a different flow distribution. As shown in Figs 10.9 and 10.10, most flow maldistribution occurs in the first group. This is so because the lower hydraulic resistance in the channels increases the flow rate, in addition to the decrease of the channel pressure drop for the U arrangement. These two effects reinforce each other, leading to higher maldistribution for this arrangement type than for the first arrangement type. Therefore, a U arrangement should place the plates with higher chevron angle in the first group. The reverse is true for Z arrangement, i.e. the plates with lower chevron angles should be placed in the first group.

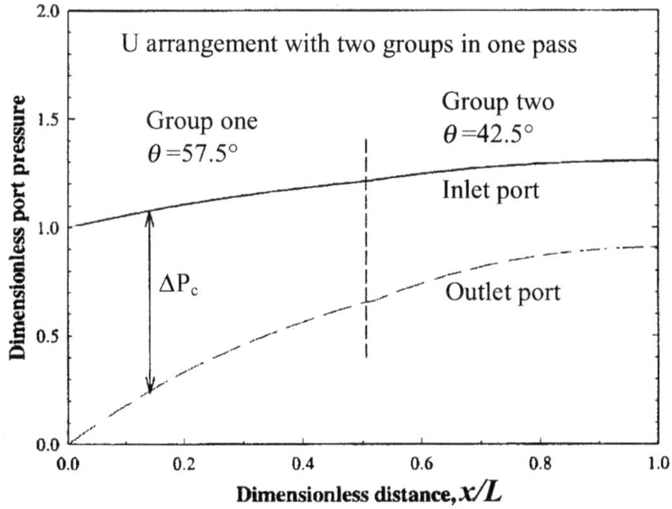

Figure 10.7: Port pressure distribution in two plate groups [1].

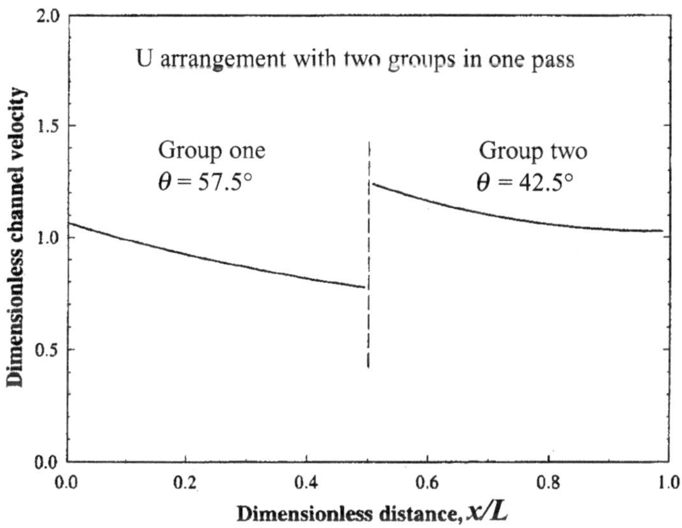

Figure 10.8: Channel velocity distribution in two plate groups [1].

The flow maldistribution certainly has an effect on the overall thermal and hydraulic performance of PHEs. However, it may surprise readers that the effect on the overall thermal performance is generally very small except for a few extreme cases [10, 12]. This can be explained as follows. Under the condition of maldistribution, the fluid temperatures of some channels become larger, and the others become smaller. Because the exit temperature (corresponding to overall thermal

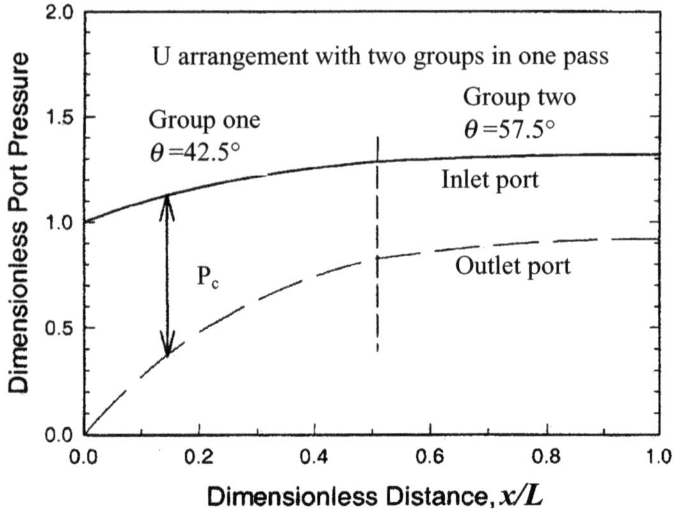

Figure 10.9: Port pressure distribution in two plate groups [1].

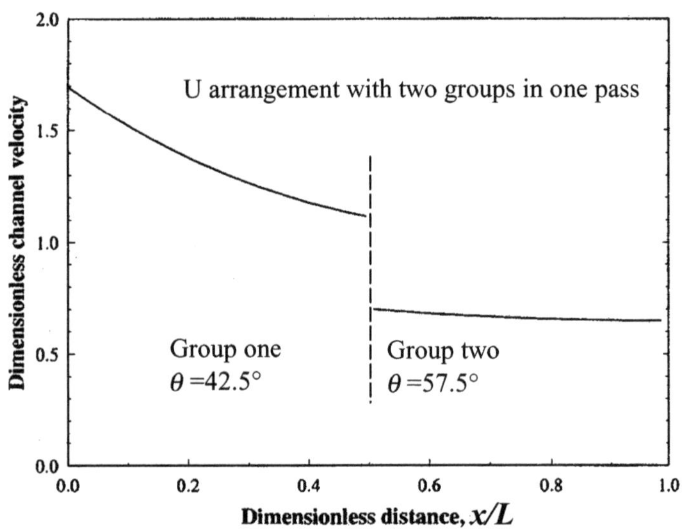

Figure 10.10: Channel velocity distribution in two plate groups [1].

performance) is the average value of the exit fluid temperatures of all channels, the difference in this value between the situations with and without maldistribution is usually very small. Therefore, the overall thermal performance may change little under flow maldistribution. However, one cannot under-estimate the negative impact of maldistribution because of this. As implied in Chapter 9, low flow velocity zones (due to maldistribution) are more prone to fouling and corrosion, which can dramatically reduce the overall performance of the exchanger. Therefore, flow maldistribution must be considered in the design of PHEs.

10.2 Numerical prediction of performance

Computational fluid dynamics (CFD) is usually referred to the broad topic encompassing the numerical solution of the governing equations for fluid flow and heat transfer by computational methods [13]. The governing equations of fluid flow are the set of the Navier–Stokes equations, continuity, and any additional conservation equations, e.g. energy or species concentrations. Due the rapid development of digital computers, CFD has been widely employed in a variety of industrial applications, including the field of fluid mechanics and heat transfer. Attempts have been carried out for PHEs in single-phase flows [2, 14–16]. Because the details of the flow and temperature fields can be provided by numerical solutions, improved understanding of the heat and momentum transport process can be achieved. Generally, these CFD calculations can be divided into two categories: whole plate calculation and 'repeated' unit calculation. However, in the CFD calculation for whole plate, calculation has to be limited to flat plates so far due to the high computation demand. Obviously, this is quite different from the real PHEs, which often have the cross-corrugated plate pattern. Therefore, this type of numerical prediction cannot reflect the real heat and flow conditions, and the influence of corrugation pitch, corrugation angle, and corrugation height cannot be investigated. However, it can give some hints to the local patterns of flow and temperature distribution, which contribute to the understanding of the local fouling formation. On the other hand, the method of 'repeated' unit calculation can give the detailed flow and temperature distribution under both laminar and turbulent conditions. The influence of different corrugation parameters can be well reflected. Therefore, the current interest of CFD application in PHEs is focused on this method. In the future, calculation in such a 'repeated' unit can be extended to the whole plate, if better turbulence modelling, faster computers, and more efficient algorithms are available. When this is realized, the thermal and hydraulic design of PHEs can be carried out solely by CFD.

10.2.1 Problem statement

As indicated in Chapter 3, the chevron wave pattern plate is the most common design offered by the majority of manufacturers. The important geometrical dimensions of the chevron plate are already shown in Figs 2.12 and 2.13. To calculate the fluid flow and heat transfer over the whole plate is too demanding in computation for the current computer technology. In order to circumvent this problem, a small segment of the exchanger is selected. The smallest volume that is worthwhile resolving is the one representing a 'repeated' building block. Periodic conditions for such a small cell unit are usually assumed, and the whole plate consists of many such cells. The concept of periodic condition dates back to Patankar et al. [17]. Figure 10.11 depicts the unit or elementary cell usually considered in numerical investigation for chevron PHEs. In this figure, the most important dimensions are indicated, namely

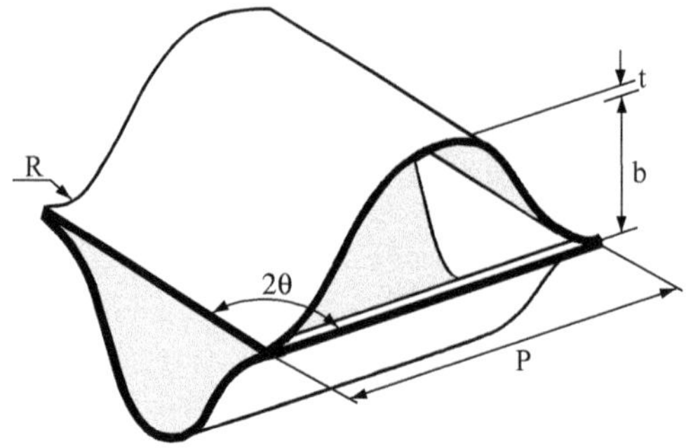

Figure 10.11: Unit cell considered in the numerical investigation.

pitch p, internal height b, angle θ, and plate thickness t. The main purpose of this numerical investigation is to determine the flow and heat transfer characteristics and provide friction factors and Nusselt numbers.

10.2.2 Mathematical modelling

The governing equations are the continuity, momentum, and energy equations. Fully developed periodic flow is prescribed. This means that the flow and temperature fields repeat themselves successively in the main flow direction. Steady-state and constant thermophysical properties are assumed. For turbulent flows, there are three major modelling approaches, namely direct numerical simulation (DNS), LES, and Reynolds averaged Navier–Stokes equations (RANS). The details of these methods are out of the scope of this book, and can be found in many related books, e.g. Versteeg and Malalasekera [13] and Wilcox [18]. Due to the high computational demand of DNS and LES methods, RANS is the most widely used approach in current engineering practice. For RANS, the governing equations in a Cartesian coordinate system are as follows:

$$\frac{\partial}{\partial x_i}(\rho U_i) = 0 \tag{10.17}$$

$$\frac{\partial}{\partial x_j}(\rho U_i U_j) = -\frac{\partial p}{\partial x_i} + \frac{\partial}{\partial x_j}\left\{\mu\left(\frac{\partial U_i}{\partial x_j} + \frac{\partial U_j}{\partial x_i}\right)\right\} + \frac{\partial}{\partial x_j}(-\rho\overline{u_i u_j}) \tag{10.18}$$

$$\frac{\partial}{\partial x_i}(\rho U_i T) = \frac{\partial}{\partial x_i}\left(\frac{\mu}{\Pr}\frac{\partial T}{\partial x_i} + (-\rho\overline{u_i t})\right) \tag{10.19}$$

In the above equations, the terms $(-\rho\overline{u_i u_j})$ and $(\rho c_p\overline{u_i t})$ are the turbulent stresses and the turbulent heat fluxes, respectively. These terms appear as a result of Reynolds decomposition and the time averaging of the turbulent fluctuations, see Versteeg and Malalasekera [13]. For laminar flows, these terms are simply equal to zero. For turbulent flows, they must be modelled by various turbulence models. The details are not further explored, and numerous related information can be found in literature.

Periodic condition is handled in the following way. Assuming linear pressure drop characteristics in the main flow direction for periodic fully developed condition, the pressure can be calculated as follows.

$$P = -\beta x_i + P^* \tag{10.20}$$

where β is a constant representing the non-periodic pressure gradient and P^* behaves in a periodic manner from cycle to cycle (unit cell to unit cell) in the main flow direction. Inserting eqn (10.20) into eqn (10.18) will add a source term like β to the right-hand side of the momentum equation in the main flow direction.

In the energy equation, either a constant heat flux or a constant temperature boundary condition is applied. For the boundary condition of constant heat flux, the periodic thermally developed regime for repeated wall heat transfer can be handled in a similar way, i.e.

$$T = \gamma x_i + T^* \tag{10.21}$$

where T^* is the periodic part of the temperature field and γ represents the temperature gradient in the main flow direction. It may be proved easily that

$$\gamma = \dot{Q}/\dot{m}c_p P \tag{10.22}$$

where $\dot{Q}$ is the rate of heat input (over one pitch length P) to the fluid. Inserting eqn (10.22) into the energy equation will add a source term like $-\rho U\gamma$ to the right-hand side of eqn (10.19). For the boundary condition of constant temperature, a periodic dimensionless temperature must be defined. This makes the calculation a little more complicated, and interested readers are referred to Patankar et al. [17].

After the above treatment, all the governing differential equations of mass conservation, momentum, and energy can be cast into the general partial differential equations as

$$\frac{\partial}{\partial x_i}(\rho U_i \phi) = \frac{\partial}{\partial x_i}\left(\Gamma_\phi \frac{\partial \phi}{\partial x_i}\right) + S_\phi \tag{10.23}$$

where ϕ denotes any of the dependent variables, Γ is the diffusivity, and S is the source term. This is a general elliptic equation form and can be solved by finite volume method and others.

10.2.3 Sample calculation

The following sample calculation was carried out by Sundén and Di Piazza [16]. In the sample, the geometry being considered in the investigation is defined by the following parameters: $2\theta = 126°$, $P = 1.07 \cdot 10^{-2}$ m, $b = 3.05 \cdot 10^{-3}$ m, and $r = 2.3 \times 10^{-3}$ m. This was chosen because experimental results are available for the average Nusselt number and the friction factor. The Reynolds number is in the range from 900 to 20,000. For turbulence modelling, the RNG k–ε model was adopted. In the numerical methods, the partial differential equations are transformed to algebraic equations by a general finite volume technique and a body-fitted mesh is adopted to obtain good resolution of the corrugations. The momentum equations are solved for the velocity components on a non-staggered mesh arrangement. The convective terms are treated by the QUICK scheme except for the k- and ε-equations for which the hybrid scheme is used. The pressure-velocity coupling is handled by the SIMPLEC algorithm. In the iterative solution procedure under-relaxation factors were chosen as U and V velocities: 0.4, W velocity: 0.3, k and ε: 0.3, enthalpy or temperature: 0.4. Wall functions are used to create the wall boundary conditions. The grid is created by using a multi-block approach and the unit cell is covered by a three-dimensional array of hexahedral control volumes in a structured body-fitted grid fashion. In this particular case 32 × 16 × 32 control volumes were used in every block.

In addition, a curvilinear coordinate transformation method is employed in the calculation to map the complex flow domain in the physical space to a rectangular domain in the computation space. This is illustrated in Fig. 10.12.

Figure 10.13 shows a typical flow pattern at a mid-plane parallel with the lower furrow axis. It is obvious that a secondary flow is set up on the flow along a corrugation when its path is crossed by streams of fluid flowing along the corrugations on

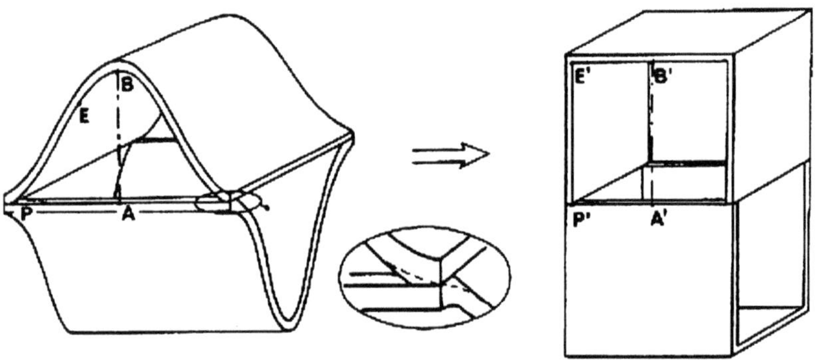

Figure 10.12: Unit cell and its mapping into cuboid.

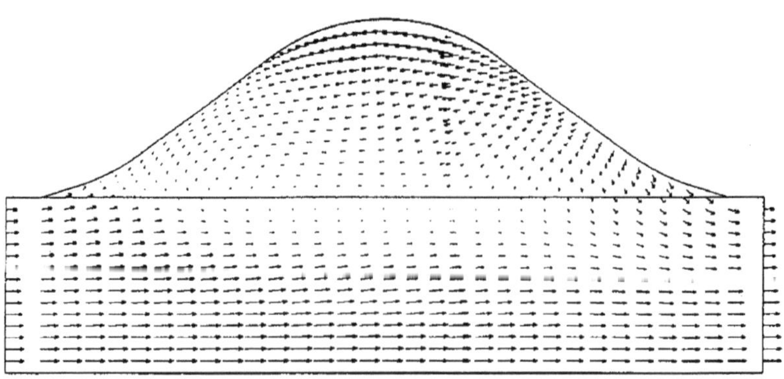

Figure 10.13: Typical flow pattern at a mid-plane parallel with the lower furrow axis.

the opposite wall. This secondary flow may be able to sustain a high temperature gradient in the fluid in regions where the secondary flow impinges on the plate surface and a low temperature gradient in regions where it detaches from the plate surface. Also, the secondary flow may be able to force the main stream fluid closer to the plate surface and thus increase the heat transfer rate. Although not shown in Fig. 10.13, it must be pointed out that secondary flow effects are present in both top and bottom furrows. It is widely believed that this secondary flow enhances the heat transfer process.

Figures 10.14 and 10.15 depict the friction factor and the average Nusselt number results, respectively. In general, the numerical values are lower than the

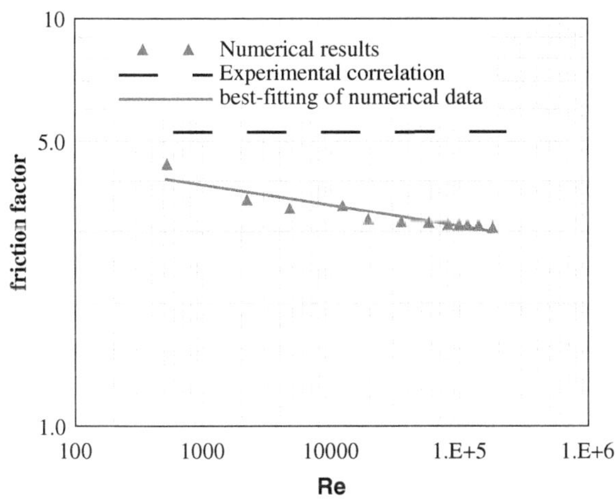

Figure 10.14: Friction factor versus Reynolds number.

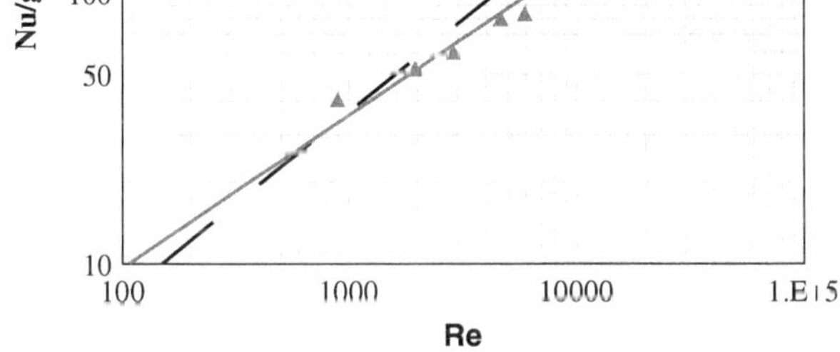

Figure 10.15: Nusselt number versus Reynolds number.

experimental ones. The deviations become bigger as the Reynolds number is increased. The maximum under-prediction of the Nusselt number is 25% while the under-prediction of the friction factor is between 17 and 40%. However, it should be noted that the experimental friction factor is independent of the Reynolds number, which may imply that the plate surfaces are rough in some way. Such effects are not included in the numerical calculations. It should also be noted that the experimental values for the unit cell were achieved from overall data from which the inlet and outlet regions and the flow distribution areas were subtracted. A certain additional degree of uncertainty is thus present in the experimental values presented in Figs 10.14 and 10.15. The experimental data for the corresponding real PHE have been provided by the manufacturer. Nevertheless, the computed results are promising and encouraging.

10.3 Multi-stream plate heat exchangers

When more than two streams are involved in a single heat exchanger, the unit is called a multi-stream heat exchanger. Common multi-stream heat exchangers are compact heat exchangers, such as multi-stream plate-fin heat exchangers and multi-stream PHEs, which are widely used in cryogenic and food industries, respectively. Because of the specific interest of this book, only multi-stream PHEs will be discussed below. General information about multi-stream heat exchangers can be found in Wang [19].

10.3.1 Practical advantages

Generally, multi-stream heat exchangers provide many unique advantages over the conventional two-stream shell-and-tube heat exchangers because of their compactness, multi-functionality, and efficiency. Due to higher heat transfer, multi-stream heat exchangers can have a significant size reduction compared to shell-and-tube heat exchangers for comparable heat duty. This is not only good for cost saving, but also very beneficial to the situation where space is a critical factor, such as off-shore platforms. Multi-functionality arises from each unit being designed to handle several streams in a single core. The high efficiency is associated with high heat transfer coefficients due to the small hydraulic diameter usually occurring in compact heat exchangers. All these advantages have made multi-stream PHEs attractive in many industrial applications.

Multi-stream heat exchangers are also attractive from the point of view of heat exchanger network optimization. The pinch technology is usually employed in the optimization of heat exchanger networks [20]. There are basically three rules in the pinch technology, i.e. heat transfer should be avoided across the pinch; external heating should be avoided below the pinch; external cooling should be avoided above the pinch. In order to satisfy these rules, a plethora of stream splits within the optimized network are often encountered, and this has been demonstrated by previous investigations [21, 22]. When the network is designed in terms of two-stream exchangers, a large number of exchangers are required. This not only increases the capital cost, but also increases the complexity of the network. Therefore, it may challenge the optimal solution, and relaxation has to be made. Using multi-stream heat exchangers might be a good way to circumvent this problem, and a number of potential benefits over the conventional two-stream heat exchangers is offered. The multi-stream heat exchangers offer large savings in capital and installation costs, such as piping costs. They also provide reduction in physical weight and space needed for the equipment. Furthermore, better integration of the process is achieved. However, the streams connected to them should not be too far away in physical space in order to save piping costs. Therefore, multi-stream heat exchangers are of interest to various industries, and there is a considerable potential for such units.

10.3.2 Design options

Multi-stream PHEs have already been used in industry for many years. The most well-known is probably the milk pasteurization application, where three-stream PHEs are typically employed (see Fig. 10.16). In addition to the general advantages of multi-stream heat exchangers, great flexibility can be obtained for the multi-stream plate-and-frame heat exchangers, because adjusting the heat transfer area is an easy task by adding or taking away some plates. Although a PHE involving more than three streams is possible in principle, the most common multi-stream PHE is the one involving three streams. Therefore, only three-stream PHEs will be considered here.

In order to incorporate more than two streams into one PHE, individual ports for these streams must be provided. There are two efficient ways to achieve this, as indicated by Heggs and Narataruksa [23]. One way is to use a dividing frame (see Fig. 10.16), which is commonly used in most multi-stream applications. In multi-stream PHEs with the dividing frame, there are two separate plate packs, which are adiabatic to each other. The dividing frame has special blocking ports and permits one stream coming out with the achieved temperature and a new stream coming in.

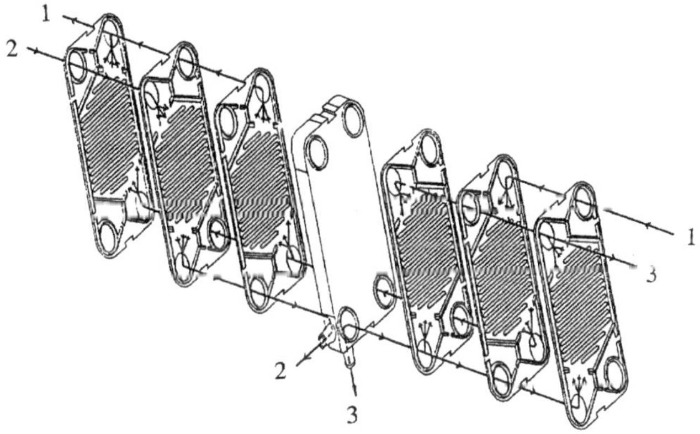

Figure 10.16: Three-stream PHE with a dividing frame [23].

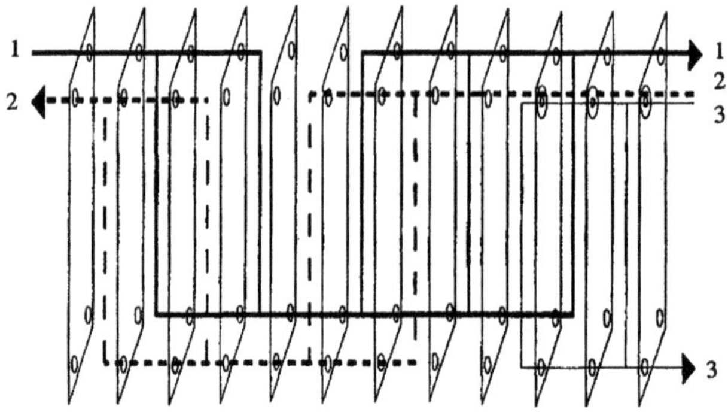

Figure 10.17: Multi-stream PHE with individual plates of annular ports [23].

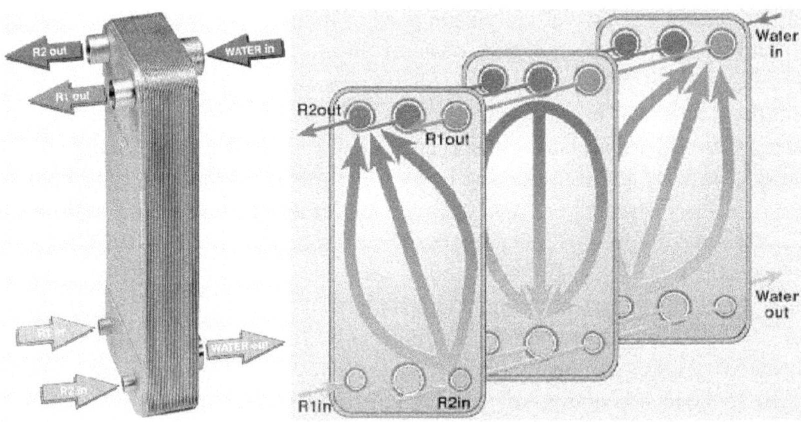

Figure 10.18: Multi-stream PHE with individual plates having separate ports (from Alfa Laval).

The alternative way is to use individual plates, which can create an extra path for the third stream to pass through. One option is to use annular ports (see Fig. 10.17), which was suggested by Heggs and Narataruksa [23]. The other option is to use another separate port for the third stream, as shown in Fig. 10.18. This is a three-stream PHE being used as an evaporator or a condenser. The design of these configurations is complicated, and all streams are thermally interconnected. It should be stressed that the demand for extra ports can dramatically increase the exchanger cost.

10.3.3 Thermal performance

The streams in three-stream PHEs can be arranged in two possible configurations, which are shown in Figs 10.19 and 10.20. In both configurations, the flow is assumed to be counter flow with single pass. In configuration 1 shown in Fig. 10.19, one

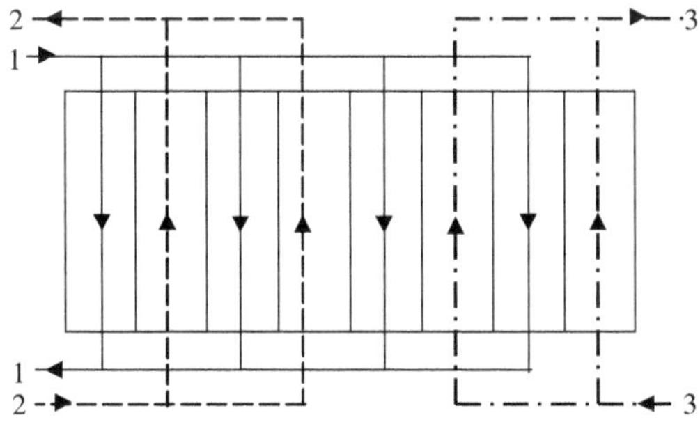

Figure 10.19: Configuration 1

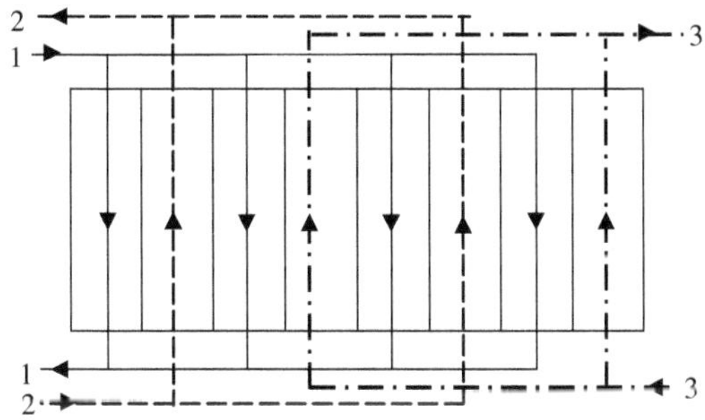

Figure 10.20: Configuration 2.

stream is divided into two parts through the port and exchanges heat with the other two streams separately. This configuration can be realized using dividing frames, and plates with conventional ports can be used. In configuration 2 shown in Fig. 10.20, all the three streams are distributed evenly inside the whole plate pack. This configuration provides simultaneous heat exchange between the hot and cold streams, and it is similar to a multi-stream plate-fin heat exchanger [19]. Note that this configuration can be realized only by extra ports, either through annular ports or through separate ports, shown in Figs 10.17 and 10.18, respectively. In addition, it should be pointed out that configuration 2 is much more complicated than configuration 1, and hence is much more costly in practice.

In order to determine the better configuration, analysis of performance was carried out by Wang and Sundén [24]. Their conclusion is that the overall thermal performance is very close for the two configurations provided the conditions of the two hot streams do not differ too much. The overall thermal performance of configuration 2 is much better than configuration 1 only at very extreme situations, where hot stream 1 has much higher values of initial temperature, number of transfer units, and ratio of heat capacities. It can be concluded that the difference of overall thermal performance between the two configurations is less than 0.05 in most practical cases. As indicated earlier, configuration 2 requires plates with annular ports. Hence, it is generally much more expensive to build such a unit.

On the other hand, it only provides slightly better overall thermal performance than configuration 1 over a wide operating range. Therefore, three-stream PHEs with configuration 2 should not be generally desired, and configuration 1 should be the first choice instead. As for those extreme cases, where configuration 2 has a much better overall thermal performance, it is still not certain whether configuration 2 should be selected. This is due to the cost difference between configurations 1 and 2, which cannot be gathered presently in a straightforward manner.

10.4 Dynamic behaviour

Most research work concentrates on the steady-state performance of heat exchangers, which reveals the energy balance relationship among fluid streams involved in heat exchanges and the time-independent temperature profiles of these streams. Obviously, it does not provide any information about the dynamic behaviour of heat exchangers. In reality, heat exchangers experience different transient processes, which are induced by several reasons: start-up, shutdown, deliberate variations in flow rates and temperatures of incoming fluid streams, etc. Fouling deposits on heat transfer surfaces also induce time-dependent operation of heat exchangers. Strictly speaking, heat exchangers are always in dynamic operation. Optimum operation, treatment of accidents, dynamic design, and real-time control demand more accurate description of the dynamic behaviour of heat exchangers. Therefore, the prediction of the dynamic behaviour of heat exchangers including PHEs is of vital importance. A literature survey reveals that there has been some research work carried out in the past. The group of Roetzel, exemplified by reference [25], is one of the most active research teams in this field, and a book on this subject has been published. In the following paragraph, a short description of the basic model for considering the dynamic behaviour of PHEs will be given. This description is based on the material from reference [25], where more detailed information can be found.

A general configuration of PHEs is shown in Fig. 10.21 to aid the derivation of the governing equations. In order to derive the equations for dynamic behaviour, some assumptions must be made. These assumptions are essentially similar to those in Chapter 6 (except for the steady state), and are not repeated here. From the energy balance, a general form of the transient heat transfer equation, which is valid for the stream in each channel including the end channels, is derived as follows

$$\frac{D_{\mathrm{f},i}}{L}\frac{\mathrm{d}T_{\mathrm{f},i}}{\mathrm{d}\tau}+(-1)^{\delta_i}C_{\mathrm{f},i}\frac{\partial T_{\mathrm{f},i}}{\partial x}=\frac{(\boldsymbol{hA})_{\mathrm{f},i,\mathrm{l}}}{L}(T_{\mathrm{p},i}-T_{\mathrm{f},i})+\frac{(\boldsymbol{hA})_{\mathrm{f},i,\mathrm{r}}}{L}(T_{\mathrm{p},i+1}-T_{\mathrm{f},i}) \tag{10.24}$$

where i is from 1 to N, $C_{\mathrm{f},i}$ is the heat capacity flow rate of the stream in channel i, and $D_{\mathrm{f},i}$ is the heat capacity of the stream in channel i, which is defined as

$$D=\rho c_{\mathrm{p}}V \tag{10.25}$$

where V is the volume of channel i. In eqn (10.24), δ_i is equal to 0 if the flow direction of the stream in channel i is identical to that of the coordinate and δ_i is equal to 1 if the flow direction of the stream in channel i is opposite to that of the

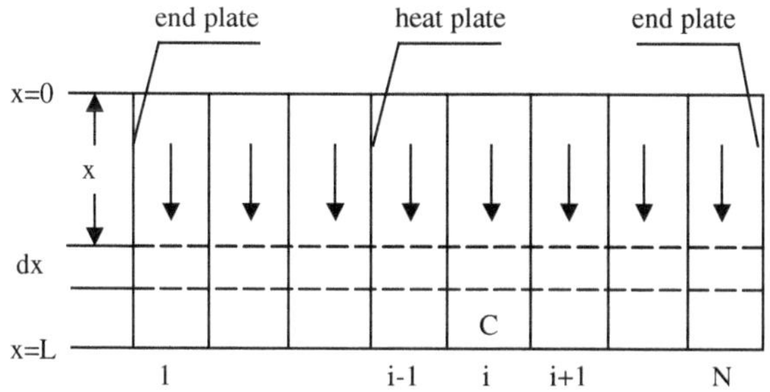

Figure 10.21: Schematic diagram of a PHE.

coordinate. $T_{p,i}$ is the temperature of the plate on the left-hand side of channel i and $T_{p,i+1}$ is the temperature of the plate on the right-hand side of channel i. $h_{f,i,l}$ is the heat transfer coefficient between the stream and the left-hand side plate of channel i and $h_{f,i,r}$ is the heat transfer coefficient between the stream and the right-hand side plate of channel i. Here the end plate on the left-hand side is counted as the first plate and the end plate on the right-hand side is plate $(N + 1)$. Note that the heat transfer coefficients are allowed to vary with channels and sides of a plate, which means eqn (10.24) is also valid for cases with maldistribution.

For plate i, the following energy equation can be derived similar to that for the channel fluid:

$$D_{p,i}\frac{dT_{p,i}}{d\tau} = (\boldsymbol{hA})_{p,i,l}(T_{f,i-1} - T_{p,i}) + (\boldsymbol{hA})_{p,i,r}(T_{f,i} - T_{p,i}) \tag{10.26}$$

where i is from 2 to N. For the first and last plates, one has the following two equations, respectively.

$$D_{p,1}\frac{dT_{p,1}}{d\tau} = (\boldsymbol{hA})_{p,1,l}(T_e - T_{p,1}) + (\boldsymbol{hA})_{p,1,r}(T_{f,1} - T_{p,1}) \tag{10.27}$$

$$D_{p,N+1}\frac{dT_{p,N+1}}{d\tau} = (\boldsymbol{hA})_{p,N+1,l}(T_{f,N} - T_{p,N+1}) + (\boldsymbol{hA})_{p,N+1,r}(T_e - T_{p,N+1}) \tag{10.28}$$

where $h_{f,1,l}$ is the heat transfer coefficient between the left-hand side of the first plate and the environmental fluid with temperature T_e; $h_{f,N+1,r}$ is the heat transfer coefficient between the right-hand side of plate $(N + 1)$ and the environmental fluid with temperature T_e. If $h_{f,1,l}$ and $h_{f,N+1,r}$ are set to zero, the end plates are considered as adiabatic.

Because the flow arrangement and the number of fluids are not involved in the derivation of the afore-mentioned governing equations, these equations are applicable to PHEs with any flow arrangements and any number of fluids. In order to solve these equations, both initial and boundary conditions must be given. The simplest initial conditions are uniform initial conditions, which correspond to the start-up case. The general forms of initial conditions can be given as

$$T_{f,i}(x,0) = g_{f,i}(x) \quad (i = 1, 2, \ldots, N) \tag{10.29}$$

$$T_{p,i}(x,0) = g_{p,i}(x) \quad (i = 1, 2, \ldots, N+1) \tag{10.30}$$

For the boundary condition, no general forms can be given. They are determined by the inlet conditions of the fluids, the flow arrangement, the number of passes, the number of channels per pass, and the coupling order of a fluid between two adjacent passes. The more detailed discussion about initial and boundary conditions can be found in reference [25].

In total, there is a set of $(2N + 1)$ equations, which form the basis for the analysis of dynamic behaviour of PHEs. Combined with initial and boundary conditions, these equations can be solved to describe the dynamic behaviour of any PHE. Some different methods can be employed, depending on different assumptions. If the flow rates of the fluids vary considerably with temperatures, dynamic simulation becomes a non-linear problem and a numerical method like the finite difference method may be preferable. If arbitrary variations of fluid temperatures and/or step disturbance of flow rates occur at inlets, dynamic simulation is considered to be linear under the assumption of constant thermal properties and the Laplace transform is applicable. No further detailed information will be given here, and interested readers are referred to reference [25].

10.5 Future developments

Although PHEs have progressed significantly since they were invented, the development will certainly continue, which will result in more industrial applications. In order to achieve this, there are still many problems to be solved. In general, these problems can be divided into two categories: construction and performance. The issue of construction includes the development of new types of PHEs, use of new materials, etc. New development in construction and material will increase the PHEs' operating limits in both temperature and pressure. New materials can also diminish the threat of corrosion and permit additional working fluids. Therefore, the development in construction is very important in order to widen the future application of PHEs. However, this part will not be explored in more detail due to the scope of this book. The current description of future developments will concentrate on the issue of both thermal and hydraulic performance. It is believed that the following three aspects will be the main research subjects in the future: more accurate prediction for single-phase and phase-change processes, consideration of maldistribution, and mitigation of fouling.

Although research in the prediction of thermal and hydraulic performance in single-phase applications has been conducted for many years, there are still some remaining problems. One issue is to develop a universal correlation for predicting the heat transfer coefficient or friction factor. In order to realize this, the influences from pitch, wavelength, corrugation angle, plate height, and width must be identified and be incorporated in the correlation. The other issue is to develop more accurate correlations for non-Newtonian fluids. Non-Newtonian fluids have significantly different behaviour from Newtonian fluids [26], and to cover this influence is never an easy task. With regard to phase-change processes, the challenge is much more ahead. The current PHEs in phase-change applications are often over-sized, which creates many problems in fouling, controlling, etc. The reason for this over-sizing is that there is still lack of correlations which can accurately predict heat transfer and pressure drop during phase-change processes. In Chapter 8, it has been clearly demonstrated that there is a great need in the future to investigate these processes, in both condensation and evaporation. In refrigeration applications, there will be certainly more new working fluids available, because the efforts for preventing the greenhouse effect will continue to push this area. In order to meet this demand, research must be carried out for these processes in PHEs. Moreover, when mixtures of fluids are used as working fluids, the prediction of thermal and hydraulic performance becomes even more difficult, which calls for more intensive research. In addition, it would be of great interest to visualize the flow in order to understand better the mechanism of condensation and evaporation in PHEs. Different measurement techniques could be employed, such as liquid crystal thermography for detailed plate surface temperature measurement, laser technique, and high speed camera together with transparent plates for velocity measurement, etc. With such efforts, the fundamental mechanism of this phase-change process can be revealed, and the correlations for predicting local heat transfer and pressure drop can be developed. Similar to single-phase applications, the influence of pitch, wavelength, corrugation angle, and plate height and width should also be investigated. For both single-phase and phase-change applications, the CFD method is another promising method in addition to the experimental work. With the rapid development of computer technology and improved models for the physical transport processes, this method will become more attractive and effective than before. However, complex geometry and turbulence modelling are certainly obstacles for CFD calculation. For phase-change processes, the tracking of the vapour–liquid interface is another obstacle. This means that effective computation algorithms must be developed, which is certainly a challenge for future research.

Research in the future must be conducted for the reduction of maldistribution in PHEs. Large PHEs working in phase-change applications are more prone to this problem due to the differences in density and velocity of the fluids in the exchanger. The prediction of maldistribution in single-phase applications has been discussed in the previous section, and the appropriate flow arrangement has been suggested for reducing this phenomenon. However, for phase-change processes, there is no effective method to predict this effect, and most effort was carried out on the development of effective inlet flow distributors to achieve better distribution. The current methods for reducing the maldistribution in phase-change applications include multiple and smaller units, feed from two sides, restrictions at inlet to each individual plate channel, manifold pipe at the inlet, etc. [27]. However, these methods are either too expensive, or too inconvenient, or associated with some other disadvantages. Therefore, models for predicting the maldistribution in phase-change applications should be established, and the methods for reducing or preventing maldistribution should be developed in the future.

Due to the high cost of fouling effects, the research in fouling formation and its prevention will continue to be important in the future. It has already been recognized that the traditional concept of fouling resistance is not enough for the accurate design and operation of PHEs. Efforts must be carried out to understand the fundamental mechanism of fouling formation and its influence on the heat transfer and pressure drop performance. Various types of fouling mechanisms should be identified, and the influence factors, e.g. temperature, pressure, velocity and concentration, must be investigated. It should be noticed that the maldistribution on the plate surface is an important factor for the local fouling formation as well as the local corrosion formation. Preventing local velocity maldistribution is beneficial to minimize the local hot or cold spots, which is good in reducing both fouling and corrosion. Except for the experimental methods, the CFD method should also be employed in this research. However, great efforts must be taken for this challenging modelling of the entire plate. After knowing the fundamental mechanism of fouling formation, design methods for reducing or preventing fouling can be developed. On the other hand, fouling is an inevitable process, and all heat exchangers will be fouled eventually. Therefore, work should be also conducted to develop cleaning methods as well as cleaning strategies.

The above discussion has presented a very big challenge for the future research in the field of PHEs. If these challenging problems can be solved, it will result in less costly but more reliable PHEs. This will certainly improve PHEs' competitiveness in the market and enlarge their industrial applications. Moreover, this is significant to the modern society with respect to conservation of limited resources and development of sustainable energy systems.

Nomenclature

$a\ a'$	coefficient
$b\ b'$	exponent
A	area, m^2
A_c	cross-sectional area, m^2
C	heat capacity flow rate, W/K
c_p	specific heat, J/(kg · K)
D	manifold diameter, m
D	heat capacity, J/K
D_e	equivalent diameter, m
f	Fanning friction factor
$g(x)$	initial temperature, K
h	heat transfer coefficient, W/$(m^2 \cdot K)$
k_1	momentum correction factor
k_2	momentum correction factor
L	manifold length, m
Nu	Nusselt number

p	pressure, Pa
P	corrugation pitch, m
p^*	periodic pressure, Pa
$\dot{Q}$	rate of heat input, W/m^2
R	radius, m
S	source term in general equation
u	velocity, m/s
T	temperature, K
U	velocity, m/s
u_0	inlet and exit velocity of manifolds, m/s
v	channel velocity, m/s
$\dot{V}$	channel flow rate, m^3/s
V	volume, m^3
x	distance along manifold from the inlet end, m
x	distance along plate, m
x_i	Cartesian co-ordinates
ΔP_c	channel pressure drop, Pa

Greek symbols

β	pressure gradient, Pa/m
δx	length of control unit, m
μ	molecular dynamic viscosity, kg/(m · s)
γ	temperature gradient, K/m
θ	corrugation angle (measured from the vertical), degree
ρ	density, kg/m^3
τ	shear stress, Pa
τ	time, s
ϕ	general dependent variable
Γ	general diffusivity
$-\rho\overline{u_i u_j}$	turbulent stresses
$\rho c_p\overline{u_i t}$	turbulent heat flux

Subscripts

1	inlet port
2	exit port
f	fluid
i	number index of plate or fluid
i,j	coordinate indices
l	left
p	plate
r	right
w	wall
ϕ	general dependent variable

References

[1] Huang, L., Port Flow Distribution in Plate Heat Exchangers, in *Compact Heat Exchangers and Enhanced Technologies for the Process Industries*, ed. R.K. Shah, A.W. Deakin, H. Honda, and T.M., Rudy, Begell House, New York, pp. 259–264, 2001.

[2] Kho, T., Effect of Flow Distribution on Scale Formation in Plate Heat Exchangers, *PhD Thesis*, University of Surrey, Surrey, UK, 1998.

[3] Rebholz, H., Heidemann, W., Hahne, E., and Müller-Steinhagen, H., Numerical Investigation of Plate Heat Exchangers, in *Compact Heat Exchangers and Enhanced Technologies for the Process Industries*, ed. R.K. Shah, A.W. Deakin, H. Honda, and T.M. Rudy, Begell House, New York, pp. 169–177, 2001.

[4] Bajura, R.A. and Jones, E.H., Flow Distribution Manifold, *ASME paper* No. 76-FE-7, pp. 654–666, 1976.

[5] Acrivos, A., Babcock, B.D., and Pigford, R.L., Flow Distribution in Manifolds, *Chemical Engineering Science*, vol. 10, no. 1/2, pp. 112–124, 1959.

[6] Wilkinson, W.L., Flow Distribution in Plate Heat Exchangers, *The Chemical Engineer*, pp. 280–293, May 1974.

[7] Bassiouny, M.K. and Martin, H., Flow Distribution and Pressure Drop in Plate Heat Exchangers - I: U-type Arrangement, *Chemical Engineering Science*, vol. 39, no. 4, pp. 693–700, 1984.

[8] Bassiouny, M.K. and Martin, H., Flow Distribution and Pressure Drop in Plate Heat Exchangers - II: Z-type Arrangement, *Chemical Engineering Science*, vol. 39, no. 4, pp. 701–704, 1984.

[9] Shen, P.I., The Effect of Friction on Flow Distribution in Dividing and Combining Flow Manifolds, *ASME Journal of Fluid Engineering*, vol. 114, pp. 121–123, 1992.

[10] Heggs, P.J. and Scheidat, H.J., Thermal Performance of Plate Heat Exchangers with Flow Maldistribution, in *Compact Heat Exchangers for Power and Process Industries*, ed. R.K. Shah, T.M. Rudy, J.M. Robertson, and K.M. Hostetler, HTD-vol. 201, ASME, New York, pp. 87–93, 1992.

[11] Marriott, J., Where and How to Use Plate Heat Exchangers, *Chemical Engineering*, vol. 78, No. 8, pp. 127–134, 1971.

[12] Yang, Q. and Wang, L., Thermal Performance and Maldistribution of Multipass Plate Heat Exchangers, in *Heat Transfer Science and Technology*, ed. B. Wang, Higher Education Press, China, pp. 621–626, 1996.

[13] Versteeg, H.K. and Malalasekera, W., *An Introduction to Computational Fluid Dynamics: The Finite Volume Method*, Longman Scientifical & Technical Publishers, London, 1995.

[14] Mehrabian, M.A., Quarini, G.L., Poulter, R., and Tierney, M.J., Effect of Herring-bone Angle on the Performance of Plate Heat Exchangers, in *Proceedings 5th UK National Conference on Heat Transfer*, IChemE, London, Industrial Session II, 1997.

[15] Ciofalo, M., Collins, M.W., and Stasiek, J.A., Flow and Heat Transfer Predictions in Flow Passages of Air Preheaters: Assessment of Alternate Modelling Approaches, in *Computer Simulations in Compact Heat Exchangers*, ed. B. Sundén and M. Faghri, Computational Mechanics Publications, Southampton, chapter 7, pp. 169–225, 1998.

[16] Sundén, B. and Di Piazza, I., Numerical Analysis of Fluid Flow and Heat Transfer in Plate-and-flame Heat Exchangers, *ASME HTD*, vol. 361–3, pp. 287–293, 1998.

[17] Patankar, S.V., Liu, C.H., and Sparrow, E.M., Fully Developed Flow and Heat Transfer in Ducts Having Streamwise-periodic Variations of Cross-sectional Area, *ASME Journal of Heat Transfer*, vol. 99, pp. 180–186, 1977.

[18] Wilcox, D.C., *Turbulence Modeling for CFD*, 2nd edn, DCW Industries CA, 2000.

[19] Wang, L., Performance Analysis and Optimal Design of Heat Exchangers and Heat Exchanger Networks, *PhD Thesis*, Lund Institute of Technology, Lund, Sweden, 2001.

[20] Linnhoff, B., Townsend, D.W., Boland, D., Hewitt G.F., Thomas, B.E.A., Guy, A.R., and Marsland, R., *User Guide on Process Integration for the Efficient Use of Energy*, Institution of Chemical Engineers, Pergamon Press, Oxford, 1982.

[21] Sundén, B., Analysis of the Heat Recovery in Two Crude Distillation Units, *Heat Recovery Systems and CHP*, vol. 8, no. 5, pp. 483–488, 1988.

[22] Farhanieh, B. and Sundén, B., Analysis of an Existing Heat Exchanger Network and Effects of Heat Pump Installations, *Heat Recovery Systems and CHP*, vol. 10, no. 3, pp. 285–296, 1990.

[23] Heggs, P.J. and Narataruksa, P., The Use of a Plate and Frame Heat Exchanger as a Three Stream Compact Exchanger, in *Compact Heat Exchangers and Enhancement Technology for the Process Industries*, ed. R.K. Shah, K.J. Bell, H. Honda, and B. Thonon, Begell House, New York, pp. 251–262, 1999.

[24] Wang, L. and Sundén, B., Thermal Performance Analysis of Multi-stream Plate Heat Exchangers, *35th National Heat Transfer Conference*, Anaheim, CA, NHTC2001–20052, 2001.

[25] Roetzel, W. and Xuan, Y., *Dynamic Behaviour of Heat Exchangers*, WIT Press, Southampton, UK, 1999.

[26] Cho, Y.I. and Hartnett, J.P., Non-Newtonian Fluids, in *Handbook of Heat Transfer Applications*, 2nd edn, W.M. Rohsenow, J.P. Hartnett, and E.N. Ganic, chapter 2, 1985.

[27] Holm, M., Nilsson, I., and Wilhelmsson, B., Reduction of Maldistribution in Large Rising Film Plate Evaporators, in *Compact Heat Exchangers and Enhanced Technologies for the Process Industries*, ed. R.K. Shah, A.W. Deakin, H. Honda, and T.M., Rudy, Begell House, New York, pp. 519–524, 2001.

Appendix

Table A.1: Effectiveness ε and correction factor F for arrangement no. 111.

		$N = 3$		$N = 7$		$N = 13$		$N = 23$		$N = 41$		$N = 71$		$N = 99$		$N = \infty$	
R	NTU	ε	F	ε	F	ε	F	ε	F	ε	F	ε	F	ε	F	ε	F
	0.2	0.1794	0.9889	0.1802	0.9937	0.1807	0.9963	0.1809	0.9978	0.1811	0.9988	0.1811	0.9993	0.1812	0.9995	0.1813	1.0000
	0.4	0.3237	0.9778	0.3263	0.9873	0.3277	0.9925	0.3285	0.9956	0.3290	0.9974	0.3293	0.9985	0.3294	0.9989	0.3297	1.0000
	0.6	0.4402	0.9669	0.4448	0.9806	0.4474	0.9885	0.4489	0.9932	0.4499	0.9960	0.4504	0.9977	0.4506	0.9983	0.4512	1.0000
	0.8	0.5346	0.9560	0.5411	0.9737	0.5450	0.9843	0.5473	0.9906	0.5487	0.9946	0.5495	0.9968	0.5498	0.9977	0.5507	1.0000
0.0	1.0	0.6115	0.9454	0.6196	0.9666	0.6246	0.9798	0.6277	0.9879	0.6295	0.9930	0.6306	0.9959	0.6310	0.9970	0.6321	1.0000
	1.5	0.7484	0.9199	0.7588	0.9482	0.7659	0.9679	0.7703	0.9806	0.7730	0.9886	0.7746	0.9933	0.7752	0.9951	0.7769	1.0000
	2.0	0.8335	0.8952	0.8440	0.9289	0.8519	0.9548	0.8569	0.9722	0.8601	0.9836	0.8620	0.9902	0.8627	0.9929	0.8647	1.0000
	3.0	0.9232	0.8554	0.9307	0.8896	0.9377	0.9254	0.9425	0.9522	0.9457	0.9711	0.9475	0.9826	0.9483	0.9873	0.9502	1.0000
	5.0	0.9815	0.7984	0.9832	0.8170	0.9864	0.8596	0.9889	0.9003	0.9907	0.9346	0.9917	0.9585	0.9921	0.9691	0.9933	1.0000
	7.0	0.9953	0.7646	0.9952	0.7621	0.9963	0.7982	0.9972	0.8413	0.9980	0.8846	0.9984	0.9202	0.9986	0.9379	0.9991	1.0000
	0.2	0.1761	0.9868	0.1771	0.9927	0.1776	0.9957	0.1778	0.9975	0.1780	0.9986	0.1781	0.9991	0.1781	0.9994	0.1782	1.0000
	0.4	0.3137	0.9738	0.3166	0.9853	0.3182	0.9914	0.3191	0.9949	0.3196	0.9971	0.3199	0.9983	0.3201	0.9988	0.3204	1.0000
	0.6	0.4229	0.9611	0.4282	0.9780	0.4310	0.9870	0.4327	0.9923	0.4337	0.9956	0.4343	0.9974	0.4345	0.9981	0.4351	1.0000
	0.8	0.5108	0.9488	0.5184	0.9706	0.5225	0.9826	0.5249	0.9897	0.5264	0.9940	0.5272	0.9965	0.5276	0.9975	0.5284	1.0000
0.2	1.0	0.5825	0.9369	0.5921	0.9632	0.5974	0.9781	0.6005	0.9869	0.6024	0.9924	0.6035	0.9955	0.6039	0.9968	0.6050	1.0000
	1.5	0.7119	0.9090	0.7248	0.9446	0.7323	0.9663	0.7369	0.9797	0.7397	0.9882	0.7413	0.9930	0.7419	0.9949	0.7436	1.0000
	2.0	0.7956	0.8839	0.8094	0.9257	0.8182	0.9539	0.8236	0.9719	0.8270	0.9835	0.8289	0.9902	0.8297	0.9929	0.8317	1.0000
	3.0	0.8912	0.8424	0.9028	0.8883	0.9117	0.9272	0.9173	0.9543	0.9210	0.9727	0.9231	0.9836	0.9239	0.9881	0.9261	1.0000
	5.0	0.9654	0.7873	0.9697	0.8204	0.9751	0.8693	0.9790	0.9109	0.9815	0.9435	0.9831	0.9649	0.9837	0.9741	0.9853	1.0000
	7.0	0.9884	0.7558	0.9891	0.7677	0.9916	0.8135	0.9935	0.8610	0.9949	0.9044	0.9958	0.9369	0.9961	0.9521	0.9970	1.0000

(*Continued*)

Table A.1: (*Continued*).

		$N = 3$		$N = 7$		$N = 13$		$N = 23$		$N = 41$		$N = 71$		$N = 99$		$N = \infty$	
R	NTU	ε	F	ε	F	ε	F	ε	F	ε	F	ε	F	ε	F	ε	F
	0.2	0.1729	0.9846	0.1740	0.9916	0.1745	0.9951	0.1748	0.9971	0.1750	0.9983	0.1751	0.9990	0.1751	0.9993	0.1753	1.0000
	0.4	0.3040	0.9698	0.3073	0.9833	0.3090	0.9903	0.3100	0.9943	0.3105	0.9967	0.3109	0.9981	0.3110	0.9986	0.3113	1.0000
	0.6	0.4062	0.9554	0.4121	0.9753	0.4151	0.9855	0.4169	0.9915	0.4179	0.9951	0.4185	0.9971	0.4187	0.9979	0.4194	1.0000
	0.8	0.4878	0.9416	0.4962	0.9674	0.5005	0.9808	0.5030	0.9886	0.5045	0.9934	0.5054	0.9961	0.5057	0.9972	0.5066	1.0000
0.4	1.0	0.5541	0.9284	0.5647	0.9595	0.5702	0.9761	0.5735	0.9858	0.5754	0.9918	0.5765	0.9952	0.5770	0.9965	0.5781	1.0000
	1.5	0.6747	0.8982	0.6892	0.9402	0.6972	0.9642	0.7019	0.9786	0.7048	0.9876	0.7064	0.9927	0.7070	0.9947	0.7087	1.0000
	2.0	0.7548	0.8718	0.7710	0.9211	0.7806	0.9521	0.7862	0.9711	0.7897	0.9831	0.7917	0.9900	0.7925	0.9928	0.7945	1.0000
	3.0	0.8519	0.8296	0.8671	0.8844	0.8777	0.9271	0.8842	0.9550	0.8882	0.9734	0.8905	0.9842	0.8914	0.9885	0.8938	1.0000
	5.0	0.9392	0.7764	0.9468	0.8193	0.9553	0.8753	0.9608	0.9184	0.9644	0.9497	0.9665	0.9693	0.9673	0.9775	0.9695	1.0000
	7.0	0.9735	0.7471	0.9759	0.7689	0.9811	0.8257	0.9848	0.8775	0.9873	0.9200	0.9888	0.9493	0.9894	0.9622	0.9909	1.0000
	0.2	0.1697	0.9825	0.1709	0.9905	0.1715	0.9944	0.1719	0.9967	0.1721	0.9981	0.1722	0.9989	0.1722	0.9992	0.1723	1.0000
	0.4	0.2947	0.9657	0.2983	0.9813	0.3000	0.9891	0.3011	0.9936	0.3017	0.9963	0.3020	0.9978	0.3022	0.9984	0.3025	1.0000
	0.6	0.3902	0.9497	0.3966	0.9725	0.3997	0.9839	0.4015	0.9905	0.4026	0.9945	0.4032	0.9968	0.4035	0.9977	0.4041	1.0000
	0.8	0.4656	0.9345	0.4746	0.9639	0.4790	0.9788	0.4816	0.9875	0.4832	0.9928	0.4840	0.9958	0.4844	0.9969	0.4853	1.0000
0.6	1.0	0.5266	0.9200	0.5378	0.9554	0.5435	0.9738	0.5468	0.9845	0.5488	0.9911	0.5499	0.9947	0.5503	0.9962	0.5515	1.0000
	1.5	0.6374	0.8876	0.6529	0.9350	0.6611	0.9614	0.6659	0.9770	0.6688	0.9867	0.6704	0.9922	0.671	0.9943	0.6727	1.0000
	2.0	0.7122	0.8600	0.7297	0.9152	0.7397	0.9492	0.7455	0.9696	0.7491	0.9823	0.7511	0.9896	0.7519	0.9924	0.7539	1.0000
	3.0	0.8065	0.8174	0.8236	0.8779	0.8356	0.9248	0.8427	0.9543	0.8470	0.9732	0.8495	0.9841	0.8504	0.9884	0.8529	1.0000
	5.0	0.9007	0.7662	0.9109	0.8137	0.9227	0.8769	0.9301	0.9221	0.9347	0.9531	0.9373	0.9717	0.9384	0.9794	0.9411	1.0000
	7.0	0.9454	0.7391	0.9495	0.7649	0.9587	0.8321	0.9650	0.8884	0.9691	0.9304	0.9714	0.9571	0.9723	0.9684	0.9748	1.0000

	0.2	0.1667	0.9804	0.1679	0.9893	0.1686	0.9938	0.1689	0.9964	0.1692	0.9979	0.1693	0.9988	0.1693	0.9991	0.1695	1.0000
	0.4	0.2857	0.9617	0.2895	0.9792	0.2914	0.9879	0.2925	0.9929	0.2931	0.9959	0.2935	0.9976	0.2936	0.9983	0.2940	1.0000
	0.6	0.3749	0.9441	0.3815	0.9695	0.3848	0.9822	0.3867	0.9895	0.3878	0.9940	0.3884	0.9965	0.3887	0.9975	0.3893	1.0000
	0.8	0.4444	0.9274	0.4536	0.9601	0.4582	0.9767	0.4608	0.9862	0.4624	0.9921	0.4633	0.9954	0.4636	0.9967	0.4645	1.0000
0.8	1.0	0.5001	0.9113	0.5116	0.9510	0.5174	0.9713	0.5207	0.9830	0.5227	0.9902	0.5238	0.9943	0.5243	0.9959	0.5254	1.0000
	1.5	0.6009	0.8773	0.6164	0.9290	0.6248	0.9581	0.6295	0.9751	0.6324	0.9856	0.6340	0.9916	0.6346	0.9939	0.6363	1.0000
	2.0	0.6690	0.8487	0.6865	0.9081	0.6968	0.9453	0.7026	0.9674	0.7061	0.9811	0.7081	0.9889	0.7089	0.9919	0.7109	1.0000
	3.0	0.7566	0.8059	0.7739	0.8691	0.7866	0.9203	0.7939	0.9520	0.7983	0.9720	0.8008	0.9834	0.8018	0.9880	0.8043	1.0000
	5.0	0.8498	0.7570	0.8606	0.8040	0.8746	0.8732	0.8833	0.9216	0.8886	0.9535	0.8915	0.9722	0.8927	0.9798	0.8957	1.0000
	7.0	0.8993	0.7321	0.9039	0.7560	0.9166	0.8306	0.9255	0.8919	0.9311	0.9347	0.9342	0.9605	0.9354	0.9711	0.9386	1.0000
	0.2	0.1636	0.9783	0.1650	0.9882	0.1657	0.9931	0.1661	0.9960	0.1663	0.9977	0.1665	0.9987	0.1665	0.9990	0.1667	1.0000
	0.4	0.2770	0.9578	0.2810	0.9771	0.2830	0.9867	0.2841	0.9922	0.2848	0.9955	0.2852	0.9974	0.2853	0.9981	0.2857	1.0000
	0.6	0.3602	0.9384	0.3670	0.9665	0.3704	0.9804	0.3723	0.9885	0.3734	0.9934	0.3741	0.9961	0.3743	0.9972	0.3750	1.0000
	0.8	0.4241	0.9204	0.4334	0.9562	0.4381	0.9744	0.4407	0.9849	0.4423	0.9913	0.4432	0.9949	0.4435	0.9963	0.4444	1.0000
1.0	1.0	0.4747	0.9037	0.4862	0.9463	0.4920	0.9685	0.4953	0.9814	0.4973	0.9893	0.4984	0.9937	0.4989	0.9955	0.5000	1.0000
	1.5	0.5654	0.8673	0.5805	0.9224	0.5887	0.9542	0.5934	0.9728	0.5962	0.9843	0.5978	0.9908	0.5984	0.9934	0.6000	1.0000
	2.0	0.6263	0.8380	0.6428	0.8998	0.6529	0.9404	0.6586	0.9645	0.6620	0.9794	0.6640	0.9879	0.6647	0.9913	0.6667	1.0000
	3.0	0.7047	0.7954	0.7203	0.8583	0.7327	0.9137	0.7409	0.9531	0.7442	0.9698	0.7466	0.9822	0.7476	0.9871	0.7500	1.0000
	5.0	0.7892	0.7490	0.7982	0.7913	0.8121	0.8643	0.8209	0.9166	0.8262	0.9508	0.8292	0.9708	0.8303	0.9788	0.8333	1.0000
	7.0	0.8356	0.7263	0.8389	0.7438	0.8518	0.8208	0.8613	0.8868	0.8671	0.9324	0.8704	0.9596	0.8717	0.9706	0.8750	1.0000

Table A.2: Effectiveness ε and correction factor F for arrangement no. 113.

		$N=3$		$N=7$		$N=13$		$N=23$		$N=41$		$N=71$		$N=99$		$N=\infty$	
R	NTU	ε	F	ε	F	ε	F	ε	F	ε	F	ε	F	ε	F	ε	F
	0.2	0.1794	0.9889	0.1802	0.9937	0.1807	0.9963	0.1809	0.9978	0.1811	0.9988	0.1811	0.9993	0.1812	0.9995	0.1813	1.0000
	0.4	0.3237	0.9778	0.3263	0.9873	0.3277	0.9925	0.3285	0.9956	0.3290	0.9974	0.3293	0.9985	0.3294	0.9989	0.3297	1.0000
	0.6	0.4402	0.9669	0.4448	0.9806	0.4474	0.9885	0.4489	0.9932	0.4499	0.9960	0.4504	0.9977	0.4506	0.9983	0.4512	1.0000
	0.8	0.5346	0.9560	0.5411	0.9737	0.5450	0.9843	0.5473	0.9906	0.5487	0.9946	0.5495	0.9968	0.5498	0.9977	0.5507	1.0000
0.0	1.0	0.6115	0.9454	0.6196	0.9666	0.6246	0.9798	0.6277	0.9879	0.6295	0.9930	0.6306	0.9959	0.6310	0.9970	0.6321	1.0000
	1.5	0.7484	0.9199	0.7588	0.9482	0.7659	0.9679	0.7703	0.9806	0.773	0.9886	0.7746	0.9933	0.7752	0.9951	0.7769	1.0000
	2.0	0.8335	0.8962	0.8440	0.9289	0.8519	0.9548	0.8569	0.9722	0.8601	0.9836	0.8620	0.9902	0.8627	0.9929	0.8647	1.0000
	3.0	0.9232	0.8554	0.9307	0.8896	0.9377	0.9254	0.9425	0.9522	0.9457	0.9711	0.9475	0.9826	0.9483	0.9873	0.9502	1.0000
	5.0	0.9815	0.7984	0.9832	0.8170	0.9864	0.8596	0.9889	0.9003	0.9907	0.9346	0.9917	0.9585	0.9921	0.9691	0.9933	0.9997
	7.0	0.9953	0.7646	0.9952	0.7621	0.9963	0.7982	0.9972	0.8413	0.998	0.8846	0.9984	0.9202	0.9986	0.9379	0.9991	0.9985
	0.2	0.1756	0.9837	0.1766	0.9897	0.1771	0.9929	0.1774	0.9947	0.1776	0.9958	0.1777	0.9965	0.1777	0.9967	0.1778	0.9973
	0.4	0.3106	0.9620	0.3137	0.9740	0.3154	0.9803	0.3163	0.9841	0.3169	0.9863	0.3172	0.9876	0.3173	0.9881	0.3177	0.9894
	0.6	0.4148	0.9357	0.4204	0.9532	0.4234	0.9627	0.4252	0.9683	0.4262	0.9717	0.4268	0.9737	0.4271	0.9744	0.4277	0.9764
	0.8	0.4954	0.9058	0.5035	0.9282	0.5079	0.9405	0.5105	0.9479	0.5121	0.9524	0.5130	0.9550	0.5133	0.956	0.5143	0.9587
0.2	1.0	0.5582	0.8732	0.5684	0.8995	0.5741	0.9144	0.5774	0.9234	0.5795	0.9289	0.5807	0.9321	0.5811	0.9333	0.5823	0.9366
	1.5	0.6620	0.7855	0.6759	0.8178	0.6838	0.8369	0.6885	0.8486	0.6915	0.8560	0.6932	0.8602	0.6939	0.8619	0.6956	0.8663
	2.0	0.7198	0.6981	0.7352	0.7310	0.7441	0.7512	0.7496	0.7639	0.7530	0.7720	0.7550	0.7767	0.7557	0.7785	0.7577	0.7834
	3.0	0.7736	0.5489	0.7880	0.5749	0.7969	0.5918	0.8023	0.6026	0.8058	0.6096	0.8077	0.6136	0.8085	0.6153	0.8106	0.6195
	5.0	0.8046	0.3643	0.8143	0.3764	0.8210	0.3852	0.8251	0.3908	0.8277	0.3943	0.8291	0.3964	0.8297	0.3972	0.8313	0.3994
	7.0	0.8137	0.2683	0.8196	0.2739	0.8251	0.2791	0.8283	0.2823	0.8304	0.2844	0.8315	0.2855	0.8320	0.2860	0.8331	0.2872

	0.2	0.1720	0.9786	0.1731	0.9858	0.1736	0.9895	0.174	0.9916	0.1742	0.9929	0.1743	0.9937	0.1743	0.9939	0.1744	0.9947
	0.4	0.2983	0.9468	0.3019	0.9611	0.3037	0.9686	0.3048	0.9729	0.3054	0.9755	0.3058	0.9770	0.3059	0.9776	0.3063	0.9791
	0.6	0.3916	0.9071	0.3979	0.9277	0.4012	0.9387	0.4031	0.9451	0.4043	0.9490	0.4050	0.9512	0.4052	0.9521	0.4059	0.9544
	0.8	0.4607	0.8621	0.4697	0.8879	0.4744	0.9018	0.4772	0.9099	0.4789	0.9150	0.4799	0.9178	0.4802	0.9189	0.4812	0.9219
0.4	1.0	0.5122	0.8143	0.5235	0.8439	0.5295	0.8599	0.5330	0.8695	0.5352	0.8754	0.5364	0.8787	0.5369	0.8801	0.5381	0.8835
	1.5	0.5917	0.6952	0.6068	0.7282	0.6149	0.7466	0.6197	0.7578	0.6227	0.7648	0.6244	0.7688	0.6251	0.7704	0.6268	0.7745
	2.0	0.6323	0.5907	0.6485	0.6211	0.6575	0.6387	0.6629	0.6494	0.6663	0.6562	0.6682	0.6601	0.6689	0.6617	0.6708	0.6657
	3.0	0.6673	0.4389	0.6817	0.4591	0.6905	0.4719	0.6958	0.4799	0.699	0.4849	0.7009	0.4878	0.7017	0.4889	0.7036	0.4919
	5.0	0.6894	0.2822	0.6965	0.2886	0.7034	0.2950	0.7076	0.2989	0.7101	0.3014	0.7116	0.3028	0.7121	0.3033	0.7136	0.3048
	7.0	0.6987	0.2076	0.7002	0.2086	0.7059	0.2124	0.7093	0.2147	0.7114	0.2161	0.7126	0.2170	0.7130	0.2173	0.7142	0.2181
	0.2	0.1684	0.9735	0.1696	0.9818	0.1703	0.9861	0.1706	0.9885	0.1709	0.9900	0.1710	0.9909	0.1710	0.9912	0.1712	0.9921
	0.4	0.2868	0.9322	0.2906	0.9486	0.2926	0.9571	0.2938	0.9620	0.2945	0.9650	0.2949	0.9667	0.2950	0.9674	0.2954	0.9691
	0.6	0.3704	0.8805	0.3772	0.9039	0.3807	0.9161	0.3828	0.9232	0.384	0.9276	0.3847	0.9301	0.3850	0.9310	0.3857	0.9336
	0.8	0.4298	0.8234	0.4393	0.8518	0.4442	0.8668	0.4471	0.8757	0.4488	0.8811	0.4498	0.8842	0.4502	0.8854	0.4512	0.8886
0.6	1.0	0.4722	0.7648	0.4839	0.7963	0.4900	0.8131	0.4936	0.8231	0.4958	0.8293	0.4971	0.8328	0.4975	0.8342	0.4988	0.8378
	1.5	0.5339	0.6287	0.5488	0.6608	0.5568	0.6785	0.5615	0.6892	0.5644	0.6958	0.5660	0.6996	0.5666	0.7011	0.5683	0.7050
	2.0	0.5632	0.5199	0.5785	0.5471	0.5870	0.5628	0.5921	0.5723	0.5952	0.5783	0.5970	0.5817	0.5977	0.5831	0.5995	0.5866
	3.0	0.5880	0.3764	0.6001	0.3918	0.6081	0.4023	0.6128	0.4088	0.6158	0.4128	0.6175	0.4151	0.6181	0.4161	0.6199	0.4184
	5.0	0.6060	0.2397	0.6095	0.2425	0.6157	0.2476	0.6194	0.2507	0.6217	0.2526	0.6230	0.2537	0.6235	0.2541	0.6248	0.2552
	7.0	0.6144	0.1761	0.6125	0.1750	0.6175	0.1779	0.6206	0.1798	0.6224	0.1809	0.6235	0.1815	0.6239	0.1818	0.6250	0.1824

(Continued)

Table A.2: (*Continued*).

		$N = 3$		$N = 7$		$N = 13$		$N = 23$		$N = 41$		$N = 71$		$N = 99$		$N = \infty$	
R	NTU	ε	F	ε	F	ε	F	ε	F	ε	F	ε	F	ε	F	ε	F
	0.2	0.1649	0.9685	0.1663	0.9779	0.1670	0.9827	0.1674	0.9855	0.1676	0.9872	0.1678	0.9881	0.1678	0.9885	0.1680	0.9895
	0.4	0.2759	0.9180	0.2800	0.9364	0.2822	0.9459	0.2834	0.9514	0.2841	0.9548	0.2845	0.9567	0.2847	0.9574	0.2851	0.9594
	0.6	0.3510	0.8558	0.3581	0.8814	0.3617	0.8947	0.3638	0.9025	0.3651	0.9073	0.3659	0.9100	0.3661	0.9111	0.3669	0.9139
	0.8	0.4021	0.7888	0.4118	0.8191	0.4168	0.8350	0.4197	0.8444	0.4215	0.8502	0.4225	0.8535	0.4229	0.8548	0.4239	0.8582
0.8	1.0	0.4373	0.7223	0.4490	0.7549	0.4550	0.7722	0.4586	0.7825	0.4608	0.7888	0.4620	0.7924	0.4625	0.7939	0.4637	0.7976
	1.5	0.4858	0.5769	0.4998	0.6073	0.5073	0.6242	0.5117	0.6344	0.5145	0.6407	0.5160	0.6443	0.5166	0.6458	0.5182	0.6495
	2.0	0.5076	0.4687	0.5212	0.4924	0.5290	0.5066	0.5336	0.5151	0.5365	0.5205	0.5381	0.5236	0.5387	0.5248	0.5404	0.5280
	3.0	0.5264	0.3345	0.5357	0.3460	0.5427	0.3550	0.5469	0.3604	0.5495	0.3638	0.5510	0.3658	0.5515	0.3665	0.5530	0.3685
	5.0	0.5415	0.2120	0.5423	0.2127	0.5477	0.2169	0.5508	0.2194	0.5528	0.2209	0.5539	0.2218	0.5543	0.2222	0.5555	0.2231
	7.0	0.5485	0.1554	0.5449	0.1533	0.5491	0.1557	0.5517	0.1572	0.5534	0.1581	0.5543	0.1586	0.5546	0.1588	0.5556	0.1594
	0.2	0.1616	0.9635	0.1630	0.9740	0.1638	0.9793	0.1642	0.9824	0.1645	0.9843	0.1646	0.9854	0.1647	0.9858	0.1648	0.9869
	0.4	0.2657	0.9044	0.2700	0.9246	0.2722	0.9350	0.2735	0.9411	0.2743	0.9448	0.2747	0.9469	0.2749	0.9477	0.2753	0.9499
	0.6	0.3331	0.8326	0.3404	0.8601	0.3441	0.8745	0.3463	0.8829	0.3476	0.8880	0.3483	0.8909	0.3486	0.8921	0.3494	0.8951
	0.8	0.3773	0.7575	0.3870	0.7891	0.3920	0.8058	0.3949	0.8156	0.3966	0.8217	0.3976	0.8251	0.3980	0.8265	0.3991	0.8300
1.0	1.0	0.4066	0.6853	0.4180	0.7181	0.4239	0.7357	0.4273	0.7462	0.4294	0.7527	0.4306	0.7564	0.4311	0.7578	0.4323	0.7616
	1.5	0.4452	0.5350	0.4580	0.5634	0.4650	0.5794	0.4691	0.5890	0.4716	0.5951	0.4731	0.5985	0.4736	0.5999	0.4751	0.6034
	2.0	0.4620	0.4293	0.4736	0.4499	0.4806	0.4627	0.4848	0.4704	0.4873	0.4753	0.4888	0.4781	0.4894	0.4792	0.4908	0.482
	3.0	0.4768	0.3038	0.4838	0.3124	0.4899	0.3201	0.4935	0.3247	0.4957	0.3276	0.4970	0.3293	0.4975	0.3300	0.4988	0.3317
	5.0	0.4896	0.1919	0.4888	0.1912	0.4933	0.1947	0.4960	0.1968	0.4977	0.1982	0.4986	0.1989	0.4990	0.1992	0.5000	0.2000
	7.0	0.4952	0.1402	0.4910	0.1378	0.4945	0.1398	0.4967	0.1410	0.4981	0.1418	0.4989	0.1422	0.4992	0.1424	0.5000	0.1429

Table A.3: Effectiveness ε and correction factor F for arrangement no. 211.

		$N = 3$		$N = 7$		$N = 15$		$N = 27$		$N = 43$		$N = 71$		$N = 99$		$N = \infty$	
R	NTU	ε	F	ε	F	ε	F	ε	F	ε	F	ε	F	ε	F	ε	F
	0.2	0.1813	1.0000	0.1809	0.998	0.1810	0.9987	0.1811	0.9992	0.1812	0.9995	0.1812	0.9997	0.1812	0.9998	0.1813	1.0000
	0.4	0.3297	1.0000	0.3286	0.9959	0.3290	0.9973	0.3292	0.9983	0.3294	0.9989	0.3295	0.9993	0.3295	0.9995	0.3297	1.0000
	0.6	0.4512	1.0000	0.4492	0.9939	0.4498	0.9959	0.4503	0.9974	0.4506	0.9983	0.4508	0.9989	0.4509	0.9992	0.4512	1.0000
	0.8	0.5507	1.0000	0.5477	0.9919	0.5487	0.9945	0.5494	0.9965	0.5498	0.9977	0.5502	0.9986	0.5503	0.9990	0.5507	1.0000
0.0	1.0	0.6321	1.0000	0.6284	0.9898	0.6296	0.9930	0.6305	0.9956	0.6311	0.9971	0.6314	0.9982	0.6316	0.9987	0.6321	1.0000
	1.5	0.7769	1.0000	0.7717	0.9848	0.7733	0.9894	0.7746	0.9932	0.7754	0.9955	0.7759	0.9972	0.7762	0.9979	0.7769	1.0000
	2.0	0.8647	1.0000	0.8591	0.9799	0.8607	0.9856	0.8621	0.9907	0.8630	0.9938	0.8636	0.9961	0.8639	0.9971	0.8647	1.0000
	3.0	0.9502	1.0000	0.9456	0.9703	0.9468	0.9776	0.9480	0.9853	0.9487	0.9900	0.9493	0.9936	0.9495	0.9953	0.9502	1.0000
	5.0	0.9933	1.0000	0.9915	0.9528	0.9918	0.9607	0.9923	0.9727	0.9926	0.9809	0.9928	0.9876	0.9929	0.9908	0.9933	1.0000
	7.0	0.9991	1.0000	0.9986	0.9382	0.9986	0.9437	0.9988	0.9585	0.9989	0.9700	0.9990	0.9799	0.9990	0.9849	0.9991	1.0000
	0.2	0.1777	0.9969	0.1775	0.9956	0.1777	0.9968	0.1778	0.9975	0.1779	0.9980	0.1780	0.9982	0.1780	0.9984	0.1780	0.9987
	0.4	0.3183	0.9919	0.3176	0.9891	0.3182	0.9913	0.3185	0.9928	0.3187	0.9936	0.3189	0.9942	0.3190	0.9944	0.3191	0.9951
	0.6	0.4304	0.9852	0.4290	0.9807	0.4300	0.9839	0.4307	0.986	0.4311	0.9872	0.4313	0.9880	0.4314	0.9884	0.4317	0.9893
	0.8	0.5205	0.9768	0.5184	0.9707	0.5198	0.9747	0.5208	0.9775	0.5213	0.9790	0.5217	0.9801	0.5218	0.9805	0.5223	0.9818
0.2	1.0	0.5933	0.9666	0.5906	0.9593	0.5924	0.9641	0.5936	0.9674	0.5942	0.9693	0.5947	0.9706	0.5949	0.9712	0.5955	0.9728
	1.5	0.7212	0.9345	0.7182	0.9263	0.7206	0.9329	0.7222	0.9375	0.7231	0.9400	0.7238	0.9419	0.7241	0.9428	0.7249	0.9450
	2.0	0.7989	0.8936	0.7972	0.8887	0.8001	0.8971	0.8020	0.9029	0.8031	0.9062	0.8039	0.9085	0.8042	0.9096	0.8051	0.9124
	3.0	0.8772	0.7934	0.8815	0.8079	0.8854	0.8214	0.8877	0.8296	0.8889	0.8341	0.8898	0.8374	0.8902	0.8389	0.8913	0.8428
	5.0	0.9222	0.5874	0.9407	0.6542	0.9474	0.6837	0.9505	0.6986	0.9520	0.7065	0.9531	0.7122	0.9536	0.7147	0.9548	0.7214
	7.0	0.9308	0.4402	0.9584	0.5298	0.9680	0.5760	0.9721	0.6003	0.9741	0.6134	0.9754	0.6229	0.9760	0.6273	0.9775	0.6388

(Continued)

Table A.3: (*Continued*).

		$N=3$		$N=7$		$N=15$		$N=27$		$N=43$		$N=71$		$N=99$		$N=\infty$	
R	NTU	ε	F	ε	F	ε	F	ε	F	ε	F	ε	F	ε	F	ε	F
	0.2	0.1743	0.9937	0.1742	0.9932	0.1745	0.995	0.1746	0.996	0.1747	0.9965	0.1748	0.9969	0.1748	0.9970	0.1749	0.9974
	0.4	0.3075	0.9840	0.3070	0.9823	0.3078	0.9855	0.3083	0.9874	0.3085	0.9884	0.3087	0.9891	0.3088	0.9894	0.3090	0.9902
	0.6	0.4109	0.9711	0.4100	0.9679	0.4112	0.9723	0.4120	0.9749	0.4124	0.9764	0.4127	0.9774	0.4129	0.9778	0.4132	0.9790
	0.8	0.4923	0.9554	0.4908	0.9508	0.4925	0.9561	0.4936	0.9594	0.4942	0.9612	0.4946	0.9625	0.4948	0.9631	0.4953	0.9646
0.4	1.0	0.5571	0.9372	0.5552	0.9316	0.5573	0.9376	0.5586	0.9415	0.5593	0.9437	0.5599	0.9452	0.5601	0.9459	0.5607	0.9477
	1.5	0.6694	0.8835	0.6673	0.8779	0.6701	0.8854	0.6719	0.8904	0.6729	0.8931	0.6736	0.8952	0.6739	0.8961	0.6747	0.8984
	2.0	0.7372	0.8226	0.7368	0.8215	0.7400	0.8302	0.7421	0.8360	0.7433	0.8392	0.7441	0.8415	0.7445	0.8426	0.7455	0.8452
	3.0	0.8070	0.6975	0.8139	0.7154	0.8183	0.7271	0.8208	0.7341	0.8222	0.7380	0.8232	0.7407	0.8236	0.7420	0.8248	0.7452
	5.0	0.8511	0.4961	0.8767	0.5537	0.8847	0.5746	0.8885	0.5849	0.8904	0.5903	0.8918	0.5941	0.8924	0.5958	0.8938	0.6000
	7.0	0.8615	0.3700	0.9010	0.4442	0.9138	0.4753	0.9193	0.4902	0.9220	0.4978	0.9238	0.5032	0.9246	0.5056	0.9266	0.5116
	0.2	0.1709	0.9906	0.1710	0.9909	0.1713	0.9931	0.1715	0.9944	0.1716	0.995	0.1717	0.9955	0.1717	0.9957	0.1718	0.9962
	0.4	0.2971	0.9763	0.2969	0.9756	0.2979	0.9797	0.2984	0.9820	0.2987	0.9832	0.2989	0.9840	0.2990	0.9844	0.2992	0.9854
	0.6	0.3925	0.9577	0.3919	0.9556	0.3934	0.9610	0.3943	0.9642	0.3947	0.9658	0.3951	0.9671	0.3952	0.9676	0.3956	0.9690
	0.8	0.4660	0.9357	0.4649	0.932	0.4668	0.9383	0.4680	0.9422	0.4686	0.9443	0.4691	0.9458	0.4693	0.9464	0.4698	0.9481
0.6	1.0	0.5236	0.9109	0.5220	0.9059	0.5242	0.9129	0.5257	0.9173	0.5264	0.9197	0.5270	0.9214	0.5273	0.9222	0.5279	0.9242
	1.5	0.6216	0.8417	0.6196	0.8362	0.6223	0.8438	0.6241	0.8489	0.6251	0.8518	0.6259	0.8538	0.6262	0.8548	0.6270	0.8571
	2.0	0.6801	0.7693	0.6794	0.7674	0.6822	0.7749	0.6842	0.7803	0.6854	0.7833	0.6862	0.7855	0.6865	0.7865	0.6875	0.7890
	3.0	0.7408	0.6353	0.7465	0.6485	0.7495	0.6559	0.7517	0.6611	0.7529	0.664	0.7537	0.6662	0.7542	0.6672	0.7551	0.6696
	5.0	0.7812	0.4435	0.8046	0.4867	0.8092	0.4960	0.8117	0.5011	0.8131	0.5039	0.8141	0.5060	0.8145	0.5070	0.8156	0.5092
	7.0	0.7916	0.3299	0.8296	0.3861	0.8374	0.3994	0.8404	0.4048	0.842	0.4077	0.8431	0.4098	0.8437	0.4107	0.8449	0.4130

	0.2	0.1677	0.9876	0.1678	0.9885	0.1682	0.9913	0.1684	0.9928	0.1685	0.9935	0.1686	0.9941	0.1687	0.9943	0.1687	0.9949
	0.4	0.2872	0.9688	0.2873	0.9691	0.2883	0.9740	0.2889	0.9767	0.2892	0.9781	0.2895	0.9791	0.2896	0.9795	0.2898	0.9807
	0.6	0.3751	0.9449	0.3748	0.9436	0.3765	0.9499	0.3774	0.9536	0.3779	0.9556	0.3783	0.9570	0.3784	0.9576	0.3788	0.9591
	0.8	0.4415	0.9172	0.4405	0.9140	0.4426	0.9212	0.4439	0.9256	0.4445	0.9279	0.4450	0.9296	0.4452	0.9304	0.4458	0.9323
0.8	1.0	0.4925	0.8869	0.4910	0.8820	0.4933	0.8896	0.4948	0.8944	0.4956	0.8970	0.4962	0.8989	0.4964	0.8997	0.4971	0.9018
	1.5	0.5777	0.8061	0.5753	0.7992	0.5777	0.8062	0.5794	0.8113	0.5804	0.8142	0.5811	0.8163	0.5814	0.8172	0.5822	0.8195
	2.0	0.6277	0.7264	0.6258	0.7214	0.5279	0.7269	0.6296	0.7315	0.6306	0.7342	0.6313	0.7362	0.6316	0.7372	0.6324	0.7394
	3.0	0.6793	0.5885	0.6820	0.5948	0.6828	0.5968	0.6842	0.5999	0.6850	0.6019	0.6857	0.6035	0.6860	0.6042	0.6867	0.6058
	5.0	0.7139	0.4049	0.7310	0.4340	0.7302	0.4326	0.7304	0.4329	0.7307	0.4336	0.7311	0.4342	0.7313	0.4345	0.7316	0.4351
	7.0	0.7229	0.3000	0.7522	0.3389	0.7515	0.3379	0.7505	0.3364	0.7502	0.3361	0.7502	0.336	0.7502	0.3361	0.7501	0.3358
	0.2	0.1645	0.9846	0.1647	0.9862	0.1652	0.9895	0.1654	0.9912	0.1656	0.9921	0.1656	0.9927	0.1657	0.9929	0.1658	0.9936
	0.4	0.2778	0.9615	0.2780	0.9626	0.2792	0.9683	0.2798	0.9714	0.2802	0.9730	0.2804	0.9742	0.2805	0.9747	0.2808	0.9759
	0.6	0.3588	0.9327	0.3586	0.9319	0.3604	0.9391	0.3614	0.9433	0.3620	0.9455	0.3623	0.9470	0.3625	0.9477	0.3629	0.9495
	0.8	0.4186	0.9000	0.4177	0.8968	0.4199	0.9047	0.4212	0.9095	0.4219	0.9121	0.4224	0.9140	0.4226	0.9148	0.4231	0.9168
1.0	1.0	0.4638	0.8650	0.4622	0.8594	0.4645	0.8673	0.4659	0.8724	0.4667	0.8752	0.4673	0.8773	0.4676	0.8782	0.4682	0.8804
	1.5	0.5376	0.7750	0.5345	0.7656	0.5365	0.7716	0.5380	0.7765	0.5389	0.7793	0.5396	0.7813	0.5399	0.7823	0.5406	0.7845
	2.0	0.5799	0.6903	0.5767	0.6811	0.5776	0.6838	0.5790	0.6876	0.5798	0.6899	0.5804	0.6917	0.5807	0.6925	0.5814	0.6944
	3.0	0.6229	0.5506	0.6224	0.5495	0.6209	0.5459	0.6213	0.5469	0.6218	0.5480	0.6222	0.5489	0.6224	0.5493	0.6227	0.5502
	5.0	0.6510	0.3731	0.6611	0.3901	0.6555	0.3806	0.6536	0.3774	0.6531	0.3765	0.6528	0.3761	0.6527	0.3759	0.6523	0.3752
	7.0	0.6580	0.2748	0.6769	0.2993	0.6696	0.2896	0.6656	0.2844	0.6641	0.2824	0.6631	0.2812	0.6628	0.2808	0.6616	0.2792

Table A.4: Effectiveness ε and correction factor F for arrangement no. 221.

		$N=3$		$N=7$		$N=15$		$N=27$		$N=43$		$N=71$		$N=99$		$N=\infty$	
R	NTU	ε	F	ε	F	ε	F	ε	F	ε	F	ε	F	ε	F	ε	F
	0.2	0.1813	1.0000	0.1809	0.9980	0.1810	0.9986	0.1811	0.9992	0.1812	0.9994	0.1812	0.9997	0.1812	0.9997	0.1813	1.0000
	0.4	0.3297	1.0000	0.3286	0.9959	0.3289	0.9973	0.3292	0.9983	0.3294	0.9989	0.3295	0.9993	0.3295	0.9995	0.3297	1.0000
	0.6	0.4512	1.0000	0.4492	0.9939	0.4498	0.9959	0.4503	0.9974	0.4506	0.9983	0.4508	0.9989	0.4509	0.9992	0.4512	1.0000
	0.8	0.5507	1.0000	0.5477	0.9918	0.5487	0.9945	0.5494	0.9965	0.5498	0.9977	0.5501	0.9985	0.5503	0.9989	0.5507	1.0000
0.0	1.0	0.6321	1.0000	0.6283	0.9898	0.6295	0.993	0.6305	0.9956	0.6310	0.9971	0.6314	0.9982	0.6316	0.9986	0.6321	1.0000
	1.5	0.7769	1.0000	0.7717	0.9848	0.7733	0.9893	0.7746	0.9932	0.7753	0.9955	0.7759	0.9971	0.7762	0.9979	0.7769	1.0000
	2.0	0.8647	1.0000	0.8591	0.9798	0.8607	0.9855	0.8621	0.9907	0.8630	0.9937	0.8636	0.9960	0.8639	0.9971	0.8647	1.0000
	3.0	0.9502	1.0000	0.9456	0.9703	0.9467	0.9775	0.9480	0.9852	0.9487	0.9900	0.9492	0.9936	0.9495	0.9953	0.9502	1.0000
	5.0	0.9933	1.0000	0.9915	0.9528	0.9918	0.9608	0.9923	0.9727	0.9926	0.9809	0.9928	0.9876	0.9929	0.9908	0.9933	1.0000
	7.0	0.9991	1.0000	0.9986	0.9382	0.9986	0.9437	0.9988	0.9585	0.9989	0.9699	0.999	0.9799	0.9990	0.9848	0.9991	1.0000
	0.2	0.1781	0.9991	0.1778	0.9972	0.1780	0.9982	0.1781	0.9989	0.1781	0.9993	0.1782	0.9996	0.1782	0.9997	0.1782	1.0000
	0.4	0.3195	0.9964	0.3188	0.9937	0.3194	0.9961	0.3198	0.9976	0.3200	0.9984	0.3201	0.9990	0.3202	0.9993	0.3204	1.0000
	0.6	0.4326	0.9920	0.4318	0.9896	0.4331	0.9937	0.4339	0.9962	0.4343	0.9975	0.4346	0.9984	0.4347	0.9989	0.4351	1.0000
	0.8	0.5236	0.9857	0.5233	0.9849	0.5254	0.9910	0.5266	0.9945	0.5272	0.9964	0.5277	0.9978	0.5279	0.9984	0.5284	1.0000
0.2	1.0	0.5972	0.9777	0.5979	0.9796	0.6008	0.9880	0.6025	0.9927	0.6034	0.9953	0.6040	0.9970	0.6043	0.9979	0.6050	1.0000
	1.5	0.7266	0.9497	0.7315	0.9638	0.7367	0.9792	0.7395	0.9875	0.7409	0.9918	0.7419	0.9949	0.7424	0.9963	0.7436	1.0000
	2.0	0.8047	0.9110	0.8153	0.9445	0.8226	0.9684	0.8263	0.9810	0.8282	0.9876	0.8295	0.9923	0.8301	0.9944	0.8317	1.0000
	3.0	0.8817	0.8086	0.9044	0.8952	0.9145	0.9407	0.9193	0.9641	0.9217	0.9765	0.9234	0.9854	0.9241	0.9894	0.9261	1.0000
	5.0	0.9222	0.5875	0.9618	0.7626	0.9737	0.8552	0.9787	0.9073	0.9811	0.9372	0.9827	0.9598	0.9834	0.9704	0.9853	1.0000
	7.0	0.9282	0.4337	0.9753	0.6218	0.9870	0.7364	0.9914	0.8105	0.9935	0.8601	0.9949	0.9031	0.9955	0.9256	0.9970	1.0000

	0.2	0.1750	0.9982	0.1747	0.9965	0.1749	0.9978	0.1750	0.9987	0.1751	0.9991	0.1752	0.9995	0.1752	0.9996	0.1753	1.0000
	0.4	0.3096	0.9929	0.3093	0.9916	0.3101	0.995	0.3106	0.9970	0.3109	0.9980	0.3110	0.9988	0.3111	0.9991	0.3113	1.0000
	0.6	0.4147	0.9842	0.4151	0.9855	0.4169	0.9916	0.4179	0.9950	0.4184	0.9967	0.4188	0.9980	0.4189	0.9985	0.4194	1.0000
	0.8	0.4978	0.9722	0.4997	0.9783	0.5027	0.9876	0.5043	0.9926	0.5051	0.9952	0.5057	0.9970	0.5059	0.9978	0.5066	1.0000
0.4	1.0	0.5639	0.9571	0.5682	0.9700	0.5726	0.9832	0.5748	0.9900	0.5760	0.9935	0.5768	0.9960	0.5771	0.9971	0.5781	1.0000
	1.5	0.6781	0.9073	0.6909	0.9451	0.6991	0.9699	0.7030	0.9822	0.7050	0.9885	0.7064	0.9929	0.7071	0.9948	0.7087	1.0000
	2.0	0.7459	0.8463	0.7691	0.9152	0.7811	0.9539	0.7867	0.9727	0.7895	0.9823	0.7914	0.9891	0.7923	0.9921	0.7945	1.0000
	3.0	0.8120	0.7103	0.8562	0.8446	0.8747	0.9146	0.8828	0.9489	0.8867	0.9667	0.8895	0.9793	0.8907	0.9850	0.8938	1.0000
	5.0	0.8467	0.4873	0.9200	0.6891	0.9458	0.8133	0.9561	0.8813	0.9610	0.9199	0.9643	0.9489	0.9658	0.9624	0.9695	1.0000
	7.0	0.8519	0.3555	0.9389	0.5533	0.9675	0.6992	0.9779	0.7896	0.9828	0.8479	0.9860	0.8969	0.9874	0.9218	0.9909	1.0000
	0.2	0.1719	0.9973	0.1717	0.9957	0.1719	0.9974	0.1721	0.9984	0.1722	0.9990	0.1722	0.9994	0.1723	0.9995	0.1723	1.0000
	0.4	0.3001	0.9894	0.3001	0.9895	0.3011	0.9939	0.3017	0.9963	0.3020	0.9976	0.3022	0.9985	0.3023	0.9989	0.3025	1.0000
	0.6	0.3977	0.9766	0.3990	0.9815	0.4012	0.9895	0.4024	0.9937	0.4030	0.9959	0.4034	0.9975	0.4036	0.9982	0.4041	1.0000
	0.8	0.4732	0.9592	0.4770	0.9719	0.4807	0.9844	0.4826	0.9907	0.4835	0.9940	0.4842	0.9963	0.4845	0.9973	0.4853	1.0000
0.6	1.0	0.5323	0.9378	0.5395	0.9608	0.5450	0.9785	0.5477	0.9873	0.5490	0.9918	0.5500	0.9949	0.5504	0.9963	0.5515	1.0000
	1.5	0.6319	0.8712	0.6507	0.9281	0.6611	0.9613	0.6659	0.9772	0.6684	0.9853	0.6700	0.9909	0.6708	0.9934	0.6727	1.0000
	2.0	0.6894	0.7941	0.7218	0.8898	0.7373	0.9409	0.7443	0.9652	0.7478	0.9775	0.7502	0.9861	0.7512	0.9900	0.7539	1.0000
	3.0	0.7437	0.6420	0.8027	0.8051	0.8277	0.8935	0.8385	0.9365	0.8437	0.9587	0.8473	0.9743	0.8489	0.9814	0.8529	1.0000
	5.0	0.7708	0.4262	0.8665	0.6400	0.9052	0.7862	0.9207	0.8656	0.9282	0.9099	0.9332	0.9428	0.9354	0.9581	0.9411	1.0000
	7.0	0.7746	0.3089	0.8878	0.5095	0.9346	0.6801	0.9523	0.7841	0.9606	0.8483	0.9662	0.8999	0.9686	0.9251	0.9748	1.0000

(Continued)

Table A.4: (*Continued*).

		$N=3$		$N=7$		$N=15$		$N=27$		$N=43$		$N=71$		$N=99$		$N=\infty$	
R	NTU	ε	F	ε	F	ε	F	ε	F	ε	F	ε	F	ε	F	ε	F
	0.2	0.1690	0.9965	0.1687	0.9950	0.1690	0.9970	0.1692	0.9982	0.1693	0.9988	0.1694	0.9993	0.1694	0.9995	0.1695	1.0000
	0.4	0.2910	0.986	0.2913	0.9874	0.2924	0.9927	0.2931	0.9957	0.2934	0.9972	0.2936	0.9983	0.2937	0.9987	0.2940	1.0000
	0.6	0.3814	0.9691	0.3836	0.9775	0.3861	0.9874	0.3874	0.9925	0.3881	0.9952	0.3886	0.9970	0.3888	0.9978	0.3893	1.0000
	0.8	0.4499	0.9467	0.4552	0.9656	0.4594	0.9812	0.4615	0.9889	0.4626	0.9928	0.4633	0.9956	0.4637	0.9968	0.4645	1.0000
0.8	1.0	0.5024	0.9195	0.5119	0.9520	0.5182	0.9740	0.5212	0.9847	0.5227	0.9902	0.5237	0.9939	0.5242	0.9956	0.5254	1.0000
	1.5	0.5884	0.8381	0.6114	0.9121	0.6234	0.9531	0.6288	0.9725	0.6315	0.9823	0.6333	0.9891	0.6341	0.9921	0.6363	1.0000
	2.0	0.6361	0.7495	0.6745	0.8667	0.6923	0.9288	0.7002	0.9581	0.7040	0.9730	0.7067	0.9833	0.7079	0.9879	0.7109	1.0000
	3.0	0.6790	0.5880	0.7462	0.7709	0.7753	0.8746	0.7877	0.9251	0.7937	0.9513	0.7978	0.9698	0.7997	0.9781	0.8043	1.0000
	5.0	0.6986	0.3808	0.8039	0.5989	0.8510	0.7620	0.8703	0.8511	0.8796	0.9006	0.8859	0.9371	0.8886	0.9540	0.8957	1.0000
	7.0	0.7009	0.2746	0.8240	0.4719	0.8836	0.6597	0.9075	0.7757	0.9189	0.8454	0.9266	0.8997	0.9299	0.9257	0.9386	1.0000
	0.2	0.1661	0.9956	0.1659	0.9942	0.1662	0.9965	0.1664	0.9979	0.1665	0.9986	0.1665	0.9992	0.1666	0.9994	0.1667	1.0000
	0.4	0.2821	0.9826	0.2827	0.9853	0.2840	0.9916	0.2847	0.9950	0.2851	0.9968	0.2853	0.9980	0.2854	0.9985	0.2857	1.0000
	0.6	0.3659	0.9618	0.3688	0.9736	0.3716	0.9854	0.3730	0.9913	0.3737	0.9944	0.3742	0.9966	0.3744	0.9975	0.3750	1.0000
	0.8	0.4278	0.9345	0.4343	0.9595	0.4390	0.9780	0.4412	0.9870	0.4424	0.9916	0.4432	0.9949	0.4435	0.9963	0.4444	1.0000
1.0	1.0	0.4742	0.9019	0.4854	0.9434	0.4923	0.9696	0.4955	0.9822	0.4971	0.9885	0.4982	0.9929	0.4987	0.9949	0.5000	1.0000
	1.5	0.5478	0.8075	0.5736	0.8968	0.5864	0.9451	0.5921	0.9678	0.5950	0.9793	0.5969	0.9872	0.5978	0.9908	0.6000	1.0000
	2.0	0.5866	0.7095	0.6283	0.8450	0.6472	0.9171	0.6555	0.9512	0.6595	0.9685	0.6623	0.9805	0.6635	0.9859	0.6667	1.0000
	3.0	0.6193	0.5423	0.6893	0.7394	0.7198	0.8562	0.7327	0.9138	0.7390	0.9438	0.7433	0.9651	0.7452	0.9747	0.7500	1.0000
	5.0	0.6325	0.3443	0.7371	0.5607	0.7862	0.7357	0.8065	0.8338	0.8163	0.8888	0.8229	0.9296	0.8259	0.9485	0.8333	1.0000
	7.0	0.6339	0.2473	0.7531	0.4357	0.8155	0.6315	0.8412	0.7568	0.8536	0.8326	0.8619	0.8917	0.8656	0.9199	0.8750	1.0000

Table A.5: Effectiveness ε and correction factor F for arrangement no. 311.

		$N = 6$		$N = 12$		$N = 24$		$N = 36$		$N = 48$		$N = 72$		$N = 96$		$N = \infty$	
R	NTU	ε	F	ε	F	ε	F	ε	F	ε	F	ε	F	ε	F	ε	F
	0.2	0.1813	1.0000	0.1813	1.0000	0.1813	1.0000	0.1813	1.0000	0.1813	1.0000	0.1813	1.0000	0.1813	1.0000	0.1813	1.0000
	0.4	0.3297	1.0000	0.3297	1.0000	0.3297	1.0000	0.3297	1.0000	0.3297	1.0000	0.3297	1.0000	0.3297	1.0000	0.3297	1.0000
	0.6	0.4512	1.0000	0.4512	1.0000	0.4512	1.0000	0.4512	1.0000	0.4512	1.0000	0.4512	1.0000	0.4512	1.0000	0.4512	1.0000
	0.8	0.5507	1.0000	0.5507	1.0000	0.5507	1.0000	0.5507	1.0000	0.5507	1.0000	0.5507	1.0000	0.5507	1.0000	0.5507	1.0000
0.0	1.0	0.6321	1.0000	0.6321	1.0000	0.6321	1.0000	0.6321	1.0000	0.6321	1.0000	0.6321	1.0000	0.6321	1.0000	0.6321	1.0000
	1.5	0.7769	1.0000	0.7769	1.0000	0.7769	1.0000	0.7769	1.0000	0.7769	1.0000	0.7769	1.0000	0.7769	1.0000	0.7769	1.0000
	2.0	0.8647	1.0000	0.8647	1.0000	0.8647	1.0000	0.8647	1.0000	0.8647	1.0000	0.8647	1.0000	0.8647	1.0000	0.8647	1.0000
	3.0	0.9502	1.0000	0.9502	1.0000	0.9502	1.0000	0.9502	1.0000	0.9502	1.0000	0.9502	1.0000	0.9502	1.0000	0.9502	1.0000
	5.0	0.9933	1.0000	0.9933	1.0000	0.9933	1.0000	0.9933	1.0000	0.9933	1.0000	0.9933	1.0000	0.9933	1.0000	0.9933	1.0000
	7.0	0.9991	1.0000	0.9991	1.0000	0.9991	1.0000	0.9991	1.0000	0.9991	1.0000	0.9991	1.0000	0.9991	1.0000	0.9991	1.0000
	0.2	0.1777	0.9966	0.1778	0.9975	0.1779	0.9981	0.1780	0.9984	0.1780	0.9985	0.1780	0.9986	0.178	0.9987	0.1781	0.9989
	0.4	0.3181	0.9910	0.3186	0.9928	0.3189	0.9941	0.3190	0.9946	0.3191	0.9949	0.3192	0.9952	0.3192	0.9953	0.3193	0.9957
	0.6	0.4299	0.9835	0.4308	0.9864	0.4315	0.9884	0.4317	0.9892	0.4318	0.9896	0.4320	0.9900	0.432	0.9902	0.4322	0.9908
	0.8	0.5197	0.9743	0.5211	0.9784	0.5221	0.9812	0.5224	0.9823	0.5226	0.9828	0.5228	0.9834	0.5229	0.9837	0.5232	0.9845
0.2	1.0	0.5922	0.9637	0.5942	0.9691	0.5955	0.9728	0.5960	0.9742	0.5962	0.9749	0.5965	0.9756	0.5966	0.9759	0.5970	0.9771
	1.5	0.7202	0.9319	0.7237	0.9415	0.7258	0.9476	0.7266	0.9498	0.7270	0.9510	0.7274	0.9522	0.7276	0.9527	0.7283	0.9546
	2.0	0.7994	0.8950	0.8042	0.9096	0.8072	0.9186	0.8082	0.9219	0.8087	0.9235	0.8093	0.9252	0.8095	0.9260	0.8104	0.9286
	3.0	0.8835	0.8146	0.8907	0.8405	0.8948	0.8562	0.8962	0.8617	0.8970	0.8646	0.8977	0.8675	0.8981	0.8689	0.8992	0.8734
	5.0	0.9439	0.6680	0.9527	0.7100	0.9578	0.7384	0.9596	0.7490	0.9605	0.7544	0.9614	0.7601	0.9618	0.7629	0.9631	0.7717
	7.0	0.9660	0.5652	0.9736	0.6102	0.9784	0.6456	0.9801	0.6601	0.9809	0.6679	0.9818	0.6760	0.9822	0.6802	0.9835	0.6935

(Continued)

Table A.5: (*Continued*).

		$N=6$		$N=12$		$N=24$		$N=36$		$N=48$		$N=72$		$N=96$		$N=\infty$	
R	NTU	ε	F	ε	F	ε	F	ε	F	ε	F	ε	F	ε	F	ε	F
	0.2	0.1742	0.9932	0.1745	0.9949	0.1747	0.9962	0.1747	0.9967	0.1748	0.9969	0.1748	0.9972	0.1748	0.9973	0.1749	0.9977
	0.4	0.3070	0.9822	0.3079	0.9857	0.3085	0.9883	0.3087	0.9893	0.3089	0.9898	0.3090	0.9903	0.3091	0.9906	0.3093	0.9915
	0.6	0.4099	0.9677	0.4115	0.9731	0.4127	0.9771	0.4131	0.9786	0.4133	0.9794	0.4136	0.9802	0.4137	0.9806	0.4141	0.9819
	0.8	0.4907	0.9505	0.4931	0.9579	0.4949	0.9632	0.4955	0.9652	0.4958	0.9663	0.4962	0.9674	0.4964	0.9679	0.4969	0.9696
0.4	1.0	0.5550	0.9311	0.5582	0.9405	0.5605	0.9472	0.5614	0.9497	0.5619	0.9511	0.5623	0.9524	0.5626	0.9531	0.5633	0.9553
	1.5	0.6669	0.8767	0.6722	0.8912	0.6758	0.9014	0.6772	0.9052	0.6779	0.9072	0.6786	0.9093	0.6790	0.9103	0.6801	0.9136
	2.0	0.7360	0.8192	0.7431	0.8385	0.7479	0.8520	0.7496	0.8570	0.7506	0.8596	0.7515	0.8623	0.7520	0.8637	0.7535	0.8680
	3.0	0.8123	0.7110	0.8219	0.7372	0.8285	0.7557	0.8309	0.7628	0.8321	0.7666	0.8334	0.7704	0.8341	0.7723	0.8361	0.7784
	5.0	0.8760	0.5520	0.8870	0.5807	0.8948	0.6031	0.8979	0.6124	0.8995	0.6173	0.9012	0.6226	0.9021	0.6253	0.9047	0.6337
	7.0	0.9061	0.4560	0.9153	0.4792	0.9226	0.4996	0.9258	0.5091	0.9275	0.5145	0.9293	0.5203	0.9303	0.5233	0.9332	0.5330
	0.2	0.1708	0.9898	0.1712	0.9924	0.1715	0.9943	0.1716	0.9950	0.1717	0.9954	0.1717	0.9958	0.1717	0.9960	0.1718	0.9966
	0.4	0.2965	0.9736	0.2977	0.9787	0.2985	0.9826	0.2989	0.9840	0.2991	0.9848	0.2992	0.9856	0.2993	0.9860	0.2996	0.9873
	0.6	0.3911	0.9527	0.3932	0.9603	0.3948	0.9660	0.3954	0.9682	0.3957	0.9694	0.3960	0.9706	0.3962	0.9712	0.3967	0.9731
	0.8	0.4637	0.9283	0.4668	0.9383	0.4691	0.9458	0.4700	0.9487	0.4705	0.9502	0.4709	0.9518	0.4712	0.9526	0.4719	0.9551
0.6	1.0	0.5205	0.9015	0.5245	0.9137	0.5275	0.9230	0.5286	0.9265	0.5293	0.9284	0.5299	0.9303	0.5302	0.9313	0.5312	0.9344
	1.5	0.6174	0.8301	0.6234	0.8468	0.6279	0.8596	0.6296	0.8646	0.6305	0.8673	0.6315	0.8700	0.6320	0.8714	0.6334	0.8758
	2.0	0.6764	0.7596	0.6838	0.7792	0.6895	0.7945	0.6917	0.8006	0.6928	0.8038	0.6940	0.8071	0.6946	0.8088	0.6965	0.8142
	3.0	0.7419	0.6379	0.7508	0.6590	0.7578	0.6763	0.7606	0.6834	0.7621	0.6872	0.7636	0.6912	0.7644	0.6933	0.7669	0.6998
	5.0	0.7994	0.4765	0.8075	0.4926	0.8142	0.5064	0.8173	0.5129	0.8190	0.5166	0.8208	0.5206	0.8218	0.5227	0.8248	0.5294
	7.0	0.8287	0.3846	0.8335	0.3926	0.8379	0.4003	0.8405	0.4049	0.8420	0.4078	0.8438	0.4110	0.8448	0.4128	0.8479	0.4187

	0.2	0.1675	0.9865	0.1680	0.9898	0.1684	0.9924	0.1685	0.9933	0.1686	0.9939	0.1687	0.9944	0.1687	0.9947	0.1688	0.9955
	0.4	0.2864	0.9652	0.2879	0.9718	0.2890	0.9768	0.2894	0.9788	0.2896	0.9798	0.2898	0.9809	0.2900	0.9814	0.2903	0.9831
	0.6	0.3734	0.9382	0.3759	0.9478	0.3778	0.9551	0.3786	0.9580	0.3789	0.9595	0.3794	0.9611	0.3796	0.9619	0.3802	0.9644
	0.8	0.4386	0.9074	0.4421	0.9194	0.4448	0.9289	0.4459	0.9327	0.4464	0.9346	0.4470	0.9367	0.4473	0.9377	0.4482	0.9410
0.8	1.0	0.4886	0.8744	0.4930	0.8885	0.4964	0.8998	0.4978	0.9042	0.4985	0.9066	0.4992	0.9090	0.4996	0.9103	0.5008	0.9142
	1.5	0.5720	0.7897	0.5779	0.8068	0.5828	0.8212	0.5847	0.8270	0.5857	0.8301	0.5868	0.8333	0.5873	0.8350	0.5891	0.8402
	2.0	0.6216	0.7103	0.6283	0.7280	0.6340	0.7436	0.6363	0.7501	0.6375	0.7536	0.6388	0.7572	0.6395	0.7591	0.6416	0.7651
	3.0	0.6758	0.5809	0.6825	0.5961	0.6885	0.6102	0.6912	0.6166	0.6926	0.6201	0.6942	0.6238	0.6950	0.6258	0.6975	0.6320
	5.0	0.7228	0.4193	0.7269	0.4267	0.7306	0.4333	0.7327	0.4371	0.7340	0.4395	0.7354	0.4421	0.7362	0.4435	0.7388	0.4483
	7.0	0.7470	0.3315	0.7464	0.3306	0.7464	0.3307	0.7475	0.3321	0.7483	0.3333	0.7493	0.3347	0.7499	0.3356	0.7519	0.3385
	0.2	0.1643	0.9832	0.1649	0.9873	0.1653	0.9905	0.1655	0.9917	0.1656	0.9924	0.1657	0.993	0.1657	0.9933	0.1659	0.9944
	0.4	0.2768	0.9570	0.2785	0.9650	0.2798	0.9712	0.2803	0.9736	0.2805	0.9748	0.2808	0.9762	0.2810	0.9768	0.2814	0.9789
	0.6	0.3568	0.9244	0.3595	0.9355	0.3617	0.9444	0.3626	0.9479	0.3630	0.9498	0.3635	0.9517	0.3637	0.9527	0.3645	0.9558
	0.8	0.4153	0.8878	0.4190	0.9013	0.4220	0.9125	0.4232	0.9170	0.4238	0.9193	0.4244	0.9218	0.4248	0.9231	0.4258	0.9270
1.0	1.0	0.4593	0.8493	0.4637	0.8645	0.4674	0.8774	0.4688	0.8827	0.4696	0.8855	0.4704	0.8884	0.4709	0.8899	0.4722	0.8946
	1.5	0.5307	0.7538	0.5360	0.7702	0.5409	0.7853	0.5429	0.7917	0.5439	0.7951	0.5450	0.7987	0.5456	0.8006	0.5474	0.8064
	2.0	0.5719	0.6680	0.5773	0.6828	0.5825	0.6977	0.5848	0.7042	0.5860	0.7077	0.5873	0.7115	0.5880	0.7135	0.5901	0.7197
	3.0	0.6156	0.5339	0.6199	0.5436	0.6244	0.5542	0.6266	0.5594	0.6279	0.5624	0.6292	0.5656	0.6299	0.5674	0.6322	0.5729
	5.0	0.6519	0.3745	0.6525	0.3756	0.6535	0.3772	0.6546	0.3790	0.6553	0.3803	0.6562	0.3818	0.6567	0.3826	0.6585	0.3856
	7.0	0.6701	0.2902	0.6658	0.2846	0.6631	0.2812	0.6630	0.2810	0.6631	0.2812	0.6634	0.2816	0.6637	0.2819	0.6646	0.2830

Table A.6: Effectiveness ε and correction factor F for arrangement no. 321.

		$N = 11$		$N = 23$		$N = 35$		$N = 47$		$N = 59$		$N = 71$		$N = 95$		$N = \infty$	
R	NTU	ε	F	ε	F	ε	F	ε	F	ε	F	ε	F	ε	F	ε	F
	0.2	0.1811	0.9992	0.1812	0.9994	0.1812	0.9996	0.1812	0.9997	0.1812	0.9997	0.1812	0.9998	0.1812	0.9998	0.1816	1.0000
	0.4	0.3292	0.9984	0.3294	0.9989	0.3295	0.9992	0.3295	0.9994	0.3295	0.9995	0.3296	0.9996	0.3296	0.9997	0.3299	1.0000
	0.6	0.4504	0.9975	0.4506	0.9983	0.4508	0.9987	0.4509	0.9990	0.4509	0.9992	0.4510	0.9993	0.4510	0.9995	0.4512	1.0000
	0.8	0.5495	0.9967	0.5498	0.9977	0.5501	0.9983	0.5502	0.9987	0.5503	0.9989	0.5503	0.9991	0.5504	0.9993	0.5506	1.0000
0.0	1.0	0.6306	0.9959	0.6310	0.9971	0.6313	0.9979	0.6315	0.9983	0.6316	0.9986	0.6317	0.9988	0.6318	0.9991	0.6320	1.0000
	1.5	0.7748	0.9938	0.7754	0.9956	0.7758	0.9968	0.7760	0.9975	0.7762	0.9979	0.7763	0.9983	0.7764	0.9987	0.7766	1.0000
	2.0	0.8624	0.9918	0.8630	0.9940	0.8635	0.9956	0.8637	0.9966	0.8639	0.9972	0.8640	0.9976	0.8642	0.9982	0.8645	1.0000
	3.0	0.9484	0.9878	0.9488	0.9908	0.9492	0.9932	0.9494	0.9946	0.9495	0.9956	0.9496	0.9962	0.9498	0.9971	0.9500	1.0000
	5.0	0.9926	0.9800	0.9927	0.9840	0.9928	0.9878	0.9929	0.9903	0.9930	0.9919	0.9930	0.9931	0.9931	0.9947	0.9931	1.0000
	7.0	0.9989	0.9732	0.9989	0.9772	0.9990	0.9823	0.9990	0.9857	0.9990	0.9881	0.9990	0.9899	0.9990	0.9922	0.9991	1.0000
	0.2	0.1780	0.9985	0.1780	0.9988	0.1781	0.9991	0.1781	0.9992	0.1781	0.9992	0.1781	0.9993	0.1781	0.9994	0.1782	0.9996
	0.4	0.3194	0.9963	0.3196	0.9969	0.3197	0.9973	0.3197	0.9975	0.3198	0.9977	0.3198	0.9978	0.3198	0.9979	0.3200	0.9983
	0.6	0.4330	0.9934	0.4333	0.9942	0.4334	0.9948	0.4335	0.9951	0.4336	0.9953	0.4337	0.9955	0.4337	0.9957	0.4339	0.9963
	0.8	0.5250	0.9899	0.5253	0.9909	0.5256	0.9916	0.5257	0.9920	0.5258	0.9923	0.5259	0.9925	0.5260	0.9927	0.5263	0.9936
0.2	1.0	0.600[illegible]	0.9858	0.6005	0.9869	0.6008	0.9877	0.6009	0.9883	0.6011	0.9886	0.6012	0.9889	0.6013	0.9892	0.6016	0.9902
	1.5	0.7347	0.9732	0.7351	0.9745	0.7355	0.9757	0.7358	0.9764	0.7359	0.9769	0.7360	0.9773	0.7362	0.9777	0.7367	0.9792
	2.0	0.8193	0.9574	0.8198	0.9590	0.8202	0.9606	0.8205	0.9616	0.8207	0.9622	0.8209	0.9627	0.8210	0.9633	0.8216	0.9652
	3.0	0.9097	0.9182	0.9104	0.9214	0.9109	0.9238	0.9112	0.9252	0.9114	0.9262	0.9116	0.9268	0.9118	0.9277	0.9124	0.9306
	5.0	0.9694	0.8176	0.9709	0.8306	0.9715	0.8358	0.9719	0.8387	0.9721	0.8405	0.9722	0.8418	0.9724	0.8434	0.9730	0.8487
	7.0	0.9845	0.7061	0.9870	0.7356	0.9877	0.7454	0.9880	0.7505	0.9882	0.7537	0.9884	0.7559	0.9885	0.7587	0.9891	0.7674

	0.2	0.1749	0.9978	0.1750	0.9983	0.1750	0.9985	0.1750	0.9987	0.1751	0.9987	0.1751	0.9988	0.1751	0.9989	0.1751	0.9991
	0.4	0.3099	0.9942	0.3101	0.9949	0.3102	0.9954	0.3103	0.9957	0.3103	0.9959	0.3104	0.9960	0.3104	0.9962	0.3105	0.9966
	0.6	0.4162	0.9893	0.4165	0.9902	0.4167	0.9909	0.4168	0.9913	0.4169	0.9915	0.4169	0.9917	0.4170	0.9919	0.4172	0.9927
	0.8	0.5013	0.9832	0.5016	0.9842	0.5019	0.9851	0.5020	0.9856	0.5021	0.9859	0.5022	0.9861	0.5023	0.9864	0.5026	0.9873
0.4	1.0	0.5703	0.9761	0.5706	0.9771	0.5709	0.9781	0.5711	0.9787	0.5712	0.9790	0.5713	0.9793	0.5714	0.9797	0.5718	0.9808
	1.5	0.6939	0.9541	0.6942	0.9551	0.6946	0.9564	0.6949	0.9572	0.6951	0.9577	0.6952	0.9581	0.6954	0.9586	0.6959	0.9602
	2.0	0.7730	0.9275	0.7734	0.9286	0.7739	0.9302	0.7742	0.9311	0.7744	0.9318	0.7745	0.9322	0.7747	0.9329	0.7753	0.9349
	3.0	0.8620	0.8655	0.8628	0.8684	0.8634	0.8706	0.8638	0.8719	0.8640	0.8728	0.8642	0.8734	0.8644	0.8742	0.8651	0.8770
	5.0	0.9303	0.7327	0.9331	0.7460	0.9340	0.7502	0.9345	0.7525	0.9348	0.7540	0.9350	0.7550	0.9353	0.7564	0.9361	0.7606
	7.0	0.9530	0.6139	0.9584	0.6421	0.9598	0.6496	0.9604	0.6535	0.9608	0.6559	0.9611	0.6576	0.9615	0.6596	0.9625	0.6661
	0.2	0.1719	0.997[illegible]	0.1720	0.9977	0.1720	0.9980	0.1721	0.9982	0.1721	0.9983	0.1721	0.9983	0.1721	0.9984	0.1721	0.9987
	0.4	0.3007	0.9922	0.3009	0.9930	0.3011	0.9936	0.3011	0.9939	0.3012	0.9941	0.3012	0.9943	0.3013	0.9944	0.3014	0.9950
	0.6	0.4001	0.9853	0.4004	0.9863	0.4006	0.9871	0.4007	0.9875	0.4008	0.9878	0.4008	0.9880	0.4009	0.9883	0.4011	0.9891
	0.8	0.4784	0.9768	0.4787	0.9778	0.4790	0.9787	0.4792	0.9793	0.4793	0.9797	0.4794	0.9799	0.4795	0.9803	0.4798	0.9813
0.6	1.0	0.5413	0.9667	0.5416	0.9677	0.5420	0.9688	0.5422	0.9694	0.5423	0.9699	0.5424	0.9702	0.5425	0.9706	0.5429	0.9719
	1.5	0.6533	0.9363	0.6536	0.9371	0.6540	0.9385	0.6543	0.9393	0.6544	0.9399	0.6546	0.9403	0.6547	0.9408	0.6553	0.9425
	2.0	0.7251	0.9005	0.7254	0.9015	0.7259	0.9030	0.7262	0.9040	0.7264	0.9046	0.7266	0.9051	0.7268	0.9057	0.7274	0.9078
	3.0	0.8078	0.8219	0.8087	0.8248	0.8092	0.8266	0.8096	0.8279	0.8098	0.8287	0.8100	0.8293	0.8102	0.8300	0.8110	0.8326
	5.0	0.8764	0.6721	0.8800	0.6848	0.8809	0.6881	0.8814	0.6899	0.8818	0.6911	0.8820	0.6919	0.8823	0.6929	0.8833	0.6964
	7.0	0.9026	0.5532	0.9102	0.5788	0.9118	0.5842	0.9125	0.5870	0.9130	0.5887	0.9133	0.5899	0.9137	0.5914	0.9150	0.5961

(Continued)

Table A.6: (*Continued*).

		$N = 11$		$N = 23$		$N = 35$		$N = 47$		$N = 59$		$N = 71$		$N = 95$		$N = \infty$	
R	NTU	ε	F	ε	F	ε	F	ε	F	ε	F	ε	F	ε	F	ε	F
	0.2	0.1690	0.9964	0.1690	0.9970	0.1691	0.9974	0.1691	0.9976	0.1691	0.9977	0.1692	0.9978	0.1692	0.9979	0.1692	0.9983
	0.4	0.2919	0.9901	0.2921	0.9911	0.2922	0.9917	0.2923	0.9921	0.2924	0.9924	0.2924	0.9925	0.2924	0.9927	0.2926	0.9934
	0.6	0.3846	0.9813	0.3848	0.9824	0.3851	0.9833	0.3852	0.9838	0.3853	0.9841	0.3853	0.9843	0.3854	0.9846	0.3857	0.9856
	0.8	0.4565	0.9704	0.4568	0.9715	0.4571	0.9725	0.4572	0.9731	0.4573	0.9736	0.4574	0.9738	0.4575	0.9742	0.4579	0.9755
0.8	1.0	0.5135	0.9576	0.5138	0.9586	0.5141	0.9598	0.5143	0.9605	0.5145	0.961	0.5146	0.9613	0.5147	0.9618	0.5151	0.9632
	1.5	0.6136	0.9194	0.6138	0.9202	0.6143	0.9216	0.6145	0.9225	0.6147	0.9231	0.6148	0.9235	0.6150	0.9241	0.6156	0.9260
	2.0	0.6771	0.8756	0.6774	0.8766	0.6778	0.8781	0.6781	0.8791	0.6783	0.8798	0.6785	0.8803	0.6787	0.8809	0.6793	0.8831
	3.0	0.7502	0.7839	0.7510	0.7868	0.7515	0.7885	0.7519	0.7896	0.7521	0.7904	0.7523	0.7909	0.7525	0.7917	0.7532	0.7941
	5.0	0.8121	0.6229	0.8159	0.6347	0.8167	0.6371	0.8171	0.6385	0.8174	0.6394	0.8176	0.6400	0.8179	0.6409	0.8188	0.6437
	7.0	0.8372	0.5051	0.8452	0.5272	0.8464	0.5306	0.847	0.5324	0.8474	0.5335	0.8476	0.5342	0.848	0.5352	0.8491	0.5384
	0.2	0.1661	0.9957	0.1662	0.9964	0.1662	0.9968	0.1663	0.9971	0.1663	0.9972	0.1663	0.9973	0.1663	0.9974	0.1664	0.9979
	0.4	0.2833	0.9880	0.2835	0.9892	0.2836	0.9899	0.2837	0.9903	0.2838	0.9906	0.2838	0.9907	0.2839	0.9910	0.2840	0.9918
	0.6	0.3697	0.9774	0.3700	0.9786	0.3702	0.9796	0.3703	0.9802	0.3704	0.9805	0.3705	0.9808	0.3705	0.9811	0.3708	0.9822
	0.8	0.4354	0.9641	0.4358	0.9654	0.4360	0.9665	0.4362	0.9672	0.4363	0.9677	0.4364	0.9680	0.4365	0.9684	0.4369	0.9698
1.0	1.0	0.4868	0.9487	0.4871	0.9498	0.4875	0.9511	0.4877	0.9519	0.4878	0.9524	0.4879	0.9528	0.4880	0.9533	0.4885	0.9549
	1.5	0.5753	0.9032	0.5756	0.9041	0.5760	0.9056	0.5762	0.9065	0.5764	0.9072	0.5765	0.9077	0.5767	0.9083	0.5773	0.9103
	2.0	0.6302	0.8522	0.6305	0.8533	0.6310	0.8549	0.6312	0.8559	0.6314	0.8566	0.6316	0.8572	0.6318	0.8578	0.6324	0.8602
	3.0	0.6921	0.7494	0.6930	0.7525	0.6935	0.7540	0.6938	0.7551	0.6940	0.7559	0.6941	0.7564	0.6943	0.7571	0.6950	0.7595
	5.0	0.7437	0.5803	0.7472	0.5911	0.7477	0.5926	0.7480	0.5935	0.7482	0.5941	0.7483	0.5946	0.7485	0.5953	0.7492	0.5974
	7.0	0.7645	0.4638	0.7714	0.4822	0.7720	0.4836	0.7722	0.4843	0.7724	0.4848	0.7725	0.4852	0.7727	0.4857	0.7734	0.4875

Table A.7: Effectiveness ε and correction factor F for arrangement no. 331.

		$N = 5$		$N = 11$		$N = 17$		$N = 23$		$N = 35$		$N = 71$		$N = 95$		$N = \infty$	
R	NTU	ε	F	ε	F	ε	F	ε	F	ε	F	ε	F	ε	F	ε	F
	0.2	0.1813	1.0000	0.1811	0.9992	0.1812	0.9993	0.1812	0.9994	0.1812	0.9996	0.1812	0.9998	0.1812	0.9998	0.1813	1.0000
	0.4	0.3297	1.0000	0.3292	0.9984	0.3293	0.9986	0.3294	0.9989	0.3295	0.9992	0.3296	0.9996	0.3296	0.9997	0.3297	1.0000
	0.6	0.4512	1.0000	0.4504	0.9975	0.4505	0.9979	0.4506	0.9983	0.4508	0.9987	0.4510	0.9993	0.4510	0.9995	0.4512	1.0000
	0.8	0.5507	1.0000	0.5495	0.9967	0.5497	0.9972	0.5498	0.9977	0.5501	0.9983	0.5503	0.9991	0.5504	0.9993	0.5507	1.0000
0.0	1.0	0.6321	1.0000	0.6306	0.9959	0.6308	0.9965	0.6310	0.9971	0.6313	0.9979	0.6317	0.9988	0.6318	0.9991	0.6321	1.0000
	1.5	0.7769	1.0000	0.7748	0.9938	0.7751	0.9947	0.7754	0.9956	0.7758	0.9968	0.7763	0.9983	0.7764	0.9987	0.7769	1.0000
	2.0	0.8647	1.0000	0.8624	0.9918	0.8627	0.9928	0.863	0.9940	0.8635	0.9956	0.8640	0.9976	0.8642	0.9982	0.8647	1.0000
	3.0	0.9502	1.0000	0.9484	0.9878	0.9486	0.9891	0.9488	0.9908	0.9492	0.9932	0.9496	0.9962	0.9498	0.9971	0.9502	1.0000
	5.0	0.9933	1.0000	0.9926	0.9800	0.9926	0.9815	0.9927	0.9840	0.9928	0.9878	0.9930	0.9931	0.9931	0.9947	0.9933	1.0000
	7.0	0.9991	1.0000	0.9989	0.9732	0.9989	0.9742	0.9989	0.9772	0.9990	0.9823	0.9990	0.9899	0.9990	0.9922	0.9991	1.0000
	0.2	0.1782	0.9996	0.1780	0.9988	0.1781	0.9991	0.1781	0.9993	0.1781	0.9995	0.1782	0.9997	0.1782	0.9998	0.1782	1.0000
	0.4	0.3200	0.9984	0.3197	0.9973	0.3198	0.9978	0.3199	0.9983	0.3201	0.9988	0.3202	0.9994	0.3203	0.9995	0.3204	1.0000
	0.6	0.4339	0.9964	0.4336	0.9953	0.4339	0.9964	0.4342	0.9971	0.4344	0.9980	0.4347	0.9989	0.4348	0.9992	0.4351	1.0000
	0.8	0.5262	0.9935	0.5261	0.9931	0.5266	0.9947	0.5270	0.9958	0.5274	0.9971	0.5279	0.9985	0.5281	0.9989	0.5284	1.0000
0.2	1.0	0.6015	0.9899	0.6017	0.9904	0.6025	0.9928	0.6030	0.9943	0.6037	0.9960	0.6043	0.9979	0.6045	0.9984	0.6050	1.0000
	1.5	0.7360	0.9771	0.7377	0.9823	0.7393	0.9870	0.7402	0.9898	0.7413	0.9929	0.7424	0.9963	0.7427	0.9972	0.7436	1.0000
	2.0	0.8198	0.9592	0.8237	0.9721	0.8259	0.9797	0.8272	0.9842	0.8286	0.9890	0.8301	0.9943	0.8305	0.9957	0.8317	1.0000
	3.0	0.9076	0.9089	0.9155	0.9455	0.9187	0.9610	0.9204	0.9697	0.9222	0.9791	0.9241	0.9892	0.9246	0.9918	0.9261	1.0000
	5.0	0.9626	0.7682	0.9751	0.8690	0.9785	0.9057	0.9802	0.9262	0.9819	0.9486	0.9836	0.9731	0.9840	0.9796	0.9853	1.0000
	7.0	0.9758	0.6254	0.9894	0.7727	0.9923	0.8292	0.9936	0.8624	0.9948	0.9005	0.9960	0.9455	0.9963	0.9581	0.9970	1.0000

(Continued)

Table A.7: (*Continued*).

		$N = 5$		$N = 11$		$N = 17$		$N = 23$		$N = 35$		$N = 71$		$N = 95$		$N = \infty$	
R	NTU	ε	F	ε	F	ε	F	ε	F	ε	F	ε	F	ε	F	ε	F
	0.2	0.1751	0.9992	0.1750	0.9985	0.1751	0.9988	0.1751	0.9990	0.1752	0.9993	0.1752	0.9997	0.1752	0.9998	0.1753	1.0000
	0.4	0.3106	0.9968	0.3104	0.9962	0.3106	0.9971	0.3108	0.9977	0.3109	0.9984	0.3111	0.9992	0.3112	0.9994	0.3113	1.0000
	0.6	0.4172	0.9928	0.4174	0.9932	0.4179	0.9949	0.4182	0.9960	0.4185	0.9972	0.4189	0.9986	0.4190	0.9989	0.4194	1.0000
	0.8	0.5025	0.9872	0.5033	0.9895	0.5042	0.9922	0.5047	0.9939	0.5053	0.9958	0.5059	0.9978	0.5061	0.9983	0.5066	1.0000
0.4	1.0	0.5716	0.9801	0.5732	0.9851	0.5746	0.9891	0.5753	0.9915	0.5762	0.9942	0.5771	0.9970	0.5774	0.9977	0.5781	1.0000
	1.5	0.6945	0.9560	0.6996	0.9716	0.7022	0.9797	0.7037	0.9842	0.7053	0.9892	0.7069	0.9944	0.7074	0.9958	0.7087	1.0000
	2.0	0.7719	0.9239	0.7813	0.9545	0.7853	0.9678	0.7874	0.9752	0.7897	0.9830	0.7921	0.9913	0.7927	0.9934	0.7945	1.0000
	3.0	0.8556	0.8424	0.8740	0.9118	0.8802	0.9379	0.8835	0.9522	0.8869	0.9672	0.8903	0.9832	0.8912	0.9873	0.8938	1.0000
	5.0	0.9126	0.6610	0.9445	0.8056	0.9531	0.8601	0.9573	0.8904	0.9615	0.9235	0.9655	0.9598	0.9665	0.9695	0.9695	1.0000
	7.0	0.9272	0.5135	0.9670	0.6957	0.9761	0.7712	0.9802	0.8152	0.9840	0.8659	0.9876	0.9261	0.9885	0.9430	0.9909	1.0000
	0.2	0.1722	0.9988	0.1721	0.9981	0.1721	0.9985	0.1722	0.9988	0.1722	0.9992	0.1723	0.9996	0.1723	0.9997	0.1723	1.0000
	0.4	0.3014	0.9952	0.3014	0.9951	0.3017	0.9963	0.3019	0.9971	0.3021	0.998	0.3023	0.999	0.3024	0.9992	0.3025	1.0000
	0.6	0.4012	0.9892	0.4017	0.9910	0.4023	0.9934	0.4027	0.9948	0.4031	0.9964	0.4036	0.9982	0.4037	0.9986	0.4041	1.0000
	0.8	0.4797	0.9809	0.4812	0.9860	0.4823	0.9898	0.4830	0.9921	0.4837	0.9945	0.4845	0.9972	0.4847	0.9979	0.4853	1.0000
0.6	1.0	0.5425	0.9706	0.5454	0.9800	0.5471	0.9856	0.5481	0.9889	0.5492	0.9923	0.5503	0.9961	0.5506	0.9970	0.5515	1.0000
	1.5	0.6533	0.9362	0.6611	0.9614	0.6646	0.9727	0.6665	0.9790	0.6685	0.9856	0.6705	0.9926	0.6711	0.9944	0.6727	1.0000
	2.0	0.7225	0.8922	0.7366	0.9383	0.7419	0.9567	0.7448	0.9668	0.7477	0.9773	0.7508	0.9884	0.7515	0.9912	0.7539	1.0000
	3.0	0.7976	0.7885	0.8250	0.8831	0.8339	0.9179	0.8386	0.9368	0.8433	0.9567	0.8481	0.9778	0.8493	0.9832	0.8529	1.0000
	5.0	0.8483	0.5872	0.8989	0.7584	0.9133	0.8256	0.9203	0.8632	0.9273	0.9043	0.9342	0.9496	0.9359	0.9617	0.9411	1.0000
	7.0	0.8599	0.4428	0.9264	0.6422	0.9441	0.7313	0.9522	0.7834	0.9600	0.8434	0.9676	0.9141	0.9694	0.9340	0.9748	1.0000

	0.2	0.1692	0.9984	0.1692	0.9978	0.1692	0.9983	0.1693	0.9986	0.1693	0.9991	0.1694	0.9995	0.1694	0.9997	0.1695	1.0000
	0.4	0.2926	0.9936	0.2927	0.9940	0.2930	0.9956	0.2932	0.9965	0.2935	0.9976	0.2937	0.9988	0.2938	0.9991	0.2940	1.0000
	0.6	0.3857	0.9857	0.3865	0.9889	0.3873	0.9919	0.3877	0.9937	0.3882	0.9957	0.3887	0.9978	0.3889	0.9983	0.3893	1.0000
	0.8	0.4577	0.9743	0.4598	0.9826	0.4611	0.9874	0.4619	0.9903	0.4627	0.9933	0.4636	0.9966	0.4638	0.9974	0.4645	1.0000
0.8	1.0	0.5145	0.9613	0.5184	0.9750	0.5204	0.9822	0.5216	0.9862	0.5228	0.9906	0.5241	0.9952	0.5244	0.9964	0.5254	1.0000
	1.5	0.6130	0.9173	0.6229	0.9516	0.6270	0.9660	0.6292	0.9739	0.6314	0.9822	0.6338	0.9909	0.6344	0.9931	0.6363	1.0000
	2.0	0.6734	0.863	0.6907	0.9232	0.697	0.9463	0.7003	0.9588	0.7038	0.9719	0.7073	0.9857	0.7082	0.9892	0.7109	1.0000
	3.0	0.7372	0.7421	0.7708	0.8572	0.7816	0.8997	0.7871	0.9227	0.7928	0.9470	0.7985	0.9727	0.7999	0.9794	0.8043	1.0000
	5.0	0.7775	0.5299	0.8401	0.7184	0.8588	0.7958	0.8680	0.8393	0.8772	0.8873	0.8865	0.9405	0.8888	0.9548	0.8957	1.0000
	7.0	0.7850	0.3915	0.8674	0.5977	0.8920	0.6968	0.9038	0.7555	0.9155	0.8234	0.9271	0.9035	0.9300	0.9259	0.9386	1.0000
	0.2	0.1664	0.9980	0.1663	0.9974	0.1664	0.9980	0.1664	0.9984	0.1665	0.9988	0.1666	0.9994	0.1666	0.9995	0.1667	1.0000
	0.4	0.2841	0.9920	0.2843	0.993	0.2846	0.9948	0.2849	0.9959	0.2851	0.9971	0.2854	0.9985	0.2855	0.9989	0.2857	1.0000
	0.6	0.3708	0.9822	0.3719	0.9868	0.3727	0.9904	0.3732	0.9926	0.3738	0.9948	0.3744	0.9974	0.3745	0.9980	0.3750	1.0000
	0.8	0.4366	0.9688	0.4392	0.9792	0.4407	0.9851	0.4416	0.9885	0.4425	0.9921	0.4434	0.9958	0.4437	0.9969	0.4444	1.0000
1.0	1.0	0.4877	0.9521	0.4924	0.9701	0.4946	0.9788	0.4959	0.9836	0.4972	0.9888	0.4986	0.9943	0.4989	0.9957	0.5000	1.0000
	1.5	0.5743	0.8992	0.5856	0.9422	0.5900	0.9595	0.5924	0.9689	0.5948	0.9788	0.5974	0.9892	0.5980	0.9918	0.6000	1.0000
	2.0	0.6256	0.8356	0.6451	0.9087	0.6519	0.9362	0.6554	0.9510	0.6591	0.9666	0.6628	0.9829	0.6638	0.9871	0.6667	1.0000
	3.0	0.6777	0.7008	0.7142	0.8332	0.7258	0.8824	0.7317	0.9092	0.7377	0.9376	0.7438	0.9678	0.7454	0.9757	0.7500	1.0000
	5.0	0.7073	0.4832	0.7733	0.6821	0.7933	0.7676	0.8032	0.8162	0.8131	0.8703	0.8232	0.9312	0.8257	0.9476	0.8333	1.0000
	7.0	0.7116	0.3524	0.7961	0.5579	0.8228	0.6633	0.8358	0.7271	0.8487	0.8016	0.8618	0.8908	0.8651	0.9160	0.8750	1.0000

Table A.8: Effectiveness ε and correction factor F for arrangement no. 441.

		$N = 7$		$N = 15$		$N = 23$		$N = 31$		$N = 47$		$N = 63$		$N = 95$		$N = \infty$	
R	NTU	ϵ	F	ϵ	F	ϵ	F	ϵ	F	ϵ	F	ϵ	F	ϵ	F	ϵ	F
	0.2	0.1813	1.0000	0.1812	0.9996	0.1812	0.9996	0.1812	0.9997	0.1812	0.9998	0.1812	0.9998	0.1813	0.9999	0.1813	1.0000
	0.4	0.3297	1.0000	0.3294	0.9991	0.3295	0.9993	0.3295	0.9994	0.3296	0.9996	0.3296	0.9997	0.3296	0.9998	0.3297	1.0000
	0.6	0.4512	1.0000	0.4508	0.9987	0.4508	0.9989	0.4509	0.9991	0.4510	0.9993	0.4510	0.9995	0.4511	0.9996	0.4512	1.0000
	0.8	0.5507	1.0000	0.5500	0.9982	0.5501	0.9985	0.5502	0.9987	0.5503	0.9991	0.5504	0.9993	0.5505	0.9995	0.5507	1.0000
0.0	1.0	0.6321	1.0000	0.6313	0.9978	0.6314	0.9981	0.6315	0.9984	0.6317	0.9988	0.6318	0.9991	0.6319	0.9994	0.6321	1.0000
	1.5	0.7769	1.0000	0.7758	0.9967	0.7759	0.9971	0.7761	0.9976	0.7763	0.9982	0.7764	0.9986	0.7765	0.9990	0.7769	1.0000
	2.0	0.8647	1.0000	0.8635	0.9956	0.8636	0.9961	0.8638	0.9967	0.864	0.9976	0.8642	0.9981	0.8643	0.9987	0.8647	1.0000
	3.0	0.9502	1.0000	0.9492	0.9934	0.9493	0.9941	0.9495	0.9950	0.9497	0.9963	0.9498	0.9971	0.9499	0.9980	0.9502	1.0000
	5.0	0.9933	1.0000	0.9929	0.9893	0.9929	0.9901	0.9930	0.9915	0.9930	0.9936	0.9931	0.9950	0.9931	0.9965	0.9933	1.0000
	7.0	0.9991	1.0000	0.9990	0.9861	0.9990	0.9868	0.9990	0.9886	0.9990	0.9914	0.9990	0.9933	0.9991	0.9955	0.9991	1.0000
	0.2	0.1778	0.9976	0.1779	0.9977	0.1779	0.998	0.1779	0.9981	0.1780	0.9983	0.1780	0.9984	0.1780	0.9985	0.1780	0.9987
	0.4	0.3186	0.9929	0.3186	0.9931	0.3188	0.9936	0.3188	0.9940	0.3189	0.9943	0.3190	0.9945	0.3190	0.9947	0.3191	0.9952
	0.6	0.4308	0.9862	0.4309	0.9866	0.4311	0.9874	0.4313	0.9878	0.4314	0.9884	0.4315	0.9887	0.4316	0.9890	0.4318	0.9896
	0.8	0.5209	0.9779	0.5211	0.9784	0.5214	0.9794	0.5217	0.9801	0.5219	0.9808	0.5220	0.9812	0.5222	0.9816	0.5225	0.9824
0.2	1.0	0.5938	0.9680	0.5941	0.9689	0.5945	0.9702	0.5948	0.9709	0.5951	0.9718	0.5953	0.9723	0.5955	0.9728	0.5959	0.9739
	1.5	0.7225	0.9383	0.7233	0.9405	0.7240	0.9425	0.7244	0.9437	0.7249	0.9450	0.7252	0.9457	0.7254	0.9465	0.7260	0.9481
	2.0	0.8021	0.9032	0.8036	0.9077	0.8045	0.9106	0.8051	0.9122	0.8057	0.9141	0.8060	0.9150	0.8063	0.9161	0.8070	0.9182
	3.0	0.8864	0.8249	0.8896	0.8367	0.8910	0.8417	0.8917	0.8444	0.8925	0.8474	0.8929	0.8489	0.8933	0.8505	0.8942	0.8538
	5.0	0.9445	0.6704	0.9507	0.6999	0.9527	0.7099	0.9537	0.7152	0.9547	0.7207	0.9552	0.7235	0.9557	0.7264	0.9568	0.7324
	7.0	0.9624	0.5475	0.9703	0.5893	0.9725	0.6033	0.9737	0.6108	0.9748	0.6187	0.9754	0.6227	0.9760	0.6269	0.9771	0.6357

	0.2	0.1745	0.9951	0.1746	0.9958	0.1747	0.9963	0.1747	0.9965	0.1748	0.9968	0.1748	0.997	0.1748	0.9971	0.1749	0.9975
	0.4	0.3079	0.9853	0.3082	0.9871	0.3084	0.9880	0.3086	0.9886	0.3087	0.9891	0.3088	0.9894	0.3088	0.9897	0.3090	0.9904
	0.6	0.4114	0.9723	0.4120	0.9748	0.4124	0.9761	0.4126	0.9769	0.4128	0.9777	0.4130	0.9781	0.4131	0.9785	0.4134	0.9795
	0.8	0.4928	0.9563	0.4936	0.9595	0.4942	0.9612	0.4945	0.9621	0.4949	0.9632	0.4950	0.9638	0.4952	0.9643	0.4956	0.9656
0.4	1.0	0.5575	0.9384	0.5587	0.9418	0.5594	0.9439	0.5598	0.9451	0.5603	0.9464	0.5605	0.9471	0.5607	0.9478	0.5612	0.9492
	1.5	0.6701	0.8854	0.6721	0.8911	0.6732	0.8941	0.6738	0.8958	0.6745	0.8977	0.6748	0.8986	0.6752	0.8996	0.6759	0.9017
	2.0	0.7391	0.8276	0.7422	0.8361	0.7436	0.8400	0.7444	0.8422	0.7452	0.8446	0.7457	0.8458	0.7461	0.8470	0.7470	0.8496
	3.0	0.8134	0.7139	0.8188	0.7285	0.8208	0.7341	0.8219	0.7372	0.8231	0.7403	0.8236	0.7420	0.8242	0.7436	0.8255	0.7471
	5.0	0.8688	0.5347	0.8783	0.5577	0.8812	0.5651	0.8827	0.5691	0.8842	0.5732	0.8850	0.5754	0.8859	0.5776	0.8875	0.5822
	7.0	0.8890	0.4188	0.9010	0.4442	0.9042	0.4516	0.9059	0.4556	0.9077	0.4599	0.9087	0.4622	0.9096	0.4646	0.9116	0.4696
	0.2	0.1712	0.9927	0.1714	0.9939	0.1715	0.9946	0.1716	0.995	0.1716	0.9954	0.1717	0.9956	0.1717	0.9958	0.1718	0.9962
	0.4	0.2977	0.9789	0.2982	0.9813	0.2985	0.9825	0.2987	0.9832	0.2989	0.9840	0.2990	0.9844	0.2991	0.9848	0.2992	0.9856
	0.6	0.3931	0.9598	0.3940	0.9633	0.3945	0.9651	0.3948	0.9661	0.3951	0.9672	0.3953	0.9678	0.3954	0.9683	0.3958	0.9695
	0.8	0.4663	0.9368	0.4677	0.9412	0.4684	0.9435	0.4688	0.9448	0.4692	0.9462	0.4694	0.9469	0.4697	0.9477	0.4701	0.9492
0.6	1.0	0.5235	0.9107	0.5253	0.9162	0.5262	0.9190	0.5267	0.9205	0.5272	0.9222	0.5275	0.9230	0.5278	0.9239	0.5284	0.9257
	1.5	0.6206	0.8389	0.6233	0.8467	0.6247	0.8504	0.6254	0.8524	0.6261	0.8546	0.6265	0.8558	0.6270	0.8570	0.6278	0.8594
	2.0	0.6786	0.7652	0.6823	0.7752	0.684	0.7795	0.6849	0.7819	0.6858	0.7845	0.6863	0.7858	0.6868	0.7872	0.6878	0.7900
	3.0	0.7397	0.6327	0.7453	0.6458	0.7474	0.6507	0.7485	0.6534	0.7497	0.6563	0.7503	0.6578	0.7509	0.6594	0.7523	0.6626
	5.0	0.7850	0.4501	0.7929	0.4644	0.7951	0.4684	0.7963	0.4707	0.7976	0.4732	0.7983	0.4745	0.7990	0.4759	0.8006	0.4789
	7.0	0.8021	0.3441	0.8109	0.3567	0.8124	0.3589	0.8133	0.3603	0.8144	0.3619	0.8150	0.3628	0.8157	0.3639	0.8172	0.3662

(Continued)

Table A.8: (*Continued*).

		$N=7$		$N=15$		$N=23$		$N=31$		$N=47$		$N=63$		$N=95$		$N=\infty$	
R	NTU	ϵ	F	ϵ	F	ϵ	F	ϵ	F	ϵ	F	ϵ	F	ϵ	F	ϵ	F
	0.2	0.1681	0.9902	0.1683	0.9921	0.1685	0.9929	0.1685	0.9934	0.1686	0.9939	0.1686	0.9941	0.1687	0.9944	0.1687	0.9949
	0.4	0.2879	0.9720	0.2887	0.9754	0.2890	0.9770	0.2892	0.9779	0.2894	0.9789	0.2895	0.9794	0.2896	0.9799	0.2899	0.9809
	0.6	0.3757	0.9471	0.3770	0.9519	0.3776	0.9543	0.3779	0.9555	0.3783	0.9569	0.3784	0.9576	0.3786	0.9583	0.3790	0.9598
	0.8	0.4415	0.9175	0.4433	0.9235	0.4441	0.9264	0.4445	0.9280	0.4450	0.9297	0.4453	0.9305	0.4455	0.9314	0.4461	0.9333
0.8	1.0	0.4918	0.8847	0.4940	0.8917	0.4950	0.8951	0.4956	0.8970	0.4962	0.8989	0.4965	0.8999	0.4968	0.9010	0.4975	0.9032
	1.5	0.5746	0.7972	0.5777	0.8061	0.5791	0.8102	0.5799	0.8125	0.5807	0.8149	0.5811	0.8162	0.5815	0.8175	0.5824	0.8202
	2.0	0.6223	0.7121	0.6260	0.7220	0.6276	0.7263	0.6285	0.7287	0.6295	0.7313	0.6300	0.7327	0.6305	0.7340	0.6316	0.7369
	3.0	0.6706	0.5692	0.6751	0.5793	0.6768	0.5830	0.6778	0.5852	0.6788	0.5875	0.6793	0.5888	0.6799	0.5901	0.6811	0.5928
	5.0	0.7047	0.3902	0.7094	0.3976	0.7102	0.3989	0.7108	0.3998	0.7115	0.4009	0.7119	0.4016	0.7124	0.4023	0.7134	0.4040
	7.0	0.7174	0.2933	0.7214	0.2981	0.7209	0.2975	0.7208	0.2973	0.7209	0.2974	0.7210	0.2976	0.7212	0.2979	0.7218	0.2986
	0.2	0.1650	0.9878	0.1653	0.9902	0.1655	0.9913	0.1655	0.9918	0.1656	0.9924	0.1657	0.9927	0.1657	0.9930	0.1658	0.9937
	0.4	0.2785	0.9652	0.2795	0.9696	0.2799	0.9716	0.2801	0.9727	0.2803	0.9738	0.2805	0.9744	0.2806	0.9750	0.2808	0.9762
	0.6	0.3593	0.9347	0.3608	0.9408	0.3615	0.9436	0.3619	0.9451	0.3623	0.9467	0.3625	0.9475	0.3627	0.9484	0.3631	0.9501
	0.8	0.4183	0.8989	0.4203	0.9063	0.4212	0.9097	0.4217	0.9116	0.4222	0.9135	0.4225	0.9145	0.4228	0.9156	0.4234	0.9177
1.0	1.0	0.4624	0.8600	0.4648	0.8683	0.4659	0.8722	0.4665	0.8743	0.4671	0.8765	0.4674	0.8777	0.4678	0.8788	0.4685	0.8813
	1.5	0.5325	0.7594	0.5355	0.7687	0.5369	0.7730	0.5377	0.7754	0.5385	0.7779	0.5389	0.7792	0.5394	0.7806	0.5403	0.7835
	2.0	0.5711	0.6658	0.5744	0.6747	0.5758	0.6788	0.5766	0.6810	0.5775	0.6835	0.5780	0.6848	0.5785	0.6861	0.5795	0.6890
	3.0	0.6083	0.5177	0.6113	0.5243	0.6125	0.5268	0.6132	0.5283	0.6139	0.5301	0.6144	0.5310	0.6148	0.5320	0.6158	0.5342
	5.0	0.6331	0.3450	0.6347	0.3476	0.6344	0.3471	0.6344	0.3471	0.6346	0.3473	0.6347	0.3475	0.6349	0.3478	0.6354	0.3485
	7.0	0.6419	0.2560	0.6424	0.2566	0.6406	0.2547	0.6399	0.2538	0.6393	0.2532	0.6392	0.2531	0.6391	0.2529	0.6390	0.2529

Index

www.ingramcontent.com/pod-product-compliance
Ingram Content Group UK Ltd.
Pitfield, Milton Keynes, MK11 3LW, UK
UKHW020247290926
15470UKWH00002B/793

9 781853 127373